Potato

Pinky Raigond • Brajesh Singh • Som Dutt •
Swarup Kumar Chakrabarti
Editors

Potato

Nutrition and Food Security

Springer

Editors
Pinky Raigond
Crop Physiology, Biochemistry &
Post-harvest Technology Division
ICAR-Central Potato Research Institute
Shimla, Himachal Pradesh, India

Brajesh Singh
Crop Physiology, Biochemistry &
Post Harvest Technology Division
ICAR-Central Potato Research Institute
Shimla, Himachal Pradesh, India

Som Dutt
Crop Physiology, Biochemistry &
Post Harvest Technology Division
ICAR-Central Potato Research Institute
Shimla, Himachal Pradesh, India

Swarup Kumar Chakrabarti
ICAR-Central Potato Research Institute
Shimla, Himachal Pradesh, India

ISBN 978-981-15-7661-4 ISBN 978-981-15-7662-1 (eBook)
https://doi.org/10.1007/978-981-15-7662-1

This Springer imprint is published by the registered company Springer Nature Singapore Pte Ltd.
The registered company address is: 152 Beach Road, #21-01/04 Gateway East, Singapore 189721, Singapore

Foreword

Potato (*Solanum tuberosum*) is the world's fourth most important food crop after maize, wheat and rice with 377 million tonnes of tubers produced from 19.2 million hectares of land, in over 160 countries, in 2016 (http://faostat.fao.org). Furthermore, as a result of steady increase in its demand in recent years, over 50% of production is now in Asia, with China and India the largest contributors. Harvestable potatoes are produced in 100 to 120 days, and therefore they proved suitable for double cropping and intercropping systems. Globally, its average dry weight yield (3.92 t/ha) compares favourably with the three cereals (3.88 t/ha) as a result of a higher harvest index of 0.75 compared with 0.5. Likewise, protein content at 10% of dry weight equals that of most cereals with protein quality better than most other non-animal sources. Hence, it was not entirely surprising that the United Nations named 2008 as the "International Year of the Potato", in recognition of its contribution as a major food staple to their Millennium Development Goals of providing food security and eradicating poverty. By 2015, however, out of a world population of 7.3 billion, around 800 million people still suffered from hunger and more than 2 billion from one or more micronutrient deficiencies, known as "hidden hunger", especially in women, infants and children. Ironically, about 1.9 billion people were overweight, of whom 600 million were obese. In this context, potato needs to be appreciated as more than a major supplier of carbohydrate (starch) in the human diet. It also provides significant amounts of protein, minerals, vitamins, micronutrients and phytonutrients, which include antioxidants, is high in dietary fibre and virtually free of fat and cholesterol. Hence, potato has an important role to play in the United Nations "2030 Agenda for Sustainable Development" that started on 1 January 2016. By 2030, the aim is to "ensure access by all people, in particular the poor and people in vulnerable situations, including infants, to safe, nutritious and sufficient food all year round". By then, the world population is expected to reach 8.5 billion and continue to increase to 9.7 billion in 2050, when it will also have experienced significant climate change. For potatoes, the need is to improve nutritional value and to increase sustainable production and adaptation to environmental change.

This book "Potato: Nutrition and Food Security" deals with the role of potatoes in the human diet, discusses the status of various nutritional compounds, along with the methods for their evaluation and their health benefits, and suggests genetic modifications to enhance the concentration of health-promoting compounds. The nutritional significance of current and new potato products is also discussed. I really appreciate the efforts made by the team of ICAR-Central Potato Research Institute (CPRI), Shimla, India, in compiling this book to provide the latest knowledge on the nutritional quality of potatoes to consumers and all the stakeholders. I have followed with great interest the work of CPRI since attending my first Global Conference on Potato in New Delhi in 1999 and feel a certain affinity with the organization that like me was born in 1949. I am sure that this book will help to establish the nutritional significance of potatoes internationally and hence increase the appreciation of potatoes as a highly nutritious food.

The James Hutton Institute
Dundee, UK
8 June 2020

John E. Bradshaw

Preface

Food security in a broader way may be realized when all people at all times have access to sufficient, safe and nutritious food to maintain a healthy and active life. Food security depends on the availability of food, affordability and proper utilization of food. Food and Agriculture Organization (FAO 2008) has emphatically considered and recommended potato as a potential crop for the poorest of the poor, to ensure global food, nutritional and income security in the future. Potato is a flexible crop compared to other vegetables and can be grown under conditions where other crops may fail to grow. Moreover, its short and flexible life cycle brings the yield within 100 to 120 days, and hence it is also suitable for double cropping and intercropping systems. Potato is a good option for food and is capable of producing nutritious food more quickly on lesser land compared to any other major food crops. It yields more edible energy, protein and dry matter per unit area and time compared to other crops due to its high protein–calorie ratio (17 g protein: 1000 kcal) and short life cycle. Farmers can harvest up to 80% of biomass as edible, nutritious food in case of potato, whereas in case of cereals only up to 50% can be harvested as grains. Serious food security problems are envisaged for the future due to stagnation of crop yields, exhausting soils and increasing population in the developing world. Besides, large-scale diversion of food grains to feed and bio-fuel and expected steep rise in per capita consumption of pulses, edible oil, fruits and vegetables, milk, sugar and non-vegetarian foods in the regime of steadily rising population are bound to put pressure on existing cultivable land. Since cultivable land is expected to remain more or less constant, the role of crops such as potato having higher production per unit land and time will become imperative. This way potato crop has a very high probability of making a crucial contribution to future food security.

Potato is known to everyone as a supplier of energy, but its ability to supply vital nutrients is vastly underestimated. Potato is an excellent source of complex carbohydrates, dietary fibres and vitamin C. It also contains a variety of health-promoting compounds, such as carotenoids, flavonoids, chlorogenic acid and caffeic acid, as well as unique tuber storage proteins, such as patatin, which exhibit activity against free radicals. Potato is also a substantial source of ascorbic acid, thiamine, niacin, pantothenic acid and riboflavin. Due to the nutritional value of potato, it is highly desirable in the human diet. This book in its 15 chapters elaborates the nutritional significance of potatoes. These chapters also suggest future strategies to further enhance the nutritional quality of potatoes. We sincerely believe that this book will help to establish the nutritional significance of potatoes. It will help to enhance the acceptance of potato as a staple crop due to the presence of a myriad of nutritional compounds in it. It is hoped that the contents of the book shall provide a platform for masses to understand the role of potatoes in food and nutritional security and also help in removing a few misconceptions associated with potato consumption.

Shimla, Himachal Pradesh, India Pinky Raigond
Shimla, Himachal Pradesh, India Brajesh Singh
Shimla, Himachal Pradesh, India Som Dutt
Shimla, Himachal Pradesh, India Swarup Kumar Chakrabarti

Contents

Editors and Contributors

About the Editors

Pinky Raigond Scientist, Division of CPB & PHT, ICAR-Central Potato Research Institute has specialized in the area of Plant Physiology. She has significantly contributed in the area of nutritional quality of potatoes, identification/quantification of health promoting compounds in potatoes, development of antioxidant rich, low glycaemic and flavoured potatoes for speciality sector, processing industry waste utilization for development of eco-friendly nanoparticles, disposable biodegradable tableware and biodegradable active packaging films, processing technologies for the entrepreneurs and potato post harvest management. Her research work is published in high impact journals. She is recipient of 'Young Achievers' Award' and Young Woman Scientist Award.

Brajesh Singh Principal Scientist & Head, Division of CPB & PHT, ICAR-Central Potato Research Institute has specialized in the area of Plant Physiology. He has made significant contribution in the area of development and refinement of potato storage technologies at farm and cold store level; processing technologies for the entrepreneurs; establishing the quality and nutritional laboratory; and development of protocols for nutritional profiling of potatoes. He is recipient of Associateship of National Academy of Agricultural Sciences and HSI-Dr. JC Anand Gold Medal for his work on Post-Harvest Technology. He is the Fellow of Indian Society for Plant Physiology, New Delhi; The Indian Academy of Horticultural Sciences, New Delhi; and Indian Potato Association, Shimla.

Som Dutt Principal Scientist, Division of CPB and PHT, ICAR-Central Potato Research Institute, has specialized in the area of Plant Biochemistry and Molecular Biology. During 2002–2014, he worked at CSIR-Institute of Himalayan Bioresource Technology as Scientist/Senior Scientist. He has made a significant contribution in the area of developmental biology of potato and important medicinal plants. He was awarded Junior Research Fellowship by ICAR during M.Sc. and Junior Research Fellowship/Senior Research Fellowship by CSIR during Ph.D. He was awarded Rothamsted International Fellowship for the year 2003–2004 by Rothamsted Internationals, Rothamsted Research, Harpenden, United Kingdom. He has published research work in high impact factor journals and has been associated with the development of important technologies that have been granted international patents.

Swarup Kumar Chakrabarti Former Director, ICAR-Central Potato Research Institute, Shimla and ICAR-CTCRI, Thiruvananthapuram has specialization in the area of genomics, molecular biology, and disease diagnostics. He made significant contribution in the area of potato structural and functional genomics, linkage mapping, marker assisted selection; development of transgenics; and plant disease diagnosis. He is the recipient of Shri L.C. Sikka Endowment Award of NAAS, Dr. S. Ramanujam Award of ICAR-CPRI, IPA-Kaushalaya Sikka Memorial Award, Biotechnology Overseas Associateship DBT, Dr. J.P. Verma Memorial Award, Indian Phytopathological Society and many others. He is the Fellow of National Academy of Agricultural Sciences, New Delhi; Indian Phytopathological Society, New Delhi; Indian Potato Association, Shimla; and Confederation of Horticultural Associations of India, New Delhi.

Contributors

Bindvi Arora Division of Food Science and Postharvest Technology, ICAR-Indian Agricultural Research Institute, New Delhi, India

Fiona S. Atkinson School of Life and Environmental Sciences, Faculty of Science, The University of Sydney, Sydney, NSW, Australia

Ahmad Farid Azizi Division of Food Science and Postharvest Technology, ICAR-Indian Agricultural Research Institute, New Delhi, India

Bandana ICAR-Central Potato Research Institute Campus, Meerut, Uttar Pradesh, India

Swarup Kumar Chakrabarti ICAR-Central Potato Research Institute, Shimla, Himachal Pradesh, India

Sushil Sudhakar Changan Crop Physiology, Biochemistry and Post-Harvest Technology Division, ICAR-Central Potato Research Institute, Shimla, Himachal Pradesh, India

Som Dutt Crop Physiology, Biochemistry and Post-Harvest Technology Division, ICAR-Central Potato Research Institute, Shimla, Himachal Pradesh, India

Arvind Kumar Jaiswal ICAR-Central Potato Research Station, Jalandhar, Punjab, India

Sastry S. Jayanty San Luis Valley Research Center, Department of Horticulture and LA, Colorado State University, Fort Collins, CO, USA

Rupak Jena ICAR-Directorate of Groundnut Research, Junagadh, Gujarat, India

Alka Joshi Division of Food Science and Postharvest Technology, ICAR-Indian Agricultural Research Institute, New Delhi, India

Ashok Kumar ICAR-Directorate of Onion and Garlic Research, Pune, Maharashtra, India

Awadhesh Kumar ICAR-National Rice Research Institute, Cuttack, Odisha, India

Dharmendra Kumar Crop Physiology, Biochemistry and Post-Harvest Technology Division, ICAR-Central Potato Research Institute, Shimla, Himachal Pradesh, India

Manoj Kumar ICAR-Central Potato Research Institute, Shimla, Himachal Pradesh, India

Manoj Kumar ICAR-Central Institute for Research on Cotton Technology, Mumbai, Maharashtra, India

Vinod Kumar ICAR-Central Potato Research Institute, Shimla, Himachal Pradesh, India

Milan Kumar Lal Crop Physiology, Biochemistry and Post-Harvest Technology Division, ICAR-Central Potato Research Institute, Shimla, Himachal Pradesh, India

Satish Kumar Luthra ICAR - Central Potato Research Institute, Meerut, Uttar Pradesh, India

Anshul Sharma Manjul Crop Physiology, Biochemistry and Post-Harvest Technology Division, ICAR - Central Potato Research Institute, Shimla, India

Tanuja Mishra ICAR-Central Potato Research Institute, Shimla, Himachal Pradesh, India

Augustine Okpani Oko Department of Biotechnology, Ebonyi State University, Abakaliki, Nigeria

Vandana Parmar ICAR-Central Potato Research Institute, Shimla, Himachal Pradesh, India

Pinky Raigond Crop Physiology, Biochemistry and Post-Harvest Technology Division, ICAR-Central Potato Research Institute, Shimla, Himachal Pradesh, India

Reetu ICAR-Indian Grassland and Fodder Research Institute, Jhansi, Uttar Pradesh, India

Shruti Sethi Division of Food Science and Postharvest Technology, ICAR-Indian Agricultural Research Institute, New Delhi, India

Devender Sharma ICAR-Vivekananda Parvatiya Krishi Anusandhan Sansthan, Almora, Uttarakhand, India

Vineet Sharma ICAR-Central Potato Research Institute Campus, Meerut, Uttar Pradesh, India

Brajesh Singh Crop Physiology, Biochemistry and Post-harvest Technology Division, ICAR-Central Potato Research Institute, Shimla, Himachal Pradesh, India

Mark A. Taylor Cellular and Molecular Sciences, The James Hutton Institute, Dundee, UK

Asha Thakur ICAR-Central Potato Research Institute, Shimla, Himachal Pradesh, India

Nitasha Thakur Crop Physiology, Biochemistry and Post-Harvest Technology Division, ICAR-Central Potato Research Institute, Shimla, Himachal Pradesh, India

B. Thippeswamy Division of Food Science and Postharvest Technology, ICAR-Indian Agricultural Research Institute, New Delhi, India

Maharishi Tomar ICAR-Indian Grassland and Fodder Research Institute, Jhansi, Uttar Pradesh, India

Abbreviations

3GT	Flavonoid-3-*O*-glucosyltransferase
5-UGT	5-O-Glucosyltransferase
AA	Ascorbic acid
ACE	Angiotensin-converting enzyme
ACP	Acyl carrier protein
ADP	Adenosine diphosphate
AFLP	Amplified fragment length polymorphism
AH	Arogenate dehydro
AmA1	Amaranth seed albumin
AMD	Age-related macular degeneration
APCI	Atmospheric pressure chemical ionization
APPH	A potato protein hydrolysate
APPH	Alcalase-generated potato protein hydrolysate
APPI	Atmospheric pressure photoionization
ARPs	Ama1-responsive protein spots
AS	Anthranilate synthase
ASD	Aspartate-semialdehyde
ATP	Adenosine triphosphate
B	Boron
Bf.	*Bifidobacterium* spp.
BOD	Biological oxygen demand
Br	Bromine
BRCs	Biguanide and related compounds
Ca	Calcium
$CaCl_2$	Calcium chloride
CAGR	Compound annual growth rate
CAPS	Cleaved amplified polymorphic sequence
CbL	Cystathionine-lyase
CCD	Carotenoid cleavage dioxygenase
CCK	Cholecystokinin
cDNA	Complementary DNA
CER	Ceremix 2XL
CGA	Chlorogenic acids

CGIAR	Consultative Group on International Agricultural Research
CgS	Cystathionine-synthase
CHI	Chalcone isomerase
CHS	Chalcone synthase
chuPCI	Cystine-knot metallocarboxypeptidase inhibitor
CHY-β	β-Carotene hydroxylase
CHY-ε	ε-Ring hydroxylase
CID	Collision-induced dissociation
Cl	Chlorine
CNS	Central nervous system
Co	Cobalt
CoA	Coenzyme A
CP	Crude protein
CPC	Centrifugal partition chromatography
CPRI	Central Potato Research Institute
CQAs	Caffeoylquinic acids
CRC	Colorectal cancer
CrtB	Phytoene synthase
CrtI	Phytoene desaturase/carotene isomerase
CRTISO	Carotenoid isomerase
CrtY	Lycopene beta-cyclase
Cu	Copper
CysTA	Cystathionine
DAD	Diode array detection
DAG	Diacylglycerol
DAHPS	DAHP synthase
DAP	Diaminopimelate
DAPAE	DAP epimerase
DAPAT	DAP-aminotransferase
DEP	Depol 670L
DESI-MS	Desorption electrospray ionization mass spectrometry
DF	Dietary fibre
DFR	Dihydroflavonol reductase
DGAT	Diacylglycerol acyltransferase
DHDP	Dihydrodipicolinate
DHDPS	Dihydrodipicolinate synthase
DHF	7,8-Dihydrofolate
DM	Diabetes mellitus
DOS	Days of storage
DRV	Daily requirement values
DW	Dry weight
EAE	Enzyme-assisted extraction
EAR	Estimated average requirement
ECD	Electrochemical detectors
ELISA	Enzyme-linked immunosorbent assay

ER	Endoplasmic reticulum
ESI	Electrospray ionization
ESI/MS	Electrospray ionization/mass spectrometry
EU	European Union
FAMEs	Fatty acid methyl esters
FAO	Food and Agriculture Organization
FAS	Fatty acid synthase
Fas-FADD	Fas-associated death domain
Fe	Iron
FFA	Free fatty acid
FODMAPs	Fermentable oligo-, di- and monosaccharides and polyols
FT	Fourier transform
FW	Fresh weight
GAE	Gallic acid equivalent
GALDH	L-Galactono-1,4-lactone dehydrogenase
GalUR	D-Galacturonate reductase
GBBS	Granule bound starch synthase
GBS	Genotyping by sequencing
GC/MS	Gas chromatography/mass spectrometry
GC-MS	Gas chromatography-mass spectrometry
GGPP	Geranylgeranyl diphosphate
GGPS	Geranylgeranyl pyrophosphate synthase
GI	Glycemic index
GL	Glycolipids
GPAT	*sn1*-Glycerol-3-phosphate acyltransferase
GPP	Geranyl pyrophosphate
GRAS	Generally regarded as safe
GSN	Gelsolin
GWD	α-Glucan water dikinase
H_2O_2	Hydrogen peroxide
HA	Haemagglutination
HCD	Higher energy collision-induced dissociation
HcY	Homocysteine
HDH	Homoserine dehydrogenase
HDL	Non-high-density lipoprotein
HFD	High fat diet
HG	High glucose
HM	Homocysteine methyltransferase
HNO_3	Nitric acid
HPLC	High-pressure liquid chromatography
HPLC-PDA	High-pressure liquid chromatography with a photodiode array detector
HPP	High-pressure processing
HPPD	p-Hydroxyphenylpyruvate dioxygenase

HPT	Homogentisate phytyl transferase
HPTLC	High-performance thin-layer chromatography
HS	Homoserine
HSK	Homoserine kinase
I^-	Iodide
I	Iodine
iAUC	Incremental area under curve
IBS	Irritable bowel syndrome
ICAPES	Inductively coupled argon plasma emission spectrophotometer
IGPS	Indole-3-glycerol phosphate synthase
IO^{3-}	Iodate
IPI	Isopentenyl pyrophosphate isomerise
IRT	Iron-regulated transporter
ISFET	Ion-sensitive field-effect transistor
K	Potassium
KASII	3-Ketoacyl-ACP synthases II
kDa	Kilo Dalton
kg/ha	Kilogram per hectare
KTI	Kunitz-type inhibitors
L	*Lactobacillus* spp.
LAB	Lactic acid bacteria
LC-MS	Liquid chromatography-mass spectrometry
LC-NMR	Liquid chromatography-tandem nuclear magnetic resonance
LCY-β	Lycopene β-cyclase
LCY-ε	Lycopene ε-cyclase
LD	Lipid droplet
LDAP	Lipid droplet-associated protein
LDL	Low-density lipoprotein
LPAAT	Lysophosphatidic acid acyltransferase
LWM	Low-molecular weight
MAE	Microwave-assisted extraction
MAGIC	Multiparent advanced generation intercross
MALDI-TOF	Matrix-assisted laser desorption/ionization-time of flight
MALDI-TOFMS	Matrix-assisted laser desorption ionization-time of flight mass spectrometry
MAS	Marker-assisted selection
MASL	Mean above sea level
Mg	Magnesium
mg/100 g FW	Milligram per 100-gram fresh weight
mg/kg	Milligram per kilogram
MGL	Methionine gamma-lyase
min	Minutes
Mn	Manganese
MS	Mass spectrometer
MT	Metric tons

MUFAs	Monounsaturated fatty acids
MW	Molecular weight
N	Nitrogen
Na	Sodium
NaCl	Sodium chloride
NAD	Nicotinamide adenine dinucleotide
NADP	Nicotinamide adenine dinucleotide phosphate
NADPH	Nicotinamide adenine dinucleotide phosphate
NCED	9-Cis-epoxycarotenoids dioxygenase
Ni	Nickel
NIDDK	National Institute of Diabetes and Digestive and Kidney Diseases
NIH	National Institutes of Health
NIRS	Near-infrared spectroscopy
NL	Neutral lipids
NMR	Nuclear magnetic resonance
NO	Nitric oxide
NTDs	Neural tube defects
NXS	Neoxanthin synthase
OPH	O-Phospho-homoserine
P	Phosphorus
PA	Phosphatidic acid
PAI	Phosphoribosyl anthranilate isomerase
PAL	Phenylalanine ammonia-lyase
PAT	Phosphoribosylanthranilate transferase
Pb	Lead
PC	Phosphatidylcholine
PDA	Photodiode array
PDI	Potato cathepsin D inhibitor
PDS	Phytoene desaturase
PE	Phosphatidylethanolamine
PEF	Pulsed electric field
pERK1/2	Phosphorylated extracellular signal-regulated kinases
PFE	Pressurized fluid extraction
PI	Protease inhibitors
PIN	Protease inhibitors I and II
PL	Phospholipid
PLP	Pyridoxal 5′-phosphate
PMC	Potato multicystatin
PSY	Phytoene synthase
PUFA	Polyunsaturated fatty acid
PWD	Phosphoglucan water dikinase
QSRs	Quick-service restaurants
QTLs	Quantitative trait loci
RAG	Rapidly available glucose

RDA	Recommended dietary allowances
RDS	Rapidly digestible starch
RFLP	Restriction fragment length polymorphism
RH	Relative humidity
RIL	Recombinant inbred line
ROS	Reactive oxygen species
RP-HPLC	Reverse-phase high-performance liquid chromatography
RS	Resistant starch
RSM	Response surface methodology
S	Sulphur
SAG	Slowly available glucose
SAH	S-Adenosylhomocysteine
SAM	S-Adenosyl methionine
SAMS	S-Adenosylmethionine synthetase
SAR	Structure–activity relationship
SCAR	Sequence characterized amplified region
SCFA	Short-chain fatty acids
SDSt	Slowly digestible starch
SDS	Sodium dodecyl sulphate
Se	Selenium
Si	Silicon
SNP	Single nucleotide polymorphism
STG1	Solanidine galactosyltransferase
STG2	Solanidine glucosyltransferase
STG3	Solanine rhamnosyl transferase
STZ	Streptozotocin
TAG	Triacylglycerol
TC	Tocopherol cyclase
TE	Triennium ending
TG	Triacylglycerol
THDP	Tetrahydrodipicolinate
THF	5,6,7,8-Tetrahydrofolate
TILLING	Targeting induced local lesion in genomes
TLC	Thin-layer chromatography
TPP	Thiamine pyrophosphate
TS	Thr synthase
UAE	Ultrasound-assisted extraction method
UPLC-DAD-MS	Ultra-high-performance liquid chromatography-diode array detector-tandem mass spectrometry
USA	United States of America
USDA	United States Department of Agriculture
UV/VIS	Ultraviolet/visible
VDE	Violaxanthin de-epoxidase
WHO	World Health Organization
ZDS	f-Carotene desaturase

ZEP	Zeaxanthin epoxidase
Zn	Zinc
ZRT	Zinc-regulated transporter

Potatoes for Food and Nutritional Security

1

Brajesh Singh, Pinky Raigond, Som Dutt, and Manoj Kumar

Abstract

Potatoes have been used as food since ages, however, investigations on its nutritional significance have increased its importance as alternative for both food and nutritional security. The global trend of potato production and per capita availability clearly indicates a decrease of production in developed world and accelerated growth in developing world including Asia and Africa. These two regions of the world also have higher number of undernourished, therefore, higher per capita availability of potatoes in these regions could substantiate the food and nutritional requirements of the populations. The chapter describes in brief, the nutritional significance of potatoes in terms of its energy value, carbohydrates, protein, vitamins, minerals and health-promoting compounds. The significance of all these compounds have been discussed in length under different chapters. It is hoped that the contents of the book shall provide a platform for masses to understand the role of potatoes in food and nutritional security and also help in removing few misconceptions associated with potato consumption.

Keywords

Food and nutrition · Dietary energy supply adequacy · Undernourishment · Birthweight

B. Singh (✉) · P. Raigond · S. Dutt
Crop Physiology, Biochemistry and Post-harvest Technology Division, ICAR–Central Potato Research Institute, Shimla, Himachal Pradesh, India
e-mail: birju16@gmail.com

M. Kumar
ICAR–Central Potato Research Institute, Shimla, Himachal Pradesh, India

P. Raigond et al. (eds.), *Potato*, https://doi.org/10.1007/978-981-15-7662-1_1

1.1 Introduction

Potato is a wholesome food contributing to the energy and nutritional requirements of a large population world over. It has been used as a food by humans for almost 8000 years and it is believed that cultivated potato originated in South America, most likely near Lake Titicaca on the borders of Peru and Bolivia. When Spain conquered Peru, they took potatoes from there and spread it all over the Europe by the end of the sixteenth century. Further spread of potatoes was facilitated by several European countries including Portugal and Britain. However, its significance and role in the global food system is under-appreciated. Potatoes have been known for their use as staple food, cash crop, animal feed, and also as source of starch for industrial uses in almost 160 countries world over (Ramani and Mouille 2019).

Food security depends on availability of food, affordability and proper utilization of food. Generally it is considered that surplus food grains lead to nutritional security, but it is not always true to all the population, as affordability is one of the major issues. There are few other indicators of food and nutritional security as well like average dietary energy supply adequacy, proportion of undernourished population and low birthweight. Undernourished population and <5 years old underweight children are high in several Asian countries like Bangladesh and India with the status remaining alarming over the years (Singh et al. 2020). Globally India is second in child malnutrition after Bangladesh. In India still more than 40% children are underweight, which is of great concern. However, the rate of mortality of <5 years olds has decreased over the period, but it is still up to 6.3% in India (Singh et al. 2020). The average dietary energy supply adequacy has been found to be comparatively high in North America and Europe, low in Asia and least in Africa (Fig. 1.1). The prevalence of undernourishment though has decreased over last two decades, but still is high in Africa and Asia (Fig. 1.2). Similarly, low birthweight has also decreased over the calculated period, but again is alarming in Asia (Fig. 1.3). FAO data suggests that about 11 percent of the population is undernourished, which

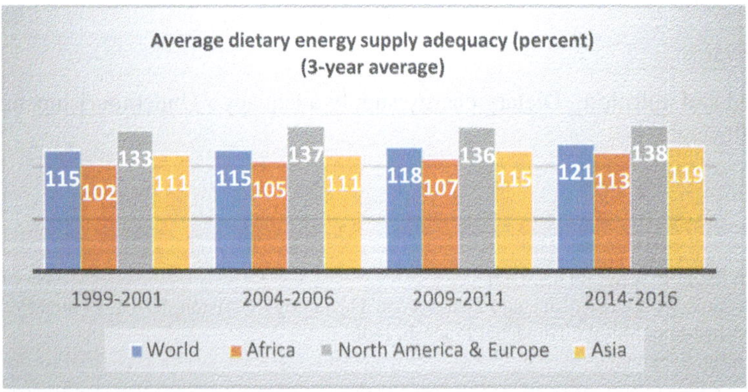

Fig. 1.1 Average dietary energy supply adequacy in selected regions and world (source: FAOSTAT 2020)

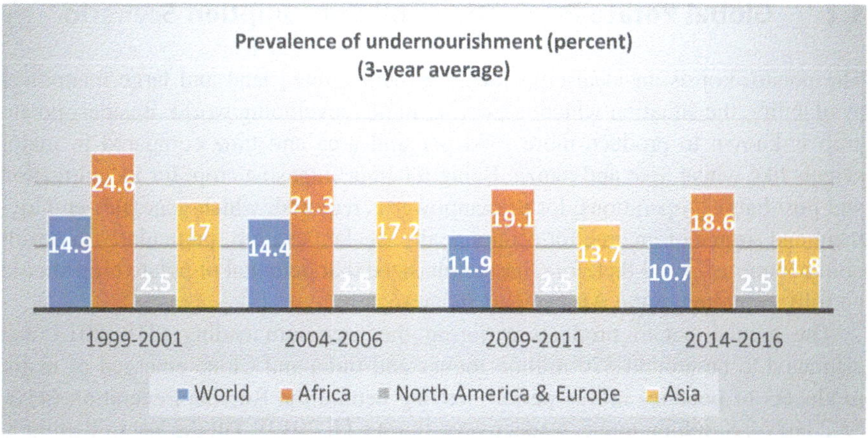

Fig. 1.2 Prevalence of undernourishment in selected regions and world (source: FAOSTAT 2020)

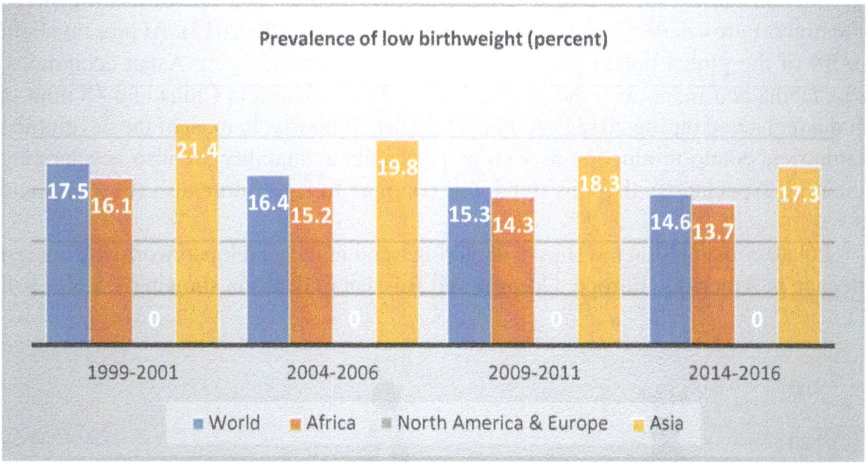

Fig. 1.3 Prevalence of low birthweight in selected regions and world (source: FAOSTAT 2020)

means that one in every nine people in the world suffers from hunger (FAOSTAT 2020).

Thus concerns of food and nutritional security need to be addressed further through nutritious and sufficient food access to populations including pregnant women and children. To overcome this situation those fruits and vegetables should be popularized, which are available throughout the regions in all the seasons and are in reach of all income groups, especially economically weaker sections.

1.2 Global Potato Production and Consumption Scenario

The potato crop is an ideal crop for areas with limited land and large manpower availability, the situation which is existing in the developing world. Besides, potato crop is known to produce more food per unit area and time compared to major cereals like wheat, rice and maize. Being a labour intensive crop, for its cultivation and post-harvest operations, lot of manpower is required, which generates employment and source of income for economically weaker sections, particularly in developing countries. FAO therefore, has emphasized that potential of potato crop should be fully exploited in the African and Asian regions.

The annual potato production during the triennium ending (TE) 2013 was estimated to be around 370 million tonnes and India and China emerged as major producers of potatoes in this period, leaving behind the Russian Federation (43.1, 88.2 and 30.8 million tonne, respectively, during TE 2013). Till the last millennium, the major producers and consumers were from developed world, whereas the scenario has changed over the period. A comparison of growth rate in potato production during TE ending in 2003 and 2013 depicts that Africa has registered the highest growth (97%) followed by Asia (Fig. 1.4) (CPRI 2015). At present about 1/3rd of the global potato production is coming from emerging Asian economies, viz. China and India, with the production levels of 90 mmt in China and 48 mmt in India registered during 2018 (FAOSTAT 2020). Similarly, in most of the developing countries, potato production as well as per capita availability are also accelerating creating expectations that this trend will continue in near future also (Figs. 1.5 and 1.6).

Potato consumption has shown declining trend in the developed world during last decade in Americas, Europe, Oceania and Russian Federation showing −8.8, −9.4,

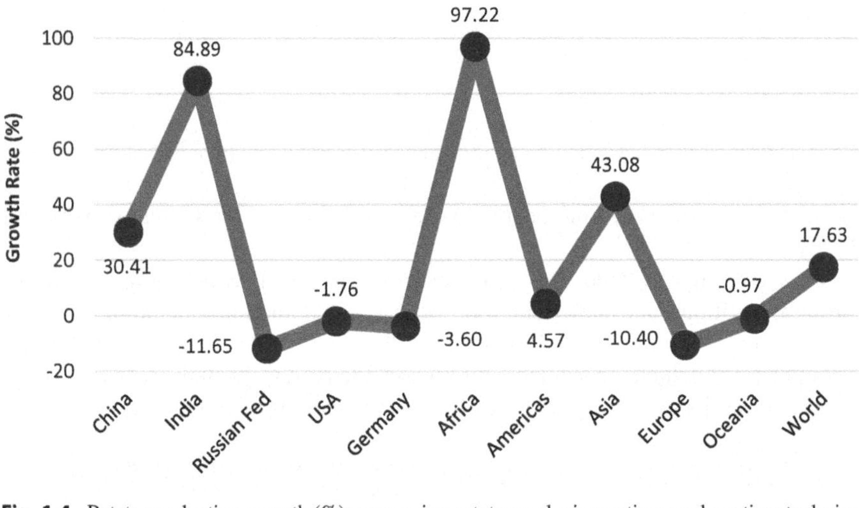

Fig. 1.4 Potato production growth (%) over major potato producing nations and continents during TE 2003 and TE 2013 (Source: CPRI 2015)

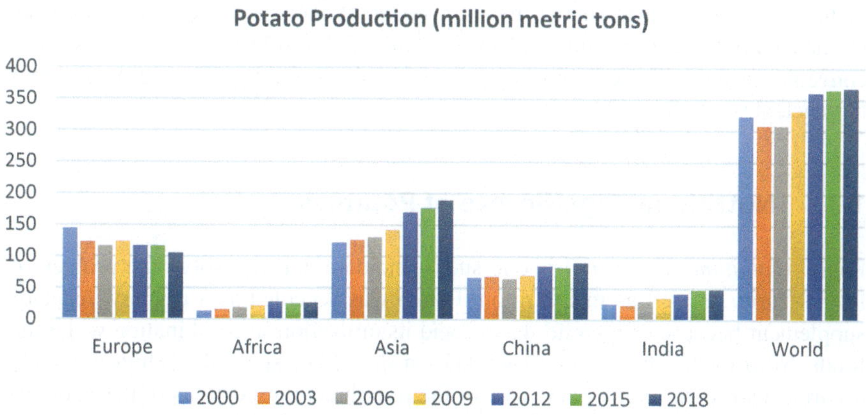

Fig. 1.5 Potato production in different parts of the world during 2000–2018 (FAOSTAT 2020)

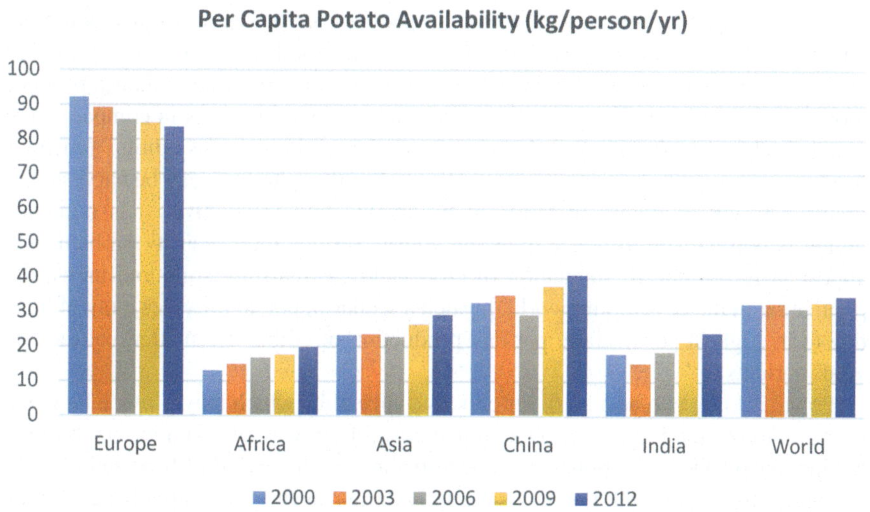

Fig. 1.6 Per capita potato availability in different parts of the world during 2000–2012 (FAOSTAT 2020)

−8.3 and −2.4% growth, respectively, whereas the developing world has shown increasing trend (Africa, Asia, India and China with 40.6, 25.6, 37.1 and 28.8% growth, respectively (FAOSTAT 2015, CPRI 2015). Thus, the growth in per capita as well as total potato consumption during this period had been the highest in Asia, though there could be issues related to productivity in several countries of Asia.

Due to the high productivity per unit area, potato is preferred in countries with high population density. The other favouring attributes include its flexibility of fitting into several prevailing cropping systems and stable yields over different environments compared to other crops. Simultaneously, the consumption of potatoes in these regions is also increasing as a result of increased industrialization and women employment creating enhanced demand for processed and ready-to-eat convenience food. There is also a perceptible shift in food preference from cereals to vegetables and fruits in such areas.

1.3 Nutritional Significance of Potatoes

Potato (*Solanum tuberosum*) is one such candidate that can solve the problem of food security as well as malnutrition. Potato was accepted as a primary vegetable supplement because of its mild flavour and its utilization in combination with other foods. Nutritional value of potato was known since long, specially its high content of ascorbic acid to prevent scurvy. One of the prominent publications of the Food and Agriculture Organization (FAO 2008) has emphatically considered and recommended potato as a potential crop for the poorest of the poor, to ensure global food, nutritional and income security in future.

Potato is a good option for food and is capable of producing nutritious food more quickly on lesser land compared to any other major food crops. Attributing to high protein-calorie ratio (17 g protein: 1000 kcal) and short life cycle, potatoes produce more edible energy, protein and dry matter yield on per unit basis in comparison to major food crops. Farmers can harvest up to 80% of biomass as edible, nutritious food in case of potato, whereas in case of cereals only up to 50% can be harvested as grains. Growth of potato in terms of production and productivity has remained higher in comparison to maize, rice and wheat. Serious food security problem will appear in future due to stagnation of crop yields, exhausting soils and increasing population in the developing world. In such a scenario, potato provides a ray of hope due to its highest per hectare, per day production of edible dry matter, calorie and vital nutrients.

Potato is known to be a highly productive vegetable, which may provide food and nutrition to bigger population. Due to its versatility in way of cooking, viz. boiling, baking, deep frying, etc. potato became popular over the period of time and is being consumed by one and all. Potato is popularly known as the 'Vegetable King'. It may be consumed in the form of snacks (chips, fries and dehydrated products) by the rich, whereas most of the low income households consume potato as primary or secondary source of food as well as nutrition. The nutritional value of potato is well acclaimed and is known as a versatile, carbohydrate rich and low-fat food. Potatoes at fresh harvest may contain approximately 80% water and 20% dry matter, out of this dry matter approximately 60–80% is constituted in the form of starch. Its content of dry matter, edible energy and edible protein makes it a good choice for nutrients availability. On dry weight basis, the protein content of potatoes is comparable to cereals and higher when compared to other roots and tubers. Potato is well known to consumers as a source of energy, but its significance of supplying vital nutrients is not well recognized. Potato is an excellent source of complex carbohydrates, dietary fibres and vitamin C. It also contains a variety of health-promoting compounds, such as phytonutrients, that have antioxidative activity. Some of the health-promoting compounds present in potatoes include carotenoids, flavonoids, and caffeic acid. Besides, unique tuber storage proteins like patatin known to exhibit activity against free radicals is also present in it. Potato is also a substantial source of ascorbic acid, thiamine, niacin, pantothenic acid and riboflavin. Due to the nutritional value of potato, it is highly desirable in human diet. The nutritive value of a potato containing food depends on the other components served with it and on the method of preparation. As it is, potato does not contain much of fat and the feeling of satiety it gives, is helpful for the people aiming weight reduction. However, the caloric value of the potato containing dish may increase if it is prepared and served with high fat ingredients. The starch of raw potato is not easily digestible, hence, potatoes have to be cooked before consumption by boiling (with or without the skin), baking or frying and depending on the cooking method, the potato composition get changed in different ways.

Raw potato (on dry weight basis) provides about 80 kcal energy, whereas a boiled potato provides about 69 kcal energy per 100 g of weight (Singh et al. 2020). Due to its low energy density, potatoes are good for weight conscious people, if they eat potatoes without adding fat in it. The energy value of potatoes is low compared to rice, wheat, maize and sorghum. The energy value is even lower than other tuber and root crops as well as food products from animal origin. Potatoes are an excellent source of complex carbohydrates. These carbohydrates take longer time for break down into glucose and result in energy that lasts longer. Complex carbohydrates are longer chains of sugars, such as starches and fibre. In potatoes, the major carbohydrate is in form of starch, whereas main sugars include sucrose, fructose and glucose. Carbohydrates are the body's primary source of fuel for energy. The energy produced through potato gets stored as glycogen in muscle and liver and functions as a readily available energy during prolonged, strenuous exercise. Sugars are the most basic carbohydrates, the building blocks of complex carbohydrates. Starch furnishes most of the energy supplied by the potato. Digestibility of starch influences the energy value of the potato and hence also the bulk of potato which must be eaten to supply a given amount of energy.

Potato is a rich source of dietary fibre. Cellulose, pectin and pectin associated substances are higher in potatoes compared to cereal bran. Dietary fibre content in raw potato tuber ranges from 1 to 2 g/100 g fresh weight. Unpeeled potatoes contain more dietary fibres than peeled potatoes. The dietary fibre from potato tuber comes mainly from its cell walls that constitute about 1.2% of the fresh weight of the tubers. To increase the dietary fibre intake, potatoes must be consumed along with peel. More than half of the dietary fibre in potato is in the form of pectic substances which improves the quality of potato dietary fibre and thus helps in lowering cholesterol levels. One medium potato may supply 8% of the daily value of fibre (about 2 g). Dietary fibre is a complex carbohydrate and it cannot be digested and absorbed in the bloodstream, though it is known to have several health benefits like improving blood lipid levels, regulating blood glucose and increasing satiety. The main constituents of dietary fibre include non-starch polysaccharides (NSP), lignin, resistant starch and non-digestible oligosaccharides. Potatoes also contain resistant starch which are 'starch and starch degradation products that escape digestion in the small intestine of healthy individuals'. Resistant starch acts in similar fashion as fibres and is found naturally in foods such as legumes, bananas, potatoes and some unprocessed whole grains. In Indian potatoes resistant starch content is approximately 1.5–2% in cooked and cooled tubers (Raigond et al. 2014). The natural resistant starch is generally insoluble and gets fermented in the large intestine and acts as a prebiotic fibre. However, some other types of resistant starch are known, which may or may not be soluble and also might not have prebiotic properties. The resistant starch concentration of potatoes gets affected by cooking and processing. Generally, cooking and cooling result in about two-fold increase in its concentration. Resistant starch is also known to be present in significant amounts in processed potatoes like flakes. It has been categorized as the third type of dietary fibre since it shows the attributes of both soluble and insoluble fibres.

Potato has been known as a very good source of high quality protein. Average protein content of potato is 2% on fresh weight basis and about 10% on dry weight basis (Sato et al. 2017). Potato protein content is lower than wheat, rice, corn, sorghum and beans but is higher than other major root and tuber crops like sweet potato, yam and cassava. The total nitrogen in potatoes is contained in the form of soluble protein, insoluble protein and soluble non-protein nitrogen. The insoluble protein fraction is mainly present in the peel. Soluble potato protein contains substantial levels of the essential amino acids. Free amino acids present in potatoes are totally available for absorption. The protein exhibits adequate ratio of essential amino acids to total amino acids and can meet the needs of infants and small children. However, the digestibility of potato protein is relatively low in infants. Potato protein has a very high biological value since all the essential amino acids are present in it are in balanced proportion. The biological value of proteins in potato is high compared to major cereals and sometimes even higher than that of animal origin like milk and beef. The high lysine content of potato is helpful in supplementing diets having limited lysine composition. Potato thus, has advantage over cereals in for vegans due to its ability to provide high quality protein. Diet which can fulfil only the energy requirement of body cannot support growth of children, if its protein content is below the recommended requirement. However, if a diet provides inadequate energy, its protein is metabolized as a source of energy rather than being used for growth. Therefore, diet should be well balanced in terms of energy and protein. Therefore, potato is a superb food which has correct balance between net protein calories and total calories adequate for all age groups.

There is a common misconception that eating potato may cause obesity due to its high fat content which is not at all a true statement, since potatoes contain very little quantity of fat. The average fat content of potato is 0.1% on fresh weight basis which is too low to have any negative nutritional significance (Ramadan 2016). Fat content in potato is lower than major cereals like rice, wheat, maize and sorghum. The little fat present in potato contributes towards potato palatability. Major proportion (i.e. nearly 60–80%) of potato fat consists of unsaturated fatty acids and linoleic acid is the predominant one and these unsaturated fatty acids, in fact increase the nutritive value of the fat. Due to its low energy density, potato is also good for weight conscious people if they consume potatoes without adding fat. Of course, if fat is added to the fried or processed potato products, it becomes rich in calorie. Especially excessive consumption of processed potato products such as chips and French fries containing up to 40% fat may not be so healthy.

Potato is also a good source of important minerals and trace elements. Hundred grams of potato contains approximately 40–65 mg of phosphorus and due to relatively small percentage of phytic acid present in potato, the assimilation of phosphorous is high. The lower phytic acid content of potatoes enhances phosphorous bioavailability to human body and also helps in increased bioavailability of calcium, iron and zinc. The potassium content of potato is also relatively high, i.e. 247–455 mg/100 g fresh weight. Since potassium content in potatoes is high, it is generally not included in the diet of patients having renal issues. The sodium content of potato is very low limited to about 11 mg/100 g fresh weight. Potatoes are a good source of iron and their iron content is comparable to most of the other vegetables. In fact, about 6 and 12% of daily iron requirement for children and adult can be met from 100g of cooked potato (O'Neill 2005). Moreover, due to high ascorbic acid content of potato, bioavailability of non-haem form of iron from potato is increased. Potatoes mixed with other food are also beneficial as it increases the bioavailability of iron from other foods also due to its high ascorbic acid content. Moreover iron availability from potato is higher compared to other foods such as kidney beans, wheat flour and bread, reason being the high proportion of iron from potato is soluble. Potatoes provide a good source of magnesium, which is up to 22 mg/100 g fresh weight. Potato can be consumed with foods low in magnesium such as milk. Magnesium content of milk is one-fifth to one-tenth of potato. Hence potato is superior to milk in terms of magnesium, but consumed together they form best combination as milk is rich in calcium and potato provides magnesium. Zinc is an important trace element though, its concentration in potato is less, but its availability is high due to low phytic acid content. Potatoes can also supply at least part of daily requirements of other trace elements like manganese, copper, molybdenum and chromium. Traces of boron, bromine, iodine, aluminium, cobalt and selenium are also present in potato. A small potato can deliver 10% daily value of folate, manganese, magnesium and phosphorus. Therefore, potato being in reach of poorest of the poor can play a vital role in eradication of 'Hidden Hunger' which is also known as micronutrient malnutrition.

Potato contains 19–58 mg of ascorbic acid per 100 g tuber. Potatoes have high quantities of vitamin C on average than other vegetables like carrots, onion and beet root. When consumed in sufficient quantities, potatoes itself can meet all the vitamin C requirements of an individual. Potato may act as an important source of several vitamins like thiamine, niacin and pyridoxine and its derivatives (vitamin B6 group). Potatoes also contain small amounts of pantothenic acid (vitamin B5), riboflavin and folic acid. These B-vitamins are essentially known for general health and growth benefits. However it is recommended that potatoes should not be washed after peeling to prevent loss of vitamins. A small potato can deliver more than 20% daily value of vitamin C and vitamin B6 (Singh et al. 2020).

Along with vitamins and minerals, potatoes contain a number of small molecules, many of which are beneficial phytonutrients such as phenols, flavonoids, kukoamines, anthocyanins and carotenoids. Coloured potatoes have natural anthocyanin pigments in them and may serve as source of antioxidant micronutrients. Potato antioxidants have potential role in immune function and disease prevention. Yellow pigmented potatoes are known to have high carotenoids content such as lutein, zeaxanthin, violaxanthin and antheraxathin. Potato carotenoids are primarily oxygenated carotenoids which are also known as xanthophylls. Purple pigmented potatoes have health benefit against cardiovascular disease while consumption of yellow pigmented potatoes enhances immune response. Some of the major carotenoids present in potatoes include lutein, zeaxanthin, violaxanthin and neoxanthin, though β-carotene has been detected only in trace amounts. The orange colour of the tuber flesh is attributed to zeaxanthin, whereas the yellow colour is due to lutein. Generally high phenolic contents such as anthocyanin are present in dark coloured potatoes (Singh et al. 2020). But white/cream fleshed potatoes also contain phenolics to some extent.

Though potatoes are consumed in large quantities compared to other vegetables in most countries, they are always being unappreciated compared to other vegetables due to misconception related to their role as contributor to obesity and diabetes development. Potatoes are high carbohydrate food and people believe it to be a high calorie and high fat food as well compared to other carbohydrate rich foods such as rice. Presence of relatively high concentrations of phytonutrients with bioactivities to combat chronic disease development has not been investigated in length in case of potatoes. However, in recent past, studies have indicated potential of potatoes to confer health benefits in human cell culture, experimental animals and human clinical studies. The studies have also associated potatoes with their anti-inflammatory, anti-obesity, anti-cancer, hypocholesterolemic and anti-diabetic properties. Pharmacological agents such as biguanides, α-glucosidase inhibitors, sulphonylureas, thiazolidinediones and meglitinide are used for treatment of diabetes due to their different mode of action (Raigond et al. 2018). Biguanide and related compounds (BRCs) include metformin, phenformin, urea, guanidine, galegine, biuret, L-arginine, which are known to increase insulin sensitivity thereby, decrease glucose absorption and finally reduce hyperglycaemic. Among all biguanide and related compounds, metformin is most popular and first line drug of choice used for the treatment of type II diabetes. Potatoes have been reported to contain compounds such as α glucosidase inhibitors and biguanide and related compounds (BRCs). Perla and Jayanty (2013) reported BRCs presence in potatoes with concentration of 7.05 μg/g. Later Raigond et al. (2018) evaluated BRCs from vegetables and fruits including potato. BRCs in range of 0.20–3.28 mg/g FW have been reported in vegetables and 0.17–1.91 mg/g FW in fruits. Potato contained 0.56 mg/g FW BRCs, and consumption of BRCs through potato is almost 29.36 mg/day when calculated on basis of 19 kg/person/year potato consumption (in India). Concentration of these compounds in potato peel is almost double than concentration present in flesh (Raigond et al. 2016).

α-glucosidase is a membrane bound enzyme present in the epithelium of small intestine. This enzyme facilitates the absorption of glucose by small intestine by catalysing the hydrolytic breakdown of oligosaccharides into absorbable monosaccharides. Inhibition of such enzymes can inhibit the breakdown of starch and hence can help in decreasing the postprandial enhancement in blood glucose. Raigond et al. (2016, 2017) reported presence of α-glucosidase inhibitors in indigenous potatoes. The in vitro inhibitory activity ranged from 0% to 52.8% and among 46 varieties these inhibitors were present in 14 potato varieties. These reports showed the potential of indigenous potato varieties to manage diabetes, particularly type II. Angiotensin converting enzyme (ACE) is known to increase the blood pressure. High ACE activity leads to increased concentration of angiotensin II which in turn leads to hypertension. ACE inhibitors are used as first line of treatment of hypertension. Plants and plant parts with high antioxidative capacity have the potential to act as ACE inhibitors. Raigond et al. (2020) reported the presence of ACE inhibitors in indigenous potatoes. Activity was recorded from 25 mg/ml water extract. The inhibitory activity ranged from 0 to 56.23% in tested samples. Out of 25 Indian varieties investigated, ACE inhibitory activity was recorded in six varieties. These reports showed that the indigenous potato varieties contain health-promoting compounds such as anti-diabetic and antihypertensive ones. Such studies will help to improve the image of potatoes and reduce the friction towards its consumption.

1.4 Conclusion

International Food Policy Research Institute report on ‘Global Hunger Index 2012’ has shown serious concerns on food security in developing world. To overcome such serious and alarming situations potato provides a way out to tackle the problem of hunger and provides food and nutritional security. Potato can be a perfect substitute for other cereals due to its high nutritional value and high production and productivity. Potato is a nourishing and wholesome food. Its low energy density is advantageous for majority of consumers. Potatoes contain high quality protein rich in essential amino acids. Vitamin C content of potatoes is high compared to many vegetables and cereals. Some of the B group vitamins are also present in sufficient quantity in potatoes. They contain many minerals and trace elements and simultaneously are low in fats. Compared to major food crops, bioavailability of minerals from potatoes is potentially high due to high ascorbate content in it. Simultaneously, low concentration of absorption inhibitors such as phytate and oxalate also improve the bioavailability of minerals from potato. Hence, potato as such is a wholesome food and anyone can live by eating potatoes alone. With ever increasing population, potatoes are destined to be very crucial for providing food and nutritional security to populations in the developing countries with higher percentage of undernourished population.

References

CPRI (2015) Vision 2050. ICAR-Central Potato Research Institute, Shimla, 33 p
FAOSTAT (2015) Statistical database. http://faostat.fao.org/

FAOSTAT (2020) Statistical database. http://faostat.fao.org/
O'Neill KR (2005) Statistical derivation of nutrition label for raw potatoes from 2001–2002 United States Department of Agriculture (USDA) Agricultural Research Service (ARS) Nutrient Data Laboratory (NDL) Data for Raw Russet, White, and Red Potatoes by Weighting by Marke. Cent Food Saf Appl Nutr FDA
Perla V, Jayanty SS (2013) Biguanide related compounds in traditional antidiabetic functional foods. Food Chem 138:1574–1580
Raigond P, Ezekiel R, Kaundal B (2014) Starch fractions of cooked potatoes at low temperature. Potato J 41(1):58–67
Raigond P, Kaundal B, Singh B et al (2016) Biguanide related compounds (anti-diabetic) in Indian potatoes. Newsletter no. 64. ICAR-CPRI, Shimla
Raigond P, Singh B, Dutt S et al (2017) Potential of Indian potatoes for the management of hyperglycemia. Indian J Hortic 74(1):103–108
Raigond P, Kaundal B, Sood A et al (2018) Quantification of biguanide and related compounds (anti-diabetic) in vegetables and fruits. J Food Compos Anal 74:82–88
Raigond P, Bhardwaj P, Raigond B et al (2020) Insulin like antigens in potatoes: A new hope for developing insulin rich potatoes for diabetics. In: Souvenir, Global Potato Conclave, 28–-31 January 2020 at Gandhinagar, Gujarat, India, pp 117–118
Ramadan MF (2016) Potato lipids. In: Advances in potato chemistry and technology, 2nd edn. Elsevier, Amsterdam
Ramani WB, Mouille B (2019) The contribution of potatoes to global food security, nutrition and healthy diets. Am J Potato Res 96:139–149
Sato H, Koizumi R, Nakazawa Y, Yamazaki M, Itoyama R, Ichisawa M, Negichi J, Sakuma R, Furusho T, Sagane Y, Takano K (2017) Data on the weights, specific gravities and chemical compositions of potato (*Solanum tuberosum*) tubers for food processing from different areas of Hokkaido, Japan. Data Brief 11:601–605
Singh B, Raigond P, Joshi A, Bandana, Dutt S, Changan SS, Lal MK, Chakrabarti SK (2020) Potatoes: a source of food and nutrition. Technical bulletin 111. ICAR-CPRI, Shimla, 73 p

Potato Carbohydrates

2

Pinky Raigond, Fiona S. Atkinson, Milan Kumar Lal, Nitasha Thakur, Brajesh Singh, and Tanuja Mishra

Abstract

Potatoes are widely consumed and are a highly popular food due to their preparation in many ways such as boiled, microwaved, fried, roasted, dehydrated, etc. Methods of processing and storage are known to affect its nutritional quality. Glycemic index of potato is affected by various factors, which include the genetic makeup of variety, amylose content, type and ratio of starches, cooking method, and the presence of other ingredients consumed with them including fiber, fat, and protein. Potatoes contain three types of starches, i.e. rapidly digestible starch, slowly digestible starch, and resistant starch. These starches affect blood glucose levels to different extents and hence the glycemic index. Processing method significantly affects the concentration of these starch types. Potato resistant starch has attracted the attention of nutritionists due to its various health benefits. Potato contains resistant starch type III in most cooked forms. Potato carbohydrates are affected by factors including cultivation practices, temperatures during crop growth, storage durations and temperatures, and processing. Various biochemical procedures are available to evaluate potato carbohydrates. This chapter deals with the types of starches present in potato, effect of cultivation practices, processing and storage on carbohydrates, methods used for evaluation, glycemic index, and genetic modifications carried out to alter the structure and function of potato carbohydrates.

P. Raigond (✉) · M. K. Lal · N. Thakur · B. Singh · T. Mishra
Crop Physiology, Biochemistry and Post-harvest Technology Division, ICAR-Central Potato Research Institute, Shimla, Himachal Pradesh, India
e-mail: jariapink@gmail.com

F. S. Atkinson
School of Life and Environmental Sciences, Faculty of Science, The University of Sydney, Sydney, NSW, Australia

P. Raigond et al. (eds.), *Potato*, https://doi.org/10.1007/978-981-15-7662-1_2

Keywords

Potato · Carbohydrates · Glycemic index · Processing · Genetic modifications

2.1 Introduction

Potato is a versatile, wholesome food highly popular worldwide. The predominant macronutrient in potatoes is carbohydrates. Potato carbohydrates are mostly present in the form of starch. The main role of carbohydrates is to provide energy to fuel the brain and the body. Freshly harvested potatoes contain approximately 80% moisture and 20% dry matter. Potato contains 60–80% starch on dry weight (DW) basis. Sugars are the basic form of carbohydrates and are the building blocks of complex carbohydrates. The energy value of potato is determined by the digestibility of potato starch. The digestibility of potato starch is low in raw state and its digestibility improves considerably after cooking/processing. Potato contains sucrose, fructose, and glucose as the main sugars, which account for 0.5–2% of total tuber weight (Mareček et al. 2013). Sugar quantity in potato influences the quality of processed potato products, such as chips and French fries. Potato starch is composed of amylose and amylopectin in 20:80 ratio. There are three types of starch present in cooked potatoes, namely rapidly digestible starch (RDS), slowly digestible starch (SDSt), and resistant starch (RS). Rapidly digestible starch is defined as a "type of starch which is readily (within 20 min) converted to glucose molecules by enzymatic digestion." Presence of a high concentration of RDS in food leads to rapid release of glucose which elevates blood glucose and insulin response, and hence is detrimental to health. Slowly digestible starch takes a longer time for digestion and is completely digested in the small intestine. It is defined as a "type of starch which is converted to glucose after 120 min of enzymatic digestion." The third type of starch, i.e. RS, is not digested in the small intestine and has attracted the attention of nutritionists. When gelatinized starch is allowed to cool, amylose and amylopectin chains recrystallize, this process is termed as "retrogradation" and amylose retrogradation leads to formation of RSIII. It is a part of starch present in the diet that escapes digestion and absorption in the small intestine and is fermented in the large intestine of humans, with the production of short chain fatty acids (SCFA). All these three types of starches are correlated. Rapidly digestible starch is inversely correlated to SDSt and RS. Slowly digestible starch and RS coexist in cooked potatoes. Resistant starch concentration is influenced by the ratio of amylose and amylopectin. RS is a linear molecule of α 1,4 D-glucan and is mainly derived from retrograded amylose, therefore, concentration of amylose influences the concentration of RS in food. Resistant starch is further classified into five types; RSI, RSII, RSIII, RSIV, and RSV (Fig. 2.1) (Raigond et al. 2015). Cooked potatoes mainly consist of RSIII which is the most common form of resistant starch formed during cooking of starchy foods in the presence of water. Resistant starch content (RSII, RSIII, RSV) in cooked potatoes can vary due to a number of factors such as amylose content, cooking methods, cooking period, cooking medium (water, oil, etc.), cooking temperatures,

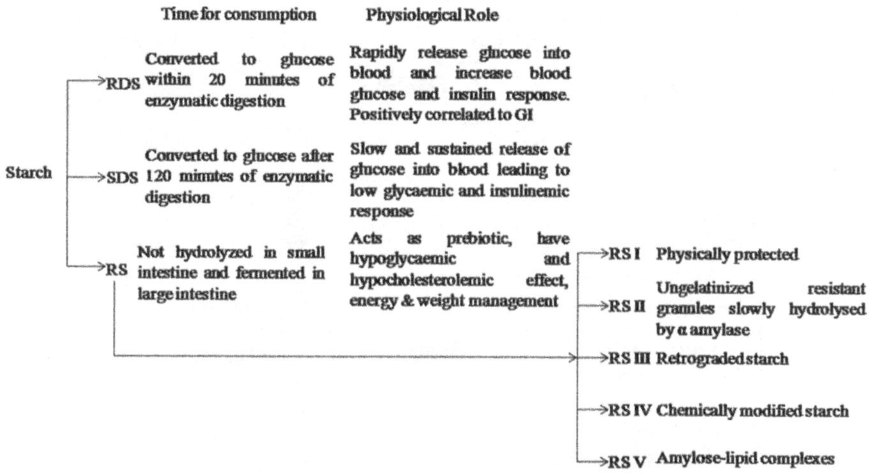

Fig. 2.1 Properties of different types of starches

and post-cooking storage (Ozturk et al. 2009; Robertson et al. 2018). Concentrations of these starches also influence the glycemic index (GI) of cooked potatoes. Carbohydrate content in potatoes is affected by various factors such as method of cultivation, combination and content of fertilizers and irrigation patterns, cooking temperature and time, storage temperatures, etc. This chapter will discuss these factors and highlight their influence on potato carbohydrates.

2.2 Effect of Cooking on Carbohydrates (Starch, Amylose, Resistant Starch, Sugars)

Potato is considered a carbohydrate-rich food. Potatoes cannot be consumed in raw form because most of the starch in raw potatoes is resistant to enzymatic hydrolysis. Therefore, cooking is necessary to convert potato starch into a digestible form. Cooking is known to enhance the gelatinization through the breakdown of starch thereby making it vulnerable to digestive enzymes (Bordoloi et al. 2012). Potatoes may be cooked in different ways such as boiled, microwaved, pressure cooked, fried, and baked. Though potatoes have high starch content, not all of the starch is digestible. In the gastrointestinal tract, starch is hydrolyzed by amylolytic enzymes leading to production of glucose, which is absorbed in the small intestine. This type of hydrolysis and digestion takes place only when starch is gelatinized during cooking and cooked food is consumed immediately after preparation (Raigond et al. 2014).

Cooking methods are known to alter the composition and ratio of starch types. Many researchers have reported the impact of cooking on starch and sugars. In raw potatoes RDS and SDSt were reported to be lower than RS (Mishra et al. 2008). When potatoes are cooked at high temperatures, starch gelatinization takes place

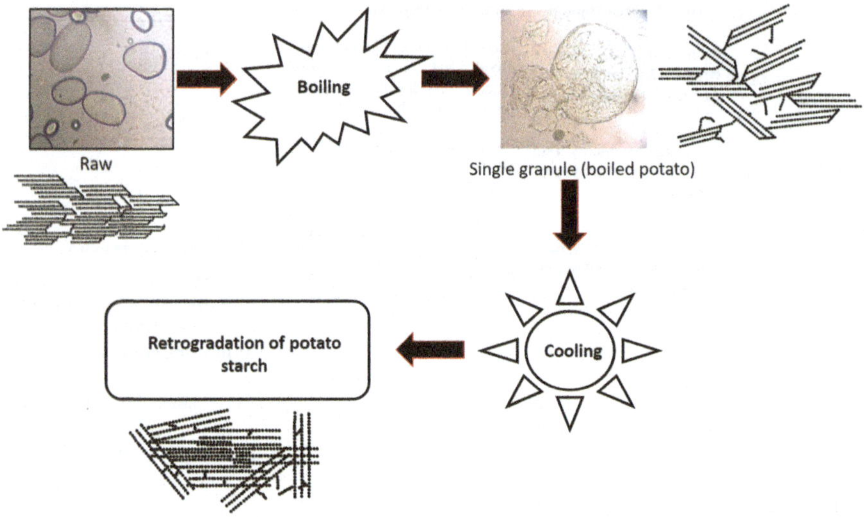

Fig. 2.2 Formation of resistant starch type 3 (RSIII)

converting most of it to RDS. However, when the same potatoes are cooled, some of RDS is converted to RS due to retrogradation (Fig. 2.2). Conversely, immediate cooling of potatoes convert starch to its SDSt form. Mulinacci et al. (2008) observed that microwaving enhanced the RS formation. Microwaving increased the RS content in all three tested cultivars, namely Kennebec, Vitelotte Noire, and Highland Burgundy Red. The extent of RS formation inversely correlated with the moisture content of the cultivar. Cultivars with low moisture content have a tightly packed microstructure that leads to reduced starch gelatinization and hence contained a higher concentration of RS. Yang et al. (2016) studied the effect of cooking on starch and sugar content in four potato cultivars; Agata, Caesar, Red Pontiac, and Kennebec. They reported total starch content in range of 644–757 g/kg lyophilized weight and RS in the range of 468–600 g/kg lyophilized weight in raw form. Their study revealed that individual sugars, specifically glucose, fructose, and sucrose, showed a six to ten-fold increase during baking or microwaving. However, 1 h baking at 250 °C resulted in the greatest sugars development. Resistant starch was reduced due to heat hydrolysis during cooking and was converted to soluble starch. Murniece et al. (2011) observed an increase in all types of sugars upon cooking of five Latvian potato cultivars. Starch content was observed to increase two to three fold upon frying, while a 1.2–1.6 fold increase was reported during roasting. Similarly, glucose content increased from 0.11–0.52 g/L to 0.17–0.74 g/L (1.4–1.5 fold) after frying and 0.27 to 0.75 g/L (1.4–2.4 fold) after roasting. Reducing sugars increased 1.3–3 folds after roasting and up to two-fold after deep frying. Monro et al. (2009) studied the effect of cooking and post-cooking low temperature storage on starch type in nine potato varieties and 37 breeding lines. Potatoes contained 62–73% RDS, 0–8.5% SDSt, and 3–6.4% RS when processed immediately after cooking. A dramatic change was observed in the starch profile after exposing the cooked potatoes to low temperature (4 °C) for 44 h. Cold storage of cooked potatoes significantly influenced the concentrations of the different starch types. Cooling changed the RDS range to 33–53%, SDSt to 15–34%, and RS to 5–16%. No significant relationship was observed between SDSt and RS of cooked and cooled potatoes. Karlsson et al. (2007) studied the effect of gelatinization treatment on RS in three genetically modified potato lines and observed a linear increase in RS with increasing amylose content. Line 527-1 contained low amylose (1%) and RS content (0.1%), while line 342 contained 64% amylose and 26% RS. These results indicated that RS content was directly proportional to the amylose content of potatoes. Raatz et al. (2016) evaluated the effect of cooking on RS content of three potato cultivars, namely Yukon Gold, Dark Red Norland, and Russet Burbank, and observed higher resistant starch with baking compared to boiling. Chilling at 4 °C resulted in high RS compared to the potatoes freshly served hot or reheated (4 °C for 6 days, then reheated).

Resistant starch in raw potatoes (RSII) is estimated to be in range of 47–59% of dry matter (DM) and is cultivar dependent (Yang et al. 2016; Zhao et al. 2018). Heating in presence of moisture reduces the RSII fraction to approximately 2–4% of DM. Resistant starch was comparatively high in baked potatoes, followed by microwaved and least in boiled potatoes. For RS formation, cooking time and intensity plays an important role as longer cooking period results in high RS formation in fried potato chips (Goni et al. 1997). Yadav (2011) compared boiled and cooled potatoes with boiled, deep or shallow fried and then cooled potatoes for RS formation. They observed maximum RS formation in boiled-cooled potatoes (1.78%), followed by shallow fried-cooled potatoes (1.11%) and least in deep fried-cooled potatoes (1.04%). The lower formation of RS in deep fried-cooled potatoes was attributed to formation of RSV that inhibited formation of RSIII and hence resulted in less total RS formation. Bembem and Sadana (2012) studied the effect of potato processing on total carbohydrates. The raw potatoes contained 20% total carbohydrates that reduced to 18.1%, 18.5%, and 16% after boiling, steaming, and pressure cooking, respectively, and increased to 25.3% after microwaving. Steam boiled, peeled, and immediately served potatoes contained 14.3% starch and 0.68% free glucose that reduced to 12.6% and 0.67% in potato cubes (peeled potatoes chopped, steamed, refrigerated, and served hot), respectively (Tahvonen et al. 2006). Eroglu and Buyuktuncer (2017) reported baking, steaming, and autoclaving increased the RS content, whereas pressure cooking decreased the RS content of food. Though frying has the potential to enhance the RS content of food, due to its detrimental effect on human health, it cannot be recommended as a suitable method to increase RS content. Bavaneethan et al. (2015) studied the effect of cooking time (15, 20 and 30 min) on RS formation in tuberous crops. Mean RS content in raw potato was 26.1% DW that decreased to 5.8% after 15 min cooking. Cooking for 20 and 30 min further decreased the RS content to 5.6% (during both cooking times). Tubers processed for <15 min contained better RS content than when cooked for 15 min, but were not palatable. Studies have indicated that the processing method has significant effects on different carbohydrate components. In general cooking decreases the starch and RS content with concomitant increase in sugars.

2.3 Glycemic Index (GI)

The concept of ranking carbohydrate-rich food based on their glycemic index was introduced by Jenkins et al. (1981). Starchy foods including rice, potato, wheat, and other cereals, and sugary foods, such as fruit, dairy products or table sugar, all produce a spike in the blood glucose (sugar) level. The GI ranks carbohydrate foods (both starches and sugars) based on the degree to which they increase blood glucose levels in the body following consumption rather than simply on their chemical structure. It is a measure of carbohydrate quality designed to compare the relative physiological impact of food and beverages on postprandial glucose responses.

Glycemic index (GI) is defined as the postprandial blood glucose response to a fixed available carbohydrate portion of a test food consumed by an individual under standard conditions expressed as a percentage of the postprandial response following consumption of a reference food consumed by the same person on a different day (FAO/WHO 1998). In other words, the GI of a food is determined as the incremental area under curve (iAUC) for blood glucose response after the consumption of a test food containing a known available carbohydrate portion, usually 50 grams of available carbohydrate, expressed as a percent of the iAUC response of an equivalent amount of available carbohydrate from a reference food (ISO 2010). Different carbohydrates can produce a wide range in GI on a scale from 0 to 100. Foods with a GI value of 55 or under are classified as low GI, between 56 and 69 are medium GI foods, and high GI foods have a GI value of 70 or more (Atkinson et al. 2008). Carbohydrate foods with low GI are either more slowly digested and absorbed or contain sugars, such as fructose or lactose, that naturally produce a lower rise in postprandial glucose response. Several different food factors influence starch digestion and the GI of foods, including the degree of starch gelatinization (Ross et al. 1987), amylose to amylopectin ratio (Granfeldt et al. 1995), particle size and physical barrier (Granfeldt et al. 1994; Liljeberg et al. 1992), viscosity and fiber (Tappy et al. 1996), acidity (Ostman et al. 2005), and the presence of antinutrients and enzyme inhibitors (Yoon et al. 1983). Glucose is used as reference food and is assigned a GI of 100 by definition. GI of fructose is quite low (15), and sucrose has an intermediate GI (65) as it is a disaccharide composed of both fructose and glucose (Atkinson et al. 2008).

Published GI data show potatoes have a broad range of GI values, depending on cultivar, cooking, and processing methods (Atkinson et al. 2008). Henry et al. (2005) evaluated the GI values of eight potato varieties in healthy volunteers and found that potatoes exhibited a wide variety of GI from 56 to 94. They reported a positive correlation between GI and potato texture, where waxy potatoes showed medium GI, whereas floury potatoes were reported to have high GI values. Ek and colleagues observed a similar broad GI range of 53–103 for seven different potato cultivars, with Carisma producing the lowest GI value of 53 (Ek et al. 2014b). Tuber size has also been shown to influence GI responses to potatoes, such that a significant positive correlation with increasing potato size producing higher GI values has been observed (Soh and Brand-Miller 1999).

The preparation and processing of potatoes influence the GI response. Tahvonen et al. (2006) observed up to a 25% decrease in GI after cold storage of cooked potatoes. In Russet Burbank potatoes, 1.25 fold reduction in GI was reported when cooking and cooling was followed by reheating (Fernandes et al. 2005). Leeman et al. (2005) studied the effect of vinegar dressing and cold storage of cooked potatoes on postprandial glycemic response and insulinemic response. They tested four meals; freshly boiled potatoes, boiled and cold-stored potatoes (8 °C, 24 h), boiled and cold-stored potatoes (8 °C, 24 h) with vinaigrette (8 g olive oil and 28 g white vinegar (6% acetic acid)), and white wheat bread as reference. Cold storage of boiled potatoes increased RS from 3.3 to 5.2% (starch basis) while vinegar dressing on cold potatoes reduced acute glycemic and insulinemic response by 43 and 31%, respectively, compared to freshly boiled potatoes (Leeman et al. 2005). The authors concluded the glycemic response of boiled potatoes could be reduced either by cold storage or with the addition of vinegar to the potatoes. However, vinegar addition had a greater GI-lowering effect (43% decrease) than low temperature exposure (26% decrease). Similarly, the beneficial influence of vinegar on glucose and insulin response to a meal in patients with diabetes mellitus was reported by Liatis et al. (2010). Postprandial glycemic and insulinemic responses of a high GI meal containing mashed potato and low-fat milk were significantly reduced after vinegar addition compared to the same meal consumed without vinegar. The addition of vinegar to a low GI meal did not affect the glucose iAUC and insulin iAUC responses. The GI-lowering effect of vinegar (acetic acid) to high GI staple carbohydrate foods such as bread, rice, and potato has been attributed to a reduced rate of gastric emptying (Liljeberg and Björck 1998; Darwiche et al. 2001). The temperature of cooked starchy food at the time of consumption can also influence the GI. To study this effect on glycemic and insulinemic indices, 9 subjects with varied insulin sensitivity consumed two potato meals (containing 50 g carbohydrates) (Najjar et al. 2004). Consumption of cooled potatoes (26.0 ± 0.6 °C) resulted in significantly lower postprandial blood glucose and lower postprandial triglyceride response producing a lower GI compared to freshly cooked potatoes served hot (83.6 ± 2 °C).

Starch properties, including amylose content, particle size and level of gelatinization, can all impact the GI of cooked potatoes. Starch extracted from Carisma potatoes (lower GI cultivar) was more thermally stable, resistant to gelatinization and showed significantly high pasting temperatures compared to other higher GI potato cultivars (Ek et al. 2014a). Pinhero et al. (2016) evaluated 14 early potato cultivars for their dietary fiber, total starch, RDS, SDSt, and RS. They reported a strong positive correlation between estimated GI and RDS compared to a strong negative correlation between estimated GI and RS. However, no correlation was observed between amylose, dietary fiber, and total starch content and the GI values of seven types of cooked potatoes (Ek et al. 2014b). Effect of processing on rapidly available glucose (RAG), slowly available glucose (SAG), and total glucose content was studied by Haase (2015). Peeling increased the relative RAG content and microwaving enhanced both the SAG and RAG contents. Processing potatoes into French fries decreased RAG significantly, with a significant increase in SAG (Haase 2015).

Polyphenols are known to inhibit α-glucosidase, an enzyme present in the brush border of the small intestine and are involved in the release of glucose from more complex carbohydrates. To establish the relationship between polyphenols, GI, and α-glucosidase, Ramdath et al. (2014) evaluated GI of white and colored potatoes in healthy adults. GI of white-fleshed potatoes was around 93, yellow was 81, while in red and purple-fleshed potatoes GI were 78 and 77, respectively. Similarly, polyphenols were 82, 108, 190, and 234 mg/100 g DW in white, yellow, red, and purple potatoes. Their study indicated a significant inverse correlation, such that colored potatoes with higher polyphenol contents produced lower GI values. The relationship is mediated through inhibition of α-glucosidase by the action of anthocyanins. Moser et al. (2018) studied the effect of potato phenolic extract (PPE) prepared from purple, red, and white-fleshed varieties on the modulation of starch digestion and intestinal glucose uptake in model systems. They observed no effect of PPE on starch digestion by α-amylase and α-glucosidase, however, a significant decrease in the rate of glucose transport was reported. Consumption of purple potato chips significantly decreased blood glucose concentrations at 30 and 60 min compared to white chips, which indicates the role of phenolics in the modulation of glycemic response.

Ramírez et al. (2015) reported the role of sodium alginate on starch hydrolysis. They used 0.5%, 1%, and 2% solution of sodium alginate in 3% starch solution and studied glucose release during in vitro digestion. In control samples, starch hydrolysis during the initial 15 min of intestinal phase was 72% that reduced to 55% during the initial 15 min of the intestinal phase when starch was mixed with 1 and 2% sodium alginate. Sodium alginate increased the viscosity and hence reduced the swelling volume and amylose leaching that resulted in a lower rate of starch hydrolysis. de Vasconcelos et al. (2015) studied the effect of blanching treatment with salts (sodium chloride (NaCl) and calcium chloride ($CaCl_2$)) on the GI of French fries. Sticks were blanched in salt solutions (3% NaCl, 3% $CaCl_2$, 3% NaCl+3% $CaCl_2$), partially fried, frozen for 24 and 30 h, and finished fried for sensory evaluation and carbohydrate analysis. RS was high in untreated fries frozen for 24 h, dietary fiber was high in French fries blanched in 3% NaCl and frozen for 30 h and the same treatment had the lowest GI and glycemic load. The results indicated that longer freezing duration leads to higher accumulation of dietary fibers that resulted in lower GI and glycemic load.

The GI of a food can be lowered by the addition of either protein or fat consumed with the food (Miller et al. 2006). A significant reduction in GI was observed after the addition of chicken breast and salad to mashed potato (GI = 54) compared to the mashed potato alone (GI = 108) (Hatonen et al. 2011). Similarly, the glucose response reduced significantly when potatoes and peas were consumed as a mixed meal compared to the consumption of potato meal alone (Schafer et al. 2003). Henry et al. (2006) studied the effect of a range of toppings on the glycemic response of baked potatoes. GI of baked potatoes reduced with the addition of toppings containing protein and fat. Cheddar cheese reduced the GI value of the baked potato by 58% and changed GI classification from high GI to low GI (39), followed by baked beans that reduced GI by 33% and changed classification from high to medium GI. Chilli con carne and tuna were found to reduce GI by 19 and 18%, respectively. Leeman et al. (2008) reported boiled potatoes to induce high satiety compared to French fries on an energy equivalent basis. French fries had a low early glycemic response and were less satiating during the first 45 min. Boiled and mashed potatoes did not differ for satiety or glycemic response. All types of potatoes regardless of cooking method contribute to dietary glycemic load (carbohydrate quality and amount) and in particular, boiled potatoes have been linked to a small increased risk of diabetes type II (Schwingshackl et al. 2019). However, it is important to acknowledge that potato cultivars can have very different GI responses and starch properties, which influence their starch digestion and absorption in the body.

2.4 Most Preferred Methods for Estimation of Total Carbohydrates, Starch, Amylose, Resistant Starch, Sugars, and Glycemic Index

2.4.1 Starch and Sugars

There is more than one procedure available for the measurement of each of the parameters; total carbohydrates, starch, sugars, amylose, and resistant starch. Anthrone method developed by Hedge and Hofreiter (1962) and phenol sulfuric acid method developed by Dubois et al. (1956) are widely accepted and used methods for estimation of total carbohydrates and total soluble sugars. Both of these methods are spectrophotometric protocols. Chen et al. (2004) suggested the simple non-destructive technique for estimation of carbohydrates using near infrared spectroscopy. For starch estimation, anthrone method developed by Mccready et al. (1958) remains popular. Method developed by Somogyi (1952) and dinitrosalicylic acid method developed by Miller (1956) are the preferred methods for reducing sugar estimation. Amylose is most commonly estimated by the method of Juliano (1971).

2.4.2 Resistant Starch

Goni et al. (1996) suggested a protocol for estimation of RS. Since then the protocol has been modified by various researchers based on food matrix. In the suggested protocol, samples are incubated with digestive enzymes (pancreatic α-amylase for 16 h at 37 °C to hydrolyze digestible starch, and amyloglucosidase for 45 min at 60 °C). Starch converted to D-Glucose with the action of digestive enzymes and is measured with glucose oxidase/peroxidase reagent (GOPOD). RS is calculated as glucose × 0.9. Other researchers (Granfeldt et al. 1992; Åkerberg et al. 1998) have suggested replacement of homogenization with chewing of food by subjects and use of dialysis tubing to imitate the small intestine to get precise results for resistant starch measurement.

2.4.3 Glycemic Index

By definition, GI ranks carbohydrate-containing foods based on their physiological response in the body rather than on their chemical structure and components. The determination of GI requires testing in at least ten healthy human participants and involves repeated testing of the reference food on different days to control for individual differences in glucose tolerance between subjects (ISO: 26642 2010). GI testing requires ethical approval and the collection and analysis of multiple blood samples (640 data points are used to generate one GI value for a food), and as such, is often considered a time- and labor-intensive process.

Various in vitro methods assessing starch digestibility have been proposed (Englyst et al. 1992; Cummings and Englyst 1995; García-Alonso and Goñi 2000; Englyst et al. 2003; Nayak et al. 2014; Kumar et al. 2018) as alternative techniques to approximate in vivo GI testing of carbohydrate foods. The biological response of carbohydrates to digestive enzyme action is an essential factor in both in vivo GI testing and in vitro methods for GI estimation. Rate of in vitro starch digestion has been suggested as a potential predictor of the physiological impact of foods. Methods to assess in vitro hydrolysis can be used to classify food carbohydrates based on the rate of digestibility, such as rapidly or slowly digestible carbohydrate (Leinonen et al. 1999; Araya et al. 2002). An early method of measuring available and non-digestible carbohydrates in food products involved the digestion of starch with enzymes such as amyloglucosidase and pullulanase (Southgate 1969a, b). However, this method does not fully hydrolyze starch. Subsequently, Englyst et al. (1982) determined non-starch polysaccharides in food by gas chromatography. They developed a quantitative method for estimation of dietary fiber and digestible starch, digesting starch using α-amylase and pullulanase followed by hydrolysis with amyloglucosidase (Englyst et al. 1982). Later methods have been proposed which simulate digestion processes in the mouth, stomach, and small intestine (Granfeldt and Björck 1991; Åkerberg et al. 1998). In vitro methods of GI estimation are typically relatively inexpensive, simple, rapid and require only small quantities of test foods. These methods aim to mimic the highly complex human gastrointestinal system, in particular the oral, gastric, and intestinal phases of in vivo digestion.

2.4.3.1 Imitation of Oral Phase

The oral phase of in vivo digestion involves significant disruption of food structure. This includes the mechanical breakdown of food by chewing followed by the commencement starch hydrolysis with salivary mixing and salivary α-amylase before swallowing (Woolnough et al. 2008). Some studies have used participants to chew foods, rather than mechanical processes, as the first part of the digestion method (Åkerberg et al. 1998; Granfeldt et al. 1992). However grinding, mincing, milling or homogenizing the sample followed by addition of α-amylase is commonly used to replicate oral digestive processes in in vitro digestion methods, for example, passing the food through a narrow gauge sieve and incubation with human α-amylase at pH 6.9 for 5 min at 37 °C (Brighenti et al. 1995).

2.4.3.2 Imitation of Gastric Phase

After completion of the oral phase, the food moves to the stomach where gastric digestion takes place. In the stomach mainly the protein component of the meal is digested within the acidic environment of pH 2.5 of gastric fluid (Turnbull et al. 2005). The in vitro method consists of a proteolysis step that is carried out by the addition of pepsin enzyme under pH 2.5 to mimic the in vivo gastric phase. This process helps fully digest the starch by disruption of the protein-starch food matrix. Pepsin degrades the protein component linked with starch and releases free starch. Granfeldt and co-workers included the digestion of a chewed food sample followed by incubation with pepsin for 30 min at pH 1.5 and 37 °C (Granfeldt et al. 1992). Other researchers have also incorporated a proteolysis step in their in vitro starch digestion methods, including protocols described by Goni et al. 1997 using pepsin incubation at 40 °C for 60 min and Kumar et al. 2018 who used pepsin incubation at 37 °C for 60 min at pH 2.5.

2.4.3.3 Imitation of Intestinal Phase

The final phase of starch digestion occurs in the small intestine, where the major starch hydrolysis enzyme is α-amylase. This enzyme and other digestive enzymes are embedded in the small intestine enterocytes, with numerous microvilli making up the brush border, and facilitate the complete breakdown of starch fragments into absorbable glucose and other monosaccharides. Monosaccharides are then absorbed in the bloodstream leading to a spike in postprandial glycemic response (Woolnough et al. 2008). In the replication of this stage in in vitro methodologies, samples are treated with α-amylase and aliquots are recovered at regular intervals over time for measurement of the extent of glucose release. Englyst et al. 1992 conducted their sample hydrolysis assay using an enzyme mixture at pH 5.2 in a 37 °C water bath with aliquots measured at 20 and 120 min, representing the RDS and SDSt fractions, respectively. The researchers used glass beads and tube shaking to help mechanical breakdown of the food and included guar gum to help homogenize the viscosity of food enzyme mixture between samples. Goni et al. (1997) in their methodology replicated the intestinal hydrolysis phase by sample incubation in a shaking 37°C water bath at pH of 6.9 including a Tris-Maleate buffer and α-amylase. Aliquots were collected every 30 min during a 3 h digestion (Goni et al. 1997). Jenkins et al. (1984) used a protocol that involved combining10 mL of human saliva with a carbohydrate food in dialysis tubing that was incubated in a stirred water bath at 37 °C for 3 h. The total concentration of sugar was measured at 3 h and the four foods were ranked based on their digestibility index relative to control food. Similarly, Kumar et al. 2018 used dialysis tubes to mimic the transportation of digested material through the small intestine wall by simple diffusion. Maltose and other simple sugars permeated out of the membrane and were released into the ambient buffer system. Aliquots were collected every 30 min for 3 h, treated with amyloglucosidase and the resulting free glucose concentration was measured.

Significant correlations have been demonstrated between in vitro starch digestion fractions and in vivo GI testing, particularly for high starch-plain foods. Rapidly available glucose has been positively correlated with both postprandial glucose response (Englyst et al. 1999) and the GI values of 39 foods (Englyst et al. 1996). Granfeldt et al. (1992) also observed a significant relationship between the rate of glucose released over 3 h, assessed as a hydrolysis index, and in vivo GI responses. A significant positive association between GI and the percentage of in vitro enzymatic starch hydrolysis in 7 types of cooked potatoes has been reported (Ek et al. 2014b). The rate of starch hydrolysis at 90 min was a reliable predictor of the GI of foods (Goni et al. 1997). In vitro starch digestion methods can produce good correlations with in vivo GI data; however, the two methods should be considered as complementary rather than interchangeable techniques.

The lack of consensus on the best in vitro starch digestion methodology to estimate GI limits the utility of these assays at this time. In vitro assays do not adequately replicate the physiological impact of gastric emptying and food bolus viscosity on starch digestion in the body. Variability in enzyme source and activity, time points for glucose appearance measurement, and differences in food processing methods hinder comparison of results between studies (Hernandez-Hernandez et al. 2019; Woolnough et al. 2008). While in vitro assays have been proposed as a cost-effective and time-efficient methodology, laboratory equipment and experienced staff are still required to achieve repeatable results. A reliable, easy to implement, standardized methodology is required. Food processing, enzyme activity and type, incubation procedures, step duration, and the use of an internal reference standard are important considerations for the development of a standardized protocol. The estimation of in vivo GI responses using in vitro starch digestion methods would be particularly useful for high-throughput screening of potato cultivars.

2.5 Effect of Growing Conditions and Storage on Carbohydrates

Not only cooking methods, but the environmental conditions during cultivation; as well as storage conditions, such as temperature and humidity, also affect the nutritional quality of potato. Lamberti et al. (2004) reported variety impacts 65% starch variability, whereas environmental conditions lead to 19% variability in starch content. The carbohydrate content of potatoes not only depends on variety but also on physiological state of tuber and tends to change during tuber development and storage (Vreugdenhil et al. 2007). To study the effect of environment on starch digestibility, Bach et al. (2013) evaluated starch profile (SDSt, RDS, and RS) of 12 potato genotypes comprising elite breeding lines and commercial varieties in six environments. They observed that genotype and environment have a significant effect on all the three types of starches. However, the interaction between genotype and environment was significant for SDSt only. Starch separated from potatoes stored at 4 °C for 14 days showed increased turbidity in gelatinized aqueous suspensions with increasing storage time period (Singh and Singh 2001). Noda et al. (2004) and Kaur et al. (2007) observed that low temperature during potato cultivation affects starch quality. Potatoes cultivated at low temperatures contained starch with higher granule size and relatively lower pasting as well as transition temperatures. Not only storage temperatures, period of storage and climatic conditions also affect the nutritional quality of potato. Late harvest date is reported to enhance the granule size, phosphorus content, peak viscosity, and breakdown and slightly decrease the amylose content, pasting temperature, and gelatinization temperature. Chryssanthopoulos et al. (2016) reported that fried potato slices prepared from two potato cultivars and fried in extra virgin olive oil cultivated under organic fertilization conditions possess similar GI in healthy male volunteers. GI of potato product is affected by maturity stage of tubers. Small and immature tubers are reported to have low GI compared to large and mature potatoes. During potato maturity, amylose content increases non-significantly, but amylopectin branching increased significantly that contributes towards high GI of mature potatoes (Soh and Brand-Miller 1999).

Method of cultivation affects the carbohydrate content of potatoes. Total carbohydrates were reported to be higher in conventionally grown potatoes (88% DW basis) compared to organic potatoes (78% DW basis). Similarly, starch, glucose, and fructose were high in inorganic potatoes and the content was 82, 2.2, and 0.3% in inorganic potatoes and 72, 1.3, and 0.2% in organic potatoes, respectively. However, sucrose was marginally higher in organic potatoes (2.6%) than inorganic potatoes (2.3%) (Carillo et al. 2012). Carbohydrates decreased with increasing dose of potassium. The decrease in carbohydrates was proposed to be due to enhanced moisture content in tubers due to high potassium (de Quadros et al. 2009). Among the organically and conventionally grown potatoes, total carbohydrates, starch, sugars (sucrose, fructose, glucose, and reducing sugars) were higher in conventionally grown potatoes, compared to organic potatoes. Irrigation level at 100% of evaporation (measured with Class A pan between two irrigations) increased the total carbohydrates, starch, fructose, glucose, and reducing sugars, but not sucrose. Total carbohydrates, fructose, glucose, and reducing sugars decreased with increasing nitrogen dose from 50 to 200 kg/ha. Starch and sucrose showed no clear trend with starch being more in 150 kg/ha and least in 50 kg/ha and nitrogen and sucrose being high in 50 kg/ha and least in 150 kg/ha nitrogen dose (Maggio et al. 2008). Total soluble carbohydrates and starch content per plant increased with increasing tuber maturity (Faruk et al. 2004). Magnesium supply at a rate of 0–100 kg/ha in field increased the dry matter by 5–8% and starch by 0.5% (Pobereżny and Wszelaczyńska 2011). High potassium concentration in tubers leads to low reducing sugar accumulation. Potassium deficiency is known to accumulate soluble sugars, organic acids, amino acids and reduce the synthesis of proteins, starch or cellulose (Wang et al. 2013). High phosphorus availability in soil leads to increased tuber phosphorous content that resulted in high amylose along with altered thermal and pasting properties of starch. High phosphorus soil also leads to higher tuber dry matter, low total sugars, and high starch and protein content (Leonel et al. 2016,

2017). Tuber yield and starch yield (g/plant) reduced significantly in potassium deficient potato plants, whereas starch content (% DW) reduced non-significantly (Koch et al. 2020). Gerendas et al. (2007) studied the effect of nitrogen and potassium fertilizer on acrylamide precursor (reducing sugars and asparagine) in potato tubers. They reported a positive relationship between nitrogen supply and reducing sugars at intermediate and high potassium supply. Reducing sugars and free amino acids increased with increasing nitrogen supply. Reducing sugars were maximum at low potassium and intermediate nitrogen supply, whereas increasing potassium supply decreased the reducing sugars. Starch content decreased with increasing potassium supply with low and medium nitrogen supply.

To keep potatoes fresh for longer durations, storage environment plays an important role. Storage at 6–10 °C, in well ventilated, dark and humid room enhances the storage life of potatoes. Storing potatoes at low temperatures inhibits sprouting but also increases the reducing sugar concentration so low temperature storage is not suitable for potatoes destined for processing. High sugar content leads to "Maillard reaction" resulting in browning of potato products along with formation of acrylamide that is a carcinogen (Robertson et al. 2018). Starch to sugar ratio in potatoes is influenced by storage duration and temperature. Fried products prepared from potatoes containing high sugars are dark in color and contain high acrylamide content (Raigond et al. 2018). Researchers are utilizing genetic modification approaches to limit cold sweetening to enable refrigerated storage of potatoes for fresh market and processing (Sowokinos 2001; Rommens et al. 2008; Furrer et al. 2018). Rivero et al. (2003) studied the change in individual carbohydrates and ascorbic acid in five potato cultivars when stored at 12 °C at 90% humidity. It was observed that starch content decreased with storage time while the amount of reducing sugars increased. Increase in amylose/amylopectin ratio was observed upon 6 weeks of storage along with significant decrease in ascorbic acid concentration. After 20 weeks the ascorbic acid concentration was half of original. Properties of starch extracted from freshly harvested tubers were compared with starch extracted from tubers stored at 5, 25, and 55 °C for 7 days before extraction. Although the rest of the properties such as amylose content, amylopectin unit chain distribution, and gelatinization temperature remained almost the same in freshly harvested tubers and tubers stored at 5 and 25 °C; storage at 55 °C resulted in in situ annealing, which can have commercial value to create new food products (Tester et al. 2005). These studies show that environmental conditions and storage have significant impact on potato carbohydrates.

2.6 Health Concern/Benefits

The scientific literature reports both positive and negative effects of potato consumption on human health. Most of the reports linked the negative effect of potato consumption to its high GI. Low GI diets have reduced rate of carbohydrate digestion and absorption and confer several health benefits, such as improved blood glucose control, reduced insulin demand, reduced blood lipid levels in healthy individuals and patients with diabetes and hypertriglyceridemia, improved satiety and increased fermentation in colon (Ludwig 2002; Venn and Green 2007). All these effects are involved in prevention of diseases such as obesity, coronary heart disease, several forms of cancer and diabetes (type II) (Nayak et al. 2014). Bang et al. (2019) tested the effect of raw starches (potato, corn, wheat, and rice) on gut microbiome and metabolome in mice. Mice were fed with different starch substitutes for 16 weeks. Potato starch fed group was observed to have lowest weight gain as well as the highest insulin sensitivity. The gut microbiota of mice fed with raw potato starch showed maximum carbohydrate and energy metabolism compared to all other starch fed mice groups. Highest propionic acid production was also observed in raw potato starch fed mice. The study concluded that among all the tested starches, raw potato starch had a positive impact on intestinal microbe composition and metabolism. The role of high intake of red meat in development of colorectal cancer (CRC) and role of RS to protect against CRC have been reported. Nielsen et al. (2019) investigated the role of high amylose starches (potato, maize, butyrylated high amylose maize starch) to oppose the negative effects caused by red meat intake in rats. They observed that high amylose potato starch and butyrylated high amylose maize starch prevented CRC. The preventive effect of these starches was because of reduced level of colonic oncogenic miR17-92 cluster miRNA expression. Balance et al. (2018) reported co-consumption of cooked broccoli with mashed potato to improve glycemic response and insulinemic response compared to consumption of mashed potato alone. Purple potatoes are rich source of polyphenolics, chlorogenic acid being the predominant polyphenol and derivatives of petunidin, peonidin, malvidin, and pelargonidin as main anthocyanin components. Purple-fleshed tuber extract has the ability to inhibit α-amylase, α-glucosidase, and aldose reductase, thereby have an impact on GI (Kalita et al. 2018). Role of carbohydrate foods such as rice, pasta, and potato was studied on satiety when consumed within standard mixed meals. Human subjects who consumed potato within a mixed meal felt less hungry, fuller and more satisfied than those who consumed rice-containing meals and pasta-containing meals (Zhang et al. 2018). The mechanism underlying this result may be found in a study by Nakajima et al. (2011) which examined the effect of a potato extract on food intake pattern in rats. Potato extract contains almost 20% protein including trypsin inhibitors along with 60% carbohydrates. Food consumption was observed for 6 h postoral administration of potato extract. Murine cholecystokinin (CCK, a satiety hormone) producing cell line STC-1 was used to study CCK-releasing activity. The results showed that oral administration of potato extract suppressed food intake for 1–6 h and showed dose dependent release of CCK in STC-1 cells. Presence of trypsin inhibitors in potatoes is considered to be responsible for the greater satiety effect. Coudray et al. (2001) studied the synergistic effect of RS and inulin (fermentable carbohydrates) on calcium (Ca) and magnesium (Mg) absorption in rat colon. Both of these carbohydrates are fermented in the large intestine and produce short chain fatty acids that help in mineral absorption. Fiber-free basal purified diet (control), diet containing 100 g inulin, diet containing 150 g pure RS (raw potato starch), and blend of inulin (50 g) and RS (75 g) per kg body weight was fed for 21 days. After 14 days, the diet containing inulin and RS increased the intestinal absorption and balance of Ca and Mg significantly compared to control diet. Blend of both fermentable carbohydrates increased the cecal soluble Ca and Mg concentrations and intestinal absorption.

2.7 Genetic Modifications to Alter Carbohydrate Composition

Factors affecting tuber quality include genetic makeup of variety, crop maturity, agronomic practices, environmental conditions, storage temperatures, and biotic and abiotic stresses. To feed a large population, increasing yield and resistance to biotic and abiotic stresses remained the primary objective of potato breeding. Nutritional quality breeding has not gained much attention in the past, however, more recently potato quality properties, such as antioxidant enrichment, high starch, low GI, micronutrient enrichment, and good taste and texture, have started gaining the attention of breeders. To enhance the process of breeding, correct selection of parents and crosses along with efficient selection procedure plays an important role. Main genetic approaches to improve quality are direct selection, traditional breeding or quantitative trait locus (QTL) mapping and subsequent marker assisted selection and transgenic approach (Kumari et al. 2018).

Most of the research on alteration of potato carbohydrates has focused on inhibition of cold induced sweetening to enhance storage life of potatoes and retain processing quality. High sugar accumulation during cold storage of potatoes is the main concern of potato processors. Researchers have utilized biotechnological tools to silence genes involved in cold induced sweetening. Vacuolar acid invertase hydrolyses sucrose (derived from starch degradation) to glucose and fructose during storage, leading to high concentration of these sugars. Menéndez et al. (2002) studied cold induced sweetening using candidate gene approach. Through Restriction Fragment Length Polymorphism (RFLP) and Amplified Fragment Length Polymorphism (AFLP) markers, they developed a QTL and linkage map of two segregating populations that were already evaluated for sugar content after cold storage. Ten potato genes with known map position and unknown role in carbon metabolism/transport were tested for their impact on sugars. Authors reported linkage between glucose, fructose, sucrose QTLs with all candidate gene loci, i.e. AGPaseS, AGPaseB, Sbel, GapC, Invap, Ppa1, Sut1, Sut2. Wiberley-Bradford et al. (2016) studied the temperature dependent regulation of sugar accumulation and expression of genes involved in carbohydrate metabolism during cold storage at 9 and 3°C in wild-type tubers and tubers with low vacuolar acid invertase expression (reduced expression with RNA interference). During cold storage, expression of vacuolar acid invertase and β-amylase increased first, followed by significant decrease in expression of AGPase and granule-bound starch synthase (GBBS). Further significant increase in β-amylase expression was observed when temperature reached 3–5°C. With higher expression of vacuolar acid invertase, large increase in sucrose, glucose, and fructose was observed. Suppression of vacuolar acid invertase did not affect the expression levels of β amylase, AGPase, and GBBS. These results suggest that temperature dependent accumulation of sucrose leads to accumulation of reducing sugars. Sucrose accumulated during storage due to degradation of starch caused by β-amylase and also reduced starch re-synthesis due to action of AGPase and GBBS. Baroja-Fernández et al. (2009) enhanced the activity of sucrose synthase in transgenic potatoes to achieve high starch content, ADPglucose, UDPglucose and total yield. Sucrose synthase, a cytosolic enzyme, is involved in the catalysis of sucrose and a nucleoside diphosphate into corresponding nucleoside diphosphate glucose and fructose. The sucrose synthase activity and sugars were measured from sucrose synthase over-expressing potato plants. Tubers with over-expressed sucrose synthase contained significantly high starch content, UDP glucose, and ADP glucose also higher tuber DW, starch content per plant and total yield compared to control plants. Their results revealed that changes in sucrose synthase activity affected the expression of genes involved in multiple biological processes, instead of those directly involved in starch metabolism. Therefore, to increase starch accumulation and yield, sucrose synthase plays an important role. Transgenic potato lines were prepared with over expression of sucrose non-fermenting-1-related protein kinase-1 gene (SnRK1). In three independent transgenic lines activity of this gene increased by 55–167%, glucose decreased by 17–56%, and starch increased by 23–30% compared to wild type. In transgenic lines expression of gene encoding sucrose synthase and ADP-glucose pyrophosphorylase (key enzymes in starch biosynthetic pathway) increased, whereas α-amylase and sucrose phosphate synthase remained unaltered. The results showed the specific role of SnRK1 in controlling starch accumulation in potato tubers (McKibbin et al. 2006). Niu et al. (2019) observed a positive correlation between starch accumulation and organelle/nuclear DNA ratios. Their study suggested that organelle/nuclear DNA ratios can act as a biomarker for development of potatoes with improved dry matter and starch content in breeding programmes. They observed an increase in dry matter and organelle/nuclear ratio during tuber bulking. Dry matter and starch content showed positive correlation with plastid/nuclear DNA ratios and mitochondrial/nuclear DNA ratios. Plastid/nuclear DNA ratios and mitochondrial/nuclear DNA ratios can be used as a biomarker for manipulation in dry matter and starch content in potatoes. Krunic et al. (2018) transferred the recessive allele IAm (increased amylose) from wild potatoes (*Solanum sandemanii*) to modern tetraploid *Solanum tuberosum* by marker assisted crossing. Amylose was 28–59% higher in IAm starch with reduced rate of digestion and higher RS content compared to control. Andersson et al. (2006) used RNA interference technique to inhibit Sbe1 and Sbe2 genes for production of high amylose potato lines. They cloned two constructs as inverted repeats controlled by potato granule-bound starch synthase promoter with 100bp segments (pHAS2) and 200bp segments (pHAS3) of both branching enzyme genes. Construct pHAS3 resulted in more than 50% transgenic lines with high amylose quality compared to antisense construct that resulted in only 3% transgenic lines with high amylose. Mitchell et al. (2017) generated transgenic potatoes that accumulated 3.3% triacylglycerol (TAG) on dry weight basis and studied the effect of TAG accumulation on nutritional and processing quality of high oil tubers. They reported positive impact of TAG accumulation by increasing energy density, total nitrogen, amino acids, organic acids, and inorganic phosphate, whereas it negatively impacted starch properties such as decreased amylose and phosphate content, reduced peak viscosity, and increased gelatinization temperature. In silico annotation of potato genome sequence has shown that 123 loci are involved in starch-sugar inter-conversion. Single nucleotide polymorphisms (SNPs) were identified in eight genes having key role in starch-sugar inter-conversion through candidate gene association mapping. Schreiber et al. (2014) cloned and characterized an allele of plastidic starch phosphorylase PHO1a that is associated with increased tuber starch content. This allele has several amino acid changes, starch/glycogen phosphorylases being the unique one. This change can be responsible for limited starch breakdown due to reduced enzyme activity caused by impaired active dimers formation (Schreiber et al. 2014). Amylopectin has been known to provide better thickening than amylose in food and textile applications. A potato variety namely "Karnicol" has been developed through starch genetic modification that leads to production of amylose free potatoes. Antisense RNA-mediated inhibition of GBBS resulted in production of amylose free potatoes (Kumari et al. 2018). Genetic engineering has been used to develop starch with desirable amylose: amylopectin ratio. Reduction in ADP-glucose pyrophosphorylase activity through antisense technology developed transgenic lines with significantly reduced amylose. In transgenic lines, along with reduced amylose, amylopectin accumulated shorter chains and starch granule size gets reduced (Lloyd et al. 1999). Schwall et al. (2000) simultaneously inhibited two isoforms of starch branching enzymes (SBE A and SBE B) using antisense technology to below 1% of wild-type activity. With this, they reported transgenic lines with increased amylose. Amylose content in one of the lines was 75% of total starch, while it was more than 40% in 19 lines. Chen et al. (2001) evaluated CAPS (cleaved amplified polymorphic sequence), SCAR (sequence characterized amplified region), and RFLP marker for 69 functional genes (already characterized) and identified 85 genetic loci that covered considerable potato genome. Their study allowed utilizing a candidate gene approach for studying starch and sugar related agronomic traits in potato. They observed correlations between map positions of 14QTLs for tuber starch content and function-related loci by comparing the QTL map for starch with molecular-function map.

2.8 Conclusion/Future Directions

Potato is a rich source of starch and starch is not only most important source of available carbohydrates but also has indigestible carbohydrates with varying health benefits. Potato RS has been shown to have prebiotic effects in animal and in vitro studies. This suggests the development of new potato varieties with enhanced formation of RS after cooking that will produce potatoes with greater health benefits. As amylose is directly proportional to RS content, breeding or molecular approaches should target the development of high amylose potato varieties. Cooking and post-cooking storage have significant effects on the GI of potatoes. A greater focus is required on the nutritional quality of the carbohydrate in potatoes, including high RS and more slowly digested lower GI starch, during development and selection of new potato cultivars that may yield benefits for overall human health.

References

Åkerberg AKE, Liljeberg HGM, Granfeldt YE et al (1998) An *in vitro* method, based on chewing, to predict resistant starch content in foods allows parallel determination of potentially available starch and dietary fiber. J Nutr 128:651–660

Andersson M, Melander M, Pojmark P et al (2006) Targeted gene suppression by RNA interference: an efficient method for production of high-amylose potato lines. J Biotechnol 123 (2):137–148

Araya H, Contreras P, Alviña M et al (2002) A comparison between an *in vitro* method to determine carbohydrate digestion rate and the glycemic response in young men. Eur J Clin Nutr 56:735–739

Atkinson FS, Foster-Powell K, Brand-Miller JC (2008) International tables of GI and GL values: 2008. Diabetes Care 31(12):2281–2283

Bach S, Yada RY, Bizimungu B et al (2013) Genotype by environment interaction effects on starch content and digestibility in potato (*Solanum tuberosum* L.). J Agric Food Chem 61 (16):3941–3948

Balance S, Knutsen SH, Fosvold ØW et al (2018) Glyceamic and insulinaemic response to mashed potato alone, or with broccoli, broccoli fibre or cellulose in healthy adults. Eur J Nutr 57:199

Bang SJ, Lee ES, Song EJ et al (2019) Effect of raw potato starch on the gut microbiome and metabolome in mice. Int J Biol Macromol 133:37–43

Baroja-Fernández E, Muñoz FJ, Montero M et al (2009) Enhancing sucrose synthase activity in transgenic potato (*Solanum tuberosum* L.) tubers results in increased levels of starch, ADPglucose and UDPglucose and total yield. Plant Cell Physiol 150(9):1651–1662

Bavaneethan Y, Vasantharuba S, Balakumar S, Thayananthan K (2015) Effect of different processing time on resistant starch content of selected tubers. World J Agric Sci 11(4):244–246

Bembem K, Sadana B (2012) Effect of cooking methods on the nutritional composition and antioxidant activity of potato tubers. Int J Food Nutr Sci 2(4):26–30

Bordoloi A, Kaur L, Singh J (2012) Parenchyma cell microstructure and textural characteristics of raw and cooked potatoes. Food Chem 133(4):1092–1100

Brighenti F, Pellegrini N, Casiraghi MC et al (1995) *In vitro* studies to predict physiological effects of dietary fibre. Eur J Clin Nutr 49(Suppl 3):S81–S88

Carillo P, Cacace D, Pascale SD et al (2012) Organic vs traditional potato powder. Food Chem 133 (4):1264–1273

Chen X, Salamini F, Gebhardt C (2001) A potato molecular-function map for carbohydrate metabolism and transport. Theor Appl Genet 102:284–295

Chen JY, Miao Y, Zhang H et al (2004) Non-destructive determination of carbohydrate content in potatoes using near infrared spectroscopy. J Near Infrared Spec 12(5):311–314

Chryssanthopoulos C, Varzakas T, Tampaki M et al (2016) Glycemic index of two different fried potato varieties cultivated under organic fertilization conditions. Curr Res Nutr Food Sci 4 (1):09–15

Coudray C, Younes H, Bellanger J et al (2001) Effect of two fermentable carbohydrates (inulin and resistant starch) and their combination on calcium and magnesium balance in rats. Br J Nutr 86 (4):479–485

Cummings JH, Englyst HN (1995) Gastrointestinal effects of food carbohydrate. Am J Clin Nutr 61:938S–945S

Darwiche G, Stman EM, Liljeberg HGM, Kallinen N, Björgell O, Björck IME, Almér L-O (2001) Measurements of the gastric emptying rate by use of ultrasonography: studies in humans using bread with added sodium propionate. Am J Clin Nutr 74:254–258 de Quadros DA, Iung MC, Ferreira SMR et al (2009) Chemical composition of potato tubers for processing, grown at different levels and sources of potassium. Food Sci Technol 29 (2):316–323

de Vasconcelos NCM, Salgado SM, Livera AVS et al (2015) Influence of heat treatment on the sensory and physical characteristics and carbohydrate fractions of french-fried potatoes (*Solanum tuberosum* L.). Food Sci Technol (Campinas) 35(3):561–569

Dubois M, Gilles KA, Hamilton JK et al (1956) Colorimetric method for determination of sugars and related substances. Anal Chem 28(3):350–356

Ek KL, Wang S, Brand-Miller JC et al (2014a) Properties of starch from potatoes differing in glycemic index. Food Funct 10:2509–2515

Ek KL, Wang S, Copeland L, Brand-Miller JC (2014b) Discovery of a low-glycaemic index potato and relationship with starch digestion *in vitro*. Br J Nutr 111:699–705

Englyst H, Wiggins HS, Cummings JH (1982) Determination of the non-starch polysaccharides in plant foods by gas-liquid chromatography of constituent sugars as alditol acetates. Analyst 107:307

Englyst HN, Kingman SM, Cummings JH (1992) Classification and measurement of nutritionally important starch fractions. Eur J Clin Nutr 46(Suppl 2):S33–S50

Englyst HN, Veenstra J, Hudson GJ (1996) Measurement of rapidly available glucose (RAG) in plant foods: a potential in vitro predictor of the glycaemic response. Brit J Nutr 75:327–337

Englyst KN, Englyst HN, Hudson GJ et al (1999) Rapidly available glucose in foods: an in vitro measurement that reflects the glycemic response. Am J Clin Nutr 69:448–454

Englyst KN, Vinoy S, Englyst HN et al (2003) Glycaemic index of cereal products explained by their content of rapidly and slowly available glucose. Br J Nutr 89:329–339

Eroglu E, Buyuktuncer Z (2017) The effect of various cooking methods on resistant starch content of foods. Nutr Food Sci 47(4):522–533

FAO (1998) Carbohydrates in human nutrition. Report of a joint FAO/WHO expert consultation. FAO Food Nutr Pap 66:1–140

Faruk AA, Kirichenko EB, Voronkova TV et al (2004) Changes in carbohydrate and polyphenolics pools during potato tuber maturation: cultivar-specific development of resistance to pathogens. Dokl Biol Sci 396:240–242

Fernandes G, Velangi A, Wolever TMS (2005) Glycemic index of potatoes commonly consumed in North America. J Am Diet Assoc 105(4):557–562

Furrer AN, Chegeni M, Ferruzzi MG (2018) Impact of potato processing on nutrients, phytochemicals and human health. Crit Rev Food Sci Nutr 58(1):146–168

García-Alonso A, Goñi I (2000) Effect of processing on potato starch: *In vitro* availability and glycaemic index. Nahrung – Food 44:19–22

Gerendas J, Heuser F, Sattelmacher B (2007) Influence of nitrogen and potassium supply on contents of acrylamide precursors in potato tubers and on acrylamide accumulation in French fries. J Plant Nutr 30(9):1499–1516

Goni L, Garcia-Diamanas E, Saura-Calixto F (1996) Analysis of resistant starch: a method for foods and food products. Food Chem 56(4):455–459

Goni I, Garcia-AIonso A, Saura-Calixto F (1997) A starch hydrolysis procedure to estimate glycemic index. Nutr Res 17(3):427437.1

Granfeldt Y, Björck I (1991) Glycemic response to starch in pasta: a study of mechanisms of limited enzyme availability. J Cereal Sci 14:47–61

Granfeldt Y, Björck I, Drews A et al (1992) An *in vitro* procedure based on chewing to predict metabolic response to starch in cereal and legume products. Eur J Clin Nutr 46:649–660

Granfeldt Y, Liljeberg H, Drews A, Newman R, Björck I (1994) Glucose and insulin responses to barley products: influence of food structure and amylose-amylopectin ratio. Am J Clin Nutr 59 (5):1075–1082

Granfeldt Y, Drews A, Björck I (1995) Arepas made from high amylose corn flour produce favorably low glucose and insulin responses in healthy humans. J Nutr 125(3):459–465

Haase NU (2015) The *in vitro* digestibility of carbohydrates in boiled and processed potatoes. Potato Res 58(2):91–102

Hatonen KA, Virtamo J, Eriksson JG et al (2011) Protein and fat modify the glycaemic and insulinaemic responses to a mashed potato-based meal. Br J Nutr 106:248–253

Hedge JE, Hofreiter BT (1962) In: Whistler RI, Be Miller JN (eds) Carbohydrate chemistry. Academic Press, New York, p 17

Henry CJK, Lightowler HJ, Strik CM et al (2005) Glycaemic index values for commercially available potatoes in Great Britain. Br J Nutr 94(6):917–921

Henry CJK, Lightowler HJ, Kendall FL et al (2006) The impact of the addition of toppings/fillings on the glycaemic response to commonly consumed carbohydrate foods. Eur J Clin Nutr 60:763–769

Hernandez-Hernandez O, Olano A, Rastall RA, Moreno FJ (2019) In vitro digestibility of dietary carbohydrates: toward a standardized methodology beyond amylolytic and microbial enzymes. Front Nutr 6:61

ISO: 26642 (2010) Food products-determination of the glycaemic index (GI) and recommendation for food classification. ISO, Geneva

Jenkins DJA, Wolever TMS, Taylor RH et al (1981) Glycemic index of foods: a physiological basis for carbohydrate exchange. Am J Clin Nutr 34:362–366

Jenkins DJA, Wolever TMS, Thorne MJ et al (1984) The relationship between glycemic response, digestibility, and factors influencing the dietary habits of diabetics. Am J Clin Nutr 40:1175–1191

Juliano BO (1971) A simplified assay for milled-rice amylose. Cereal Sci Today 16:334–360

Kalita D, Holm DG, La Barbera DV et al (2018) Inhibition of α-glucosidase, α-amylase, and aldose reductase by potato polyphenolic compounds. PLoS One 13(1):1–21

Karlsson ME, Leeman AM, Bjorck IME et al (2007) Some physical and nutritional characteristics of genetically modified potatoes varying in amylose/ amylopectin ratio. Food Chem 100:136–146

Kaur A, Singh N, Ezekiel R et al (2007) Physicochemical, thermal and pasting properties of starches separated from different potato cultivars grown at different locations. Food Chem 101 (2):643–651

Koch M, Naumann N, Pawelzik E (2020) The importance of nutrient management for potato production part I: plant nutrition and yield. Potato Res 63(1):97–119

Krunic SL, Skryhan K, Mikkelsen L et al (2018) Non-GMO potato lines with an altered starch biosynthesis pathway confer increased-amylose and resistant starch properties. Starch 70 (1–2):1600310

Kumar A, Sahoo U, Baisakha B et al (2018) Resistant starch could be decisive in determining the glycemic index of rice cultivars. J Cereal Sci 79:348–353

Kumari M, Kumar M, Solankey SS (2018) Breeding potato for quality improvement. In: Potato-From Incas to all over the world. Intech Open, London, pp 37–59

Lamberti M, Geiselmann A, Conde-Petit B et al (2004) Starch transformation and structure development in production and reconstitution of potato flakes. LWT Food Sci Technol 37 (4):417–427

Leeman M, Ostman E, Bjorck I (2005) Vinegar dressing and cold storage of potatoes lowers postprandial glycaemic and insulinaemic responses in healthy subjects. Eur J Clin Nutr 59:1266–1271

Leeman M, Ostman E, Bjork I (2008) Glycaemic and satiating properties of potato products. Eur J Clin Nutr 62:87–95

Leinonen K, Liukkonen K, Poutanen K et al (1999) Rye bread decreases postprandial insulin response but does not alter glucose response in healthy Finnish subjects. Eur J Clin Nutr 53:262–267

Leonel M, Carmo EL, Fernandes AM et al (2016) Physico-chemical properties of starches isolated from potato cultivars grown in soils with different phosphorus availability. J Sci Food Agric 96:1900–1905

Leonel M, do Carmo EL, Fernandes AM et al (2017) Chemical composition of potato tubers: the effect of cultivars and growth conditions. J Food Sci Technol 54:2372–2378
Liatis S, Grammatikou S, Poulia KA et al (2010) Vinegar reduces postprandial hyperglycaemia in patients with type II diabetes when added to a high, but not to a low, glycaemic index meal. Eur J Clin Nutr 64(7):727–732
Liljeberg HGM, Björck IME (1998) Delayed gastric emptying rate may explain improved glycaemia in healthy subjects to a starchy meal with added vinegar. Eur J Clin Nutr 52:368–371
Liljeberg H, Granfeldt Y, Björck I (1992) Metabolic responses to starch in bread containing intact kernels versus milled flour. Eur J Clin Nutr 46:561–575
Lloyd JR, Springer F, Buléon A et al (1999) The influence of alterations in ADP-glucose pyrophosphorylase activities on starch structure and composition in potato tubers. Planta 209 (2):230–238
Ludwig DDS (2002) The glycemic index - physiological mechanisms relating to obesity, diabetes, and cardiovascular disease. JAMA 287:2414–2423
Maggio A, Carillo P, Bulmetti GS et al (2008) Potato yield and metabolic profiling under conventional and organic farming. J Agron 28:343–350
Mareček J, Frančáková H, Bojňanská T et al (2013) Carbohydrates in varieties of stored potatoes and influence of storage on quality of fried products. J Microbiol. Biotechnol Food Sci 2(Special issue 1):1744–1753
Mccready RM, Guggolz J, Silviera V et al (1958) Determination of starch and amylose in vegetables. Anal Chem 22:1156
McKibbin RS, Muttucumaru N, Paul MJ et al (2006) Production of high-starch, low-glucose potatoes through over-expression of the metabolic regulator SnRK1. Plant Biotechnol J 4 (4):409–418
Menéndez CM, Ritter E, Schäfer-Pregl R et al (2002) Cold sweetening in diploid potato: mapping quantitative trait loci and candidate genes. Genetics 162(3):1423–1434
Miller GL (1956) Use of Dinitrosalicylic acid reagent for determination of reducing sugars. Anal Chem 31(3):426–428
Miller CK, Gabbay RA, Dillon J et al (2006) The effect of three snack bars on glycemic response in healthy adults. J Am Diet Assoc 106:745–748
Mishra S, Monro J, Hedderley D (2008) Effect of processing on slowly digestible starch and resistant starch in potato. Starch-Starke 60(9):500–507
Mitchell M, Pritchard J, Okada S et al (2017) Oil accumulation in transgenic potato tubers alters starch quality and nutritional profile. Front Plant Sci 8:554
Monro J, Mishra S, Blandford E et al (2009) Potato genotype differences in nutritionally distinct starch fractions after cooking, and cooking plus storing cool. J Food Compos Anal 22:539–545
Moser S, Aragona I, Furrer A et al (2018) Potato phenolics impact starch digestion and glucose transport in model systems but translation to phenolic rich potato chips results in only modest modification of glycemic response in humans. Nutr Res 52:57–70
Mulinacci N, Ieri F, Giaccherinic C et al (2008) Effect of cooking on the anthocyanins, phenolic acids, glycoalkaloids and resistant starch content in two pigmented cultivars of *Solanum tuberosum* L. J Agric Food Chem 56:11830–11837
Murniece I, Karlina D, Galoburda R et al (2011) Nutritional composition of freshly harvested and stored Latvian potato (*Solanum tuberosum* L.) varieties depending on traditional cooking methods. J Food Compos Anal 24:699–610
Najjar N, Adra N, Hwalla N (2004) Glycemic and insulinemic responses to hot vs cooled potato in males with varied insulin sensitivity. Nutr Res 24(12):993–1004
Nakajima S, Hira T, Tsubata M et al (2011) Potato extract (Potein) suppresses food intake in rats through inhibition of luminal trypsin activity and direct stimulation of cholecystokinin secretion from enteroendocrine cells. Agric Food Chem 59(17):9491–9496
Nayak B, De J, Berrios J, Tang J (2014) Impact of food processing on the glycemic index (GI) of potato products. Food Res Int 56:35–46

Nielsen TS, Bendiks Z, al TB (2019) High-amylose maize, potato, and butyrylated starch modulate large intestinal fermentation, microbial composition, and oncogenic miRNA expression in rats fed a high-protein meat diet. Int J Mol Sci 20(9):2137

Niu S, Zhang G, Li X et al (2019) Organelle DNA contents and starch accumulation in potato tubers. Theor Appl Genet 132:205–216

Noda T, Tsuda S, Mori M et al (2004) The effect of harvest dates on the starch properties of various potato cultivars. Food Chem 86(1):119–125

Ostman E, Granfeldt Y, Persson L, Björck I (2005) Vinegar supplementation lowers glucose and insulin responses and increases satiety after a bread meal in healthy subjects. Eur J Clin Nutr 59 (9):983–988

Ozturk S, Koksel H, Ng PKW (2009) Characterization of resistant starch samples prepared from two high amylose maize starches through debranching and heat treatments. Cereal Chem 86:503–510

Pinhero RG, Waduge RN, Liu Q et al (2016) Evaluation of nutritional profiles of starch and dry matter from early potato varieties and its estimated glycemic impact. Food Chem 203:356–366

Pobereżny J, Wszelaczyńska E (2011) Effect of bioelements (N, K, Mg) and long-term storage of potato tubers on quantitative and qualitative losses. Part II. Content of dry matter and starch. J Elementol 16:237–246

Raatz SK, Idso L, Johnson LK et al (2016) Resistant starch analysis of commonly consumed potatoes: content varies by cooking method and service temperature but not by variety. Food Chem 208:297–300

Raigond P, Ezekiel R, Kaundal B (2014) Starch fractions of cooked potatoes at low temperature. Potato J 41(1):58–67

Raigond P, Ezekiel R, Raigond B (2015) Resistant starch in food: a review. J Sci Food Agric 95:1968–1978

Raigond P, Mehta A, Singh B (2018) Sweetening during low temperature and long term storage of Indian potatoes. Potato Res 61:207–217

Ramdath DD, Padhi E, Hawke A et al (2014) The glycemic index of pigmented potatoes is related to their polyphenol content. Food Funct 5(5):909–915

Ramírez C, Millon C, Nuñez H et al (2015) Study of effect of sodium alginate on potato starch digestibility during *in vitro* digestion. Food Hydrocol 44:328–332

Rivero RC, Rodríguez ER, Romero CD (2003) Effects of current storage conditions on nutrient retention in several varieties of potatoes from Tenerife. Food Chem 80(4):445–450

Robertson TM, Alzaabi AZ, Robertson MD et al (2018) Starchy carbohydrates in a healthy diet: the role of the humble potato. Nutrients 10:1764

Rommens CM, Yan H, Swords K et al (2008) Low acrylamide French fries and potato chips. Plant Biotechnol J 6:843–853

Ross SW, Brand JC, Thorburn AW, Truswell AS (1987) Glycemic index of processed wheat products. Am J Clin Nutr 46(4):631–635

Schafer G, Schenk U, Ritzel U et al (2003) Comparison of the effects of dried peas with those of potatoes in mixed meals on postprandial glucose and insulin concentrations in patients with type 2 diabetes. Am J Clin Nutr 78:99–103

Schreiber L, Nader-Nieto AC, Schönhals EM et al (2014) SNPs in genes functional in starch-sugar interconversion associate with natural variation of tuber starch and sugar content of potato (*Solanum tuberosum* L.). G3 Genes Genom Genet 4:1797–1811

Schwall GP, Safford R, Westcott RJ et al (2000) Production of very-high-amylose potato starch by inhibition of SBE A and B. Nat Biotech 18:551–554

Schwingshackl L, Schwedhelm C, Hoffmann G et al (2019) Potatoes and risk of chronic disease: a systematic review and dose–response meta-analysis. Eur J Nutr 58:2243–2251

Singh J, Singh N (2001) Studies on the morphological, thermal and rheological properties of starch separated from some Indian potato cultivars. Food Chem 75(1):67–77

Soh NL, Brand-Miller J (1999) The glycemic index of potatoes: the effect of variety, cooking method and maturity. Eur J Clin Nutr 53:249–254

Somogyi M (1952) Notes on sugar determination. J Biol Chem 195:19
Southgate DAT (1969a) Determination of carbohydrates in foods. I.—available carbohydrate. J Sci Food Agric 20:326–330
Southgate DAT (1969b) Determination of carbohydrates in foods. II.—unavailable carbohydrate. J Sci Food Agric 20:331–335
Sowokinos JR (2001) Biochemical and molecular control of cold induced sweetening in potatoes. Am J Potato Res 78:221–236
Tahvonen R, Hietanena RM, Sihvonena J et al (2006) Influence of different processing methods on the glycemic index of potato (Nicola). J Food Compos Anal 19:372–378
Tappy L, Gügolz E, Würsch P (1996) Effects of breakfast cereals containing various amounts of beta-glucan fibers on plasma glucose and insulin responses in NIDDM subjects. Diabetes Care 19(8):831–834
Tester RF, Ansell R, Snape CE et al (2005) Effects of storage temperatures and annealing conditions on the structure and properties of potato (*Solanum tuberosum*) starch. Int J Biol Macromol 36(1–2):1–8
Turnbull CM, Baxter AL, Johnson SK (2005) Water-binding capacity and viscosity of Australian sweet lupin kernel fibre under *in vitro* conditions simulating the human upper gastrointestinal tract. Int J Food Sci Nutr 56:87–94
Venn BJ, Green TJ (2007) Glycemic index and glycemic load: measurement issues and their effect on diet–disease relationships. Eur J Clin Nutr 61:S122–S131
Vreugdenhil D et al (2007) Potato biology and biotechnology: advances and perspectives. Elsevier, Amsterdam, p 857
Wang M, Zheng Q, Shen Q et al (2013) The critical role of potassium in plant stress response. Int J Mol Sci 14:7370–7390
Wiberley-Bradford AE, Busse JS, Bethke PC (2016) Temperature-dependent regulation of sugar metabolism in wild-type and low-invertase transgenic chipping potatoes during and after cooling for low-temperature storage. Postharvest Biol Technol 115:60–71
Woolnough JW, Monro JA, Brennan CS et al (2008) Simulating human carbohydrate digestion *in vitro*: a review of methods and the need for standardisation. Int J Food Sci Technol 43:2245–2256
Yadav BS (2011) Effect of frying, baking and storage conditions on resistant starch content of foods. Br Food J 113:710–719
Yang Y, Achaerandio I, Pujolà M (2016) Effect of the intensity of cooking methods on the nutritional and physical properties of potato tubers. Food Chem 197:1301–1310
Yoon JH, Thompson LU, Jenkins DJA (1983) The effect of phytic acid on in vitro rate of starch digestibility and blood glucose response. Am J Chin Nutr 38:835–842
Zhang Z, Venn BJ, Monro J et al (2018) Subjective satiety following meals incorporating rice, pasta and potato. Nutrients 10(11):1739
Zhao X, Andersson M, Andersson R (2018) Resistant starch and other dietary fiber components in tubers from a high-amylose potato. Food Chem 251:58–63

Dietary Fibres in Potato

3

Milan Kumar Lal, Awadhesh Kumar, Ashok Kumar, Pinky Raigond, Augustine Okpani Oko, Nitasha Thakur, Vandana Parmar, Asha Thakur, and Brajesh Singh

Abstract

The dietary fibre (DF) is both starch and non-starch polysaccharide of our diet having a significant effect on gut physiology. It enhances the fermentation process, nutrifies gut microflora, enriches gut microbiome, and improves large bowel function. The recommended dose of DF in our diet is 20–35 g/day, which shows that how much important is the role of DF in our diet. The DF consumption not only lowers the effect of glycaemic response but also provides a protective role in large bowel cancer, diabetes, and coronary heart diseases. The DF can be classified majorly into two types, viz., water-soluble DF and water-insoluble DF. Water-soluble DF includes gum, pectin, and mucilage, whereas water-insoluble DF includes cellulose, hemicellulose, and lignin. Cooked potatoes are a good dietary source of carbohydrates which contains about 1.8% of total dry weight. Consumption of dietary fibre can confer benefits to heart health by influencing both lipid and glucose metabolism.

M. K. Lal (✉) · P. Raigond · N. Thakur · V. Parmar · A. Thakur · B. Singh
ICAR-Central Potato Research Institute, Shimla, Himachal Pradesh, India
e-mail: milan2925@gmail.com

A. Kumar
ICAR-National Rice Research Institute, Cuttack, Odisha, India

A. Kumar
ICAR-Directorate of Onion and Garlic Research, Pune, Maharashtra, India

A. O. Oko
Department of Biotechnology, Ebonyi State University, Abakaliki, Nigeria

P. Raigond et al. (eds.), *Potato*, https://doi.org/10.1007/978-981-15-7662-1_3

Keywords

Dietary fibre · Types · Cooking · Storage · Health benefits · Genetic modification

3.1 Introduction

Potato is an important vegetable crop that is gaining importance in terms of consumption and industrial uses. Due to the presence of diverse nutritional components and a rich source of carbohydrate, it is consumed as a staple food in European countries and as a staple vegetable in the Indian subcontinent. It is the third important and popular food crop after rice and wheat, grown in every continent of the world except Antarctica (Birch et al. 2012). Potato contains around 75–80% starch on a dry weight basis (Liu et al. 2003). Besides being a rich source of carbohydrate, it also contains a substantial amount of protein, ascorbic acid, phenolic substances, vitamins, and minerals. It is highly rich in minerals like potassium, phosphorus, and calcium (Tian et al. 2016). The consumers believe that potatoes are rich in calories and fat as compared to other carbohydrate sources. But this assumption is incorrect since potato is very negligible in fat and have very low energy density similar to legumes (Priestley 2006). In earlier days, the sailors in the sea used to take potato with them and consume it to prevent scurvy due to its high vitamin C content (Camire et al. 2009; Camire 2016). Despite being a high starch crop, it synthesizes more dry matter and protein per unit growing area as compared to the cereals (Razdan 2005).

Potato is rich in starch and its consumption provides carbohydrates, which are the primary source of energy for the human body. Its starch quality and quantity is different from cereals. Starch is composed of two polysaccharides, viz., amylose (linear chain of α-D-glucose) and amylopectin (highly branched chain of α-D-glucose) (Jansen et al. 2001; Singh et al. 2010; Dhital et al. 2011; Li et al. 2019; Chen et al. 2019). Amylose and amylopectin are packed in a specific pattern and form starch granules that possess a unique internal hierarchical structure. This structure may range from 20 μm to 400 μm in size (Wang et al. 2015). Amylose content in potato ranges from 15 to 33% on the total starch basis and balance is amylopectin (Dhital et al. 2011; Lin Ek et al. 2014; Lovedeep and Jaspreet 2016; Dupuis and Liu 2019; Chen et al. 2019). Depending upon the quality of starch, it is used for consumption and industrial purposes. For instance, potato starch rich in phosphate content with very large granules is used for the manufacturing of high-quality paper and generation of viscous hydrocolloid system (Wiesenborn et al. 1994; Blennow et al. 2003). The starch granules and its hydrolysis may be affected by physiochemical treatments, chemical modification, thermal processing, enzymatic action, and hydrolysis in the gastrointestinal tract (Wani et al. 2013; Chen et al. 2019). The raw potato is highly resistant to digestion in the human stomach due to the abundance of resistant starch (RS). Cooking/processing alter the structure of starch making it accessible to digestive enzymes that lead to a reduction in RS content. During cooking the starch granule structure is destroyed that results in exposure of starch component and hence increase the glycaemic index (GI) of cooked potato. The GI of cooked potato ranges from 80 to 150 (white bread is taken as reference) (Dupuis et al. 2016).

Dietary fibre (DF) is the fraction of plant material that resists digestion by the hydrolytic action of enzymes present in the human gut. The DF is categorized as cellulose, hemicellulose, fructooligosaccharides, mucilages, natural gum, pectin, and non-carbohydrates. However, the most consistent definition of DF was given by Trowell et al. (1985) that "DF consists of remnants of plant cells resistant to hydrolysis (digestion) by the alimentary enzymes of man", whose components are cellulose, hemicellulose, lignin, oligosaccharides, pectins, gum, and waxes. The DF is necessary for the digestion process that enhances the bowel movement and reduces the chance of constipation. American Association of Cereal Chemists (AACC) defined DF as "the edible form of a plant part or analogous carbohydrate which is completely or partially fermented in the large intestine and that are resistant to digestion in the small intestine". The recommendation of DF for men is higher 38 g/day than 25 g/day for women (Slavin 2005).

The RS is the fraction of starch which is resistant to the action of amylolytic enzymes present in the digestive system. This RS may be produced by subjecting food to heat, dehydration or some treatment where the ordered structure of the starch molecule is changed, making it less susceptible to digestive enzymes (Thed and Phillips 1995). The starch granules in the native state (raw) contain a high level of SDSt and RS which is rapidly converted to RDS after cooking or processing (Zhang et al. 2006). The digestibility, gelatinization temperature, and viscosity of potato depend upon the ratio of amylose and amylopectin. These properties also depend upon the distribution of amylose chain length, degree of branching, and size of amylopectin branching (Leeman et al. 2008). It acts as a functional food that provides physiological benefit for humans and also reduces the risk of chronic diseases (Raigond et al. 2015).

Food must be healthy, tasty nutritious and should have health-promoting attributes. These kinds of foods are also known as a functional food (García-Segovia et al. 2007; Sanz et al. 2008). Functional food is of great importance to humans due to its potential physiological and therapeutic benefit to reduce the risk of chronic diseases that are caused by other food product and sedentary lifestyle (Fuentes-Zaragoza et al. 2010). It may be the whole or a component of food that has a positive health benefit or can eliminate the potential disease-causing effect of a given food product. The DF is one of the components of functional food which have a profound positive impact on human health. The non-soluble fibre like RS is widely used as a functional ingredient or used as functional food mainly in the products aimed to produce DF rich food (Mikulíová et al. 2008).

3.2 Types of DF

The DF is the component of the plant which is not easily digested by the enzymatic action of the intestinal enzyme and escapes during the process of digestion. It includes molecule like starch (RS), cellulose, hemicellulose, oligosaccharides,

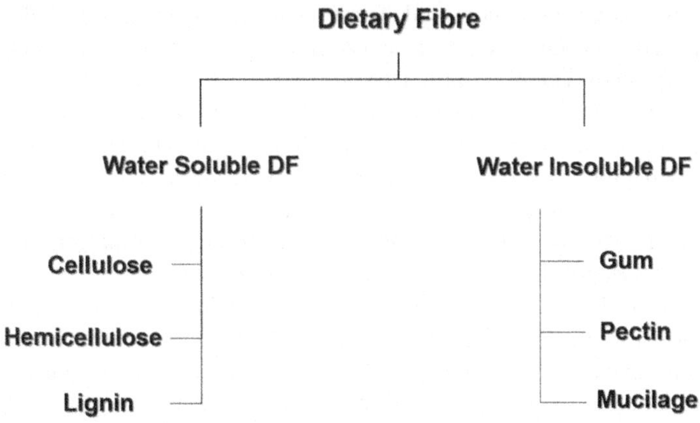

Fig. 3.1 Classification of dietary fibres based on solubility

gum, mucilages, pectin, and non-carbohydrate components of lignin (Theander et al. 1995). Along with potato flesh, potato peels are also a potential source of DF (Lazarov and Werman 1996; Camire 2016). The DF is categorized in soluble and insoluble fibres. The categorization of DF was done based on water solubility and fermentability. Water-insoluble fibres include cellulose, hemicellulose, and lignin (Fig. 3.1). These fibres are less fermented in the human digestive system. The DF like oligosaccharides, pectin, gums, and mucilages is water-soluble and can be well fermented in the small intestine (Dhingra et al. 2012). However, DF has some physicochemical properties like hydration, oil holding capacity, swelling capacity, stability, and emulsifying activity (Xie et al. 2016; 2017).

3.2.1 Resistant Starch

Potato's starch is categorized on their pace of digestion as rapidly digestible starch (RDS), slowly digestible starch (SDSt), and RS. RDS is highly digestible and hydrolyzes within 20 min. The SDSt is medium in digestion and hydrolyses within 20–120 min, whereas RS does not hydrolyse within 120 min (Englyst et al. 1992; Araya et al. 2002; Dupuis et al. 2016). The high concentration of SDSt and RS in potatoes may be helpful to diabetic persons as it acts as a functional food. The RS resists the enzyme action in the intestine and thus passes undigested from small intestine to large intestine and decrease the glycaemic index as well as glycaemic load (Dupuis et al. 2014; Nayak et al. 2014). Recently, SDSt and RS have gained attention due to its impact on health (Raigond et al. 2014; Chen et al. 2019). The higher amount of RS and SDSt and a lower amount of RDS in food and food products reduce the glycaemic index of food and have profound preventive potential against lifestyle diseases. The amount of RS present in raw potato is around 75%, whereas during cooking its RS decreases to 5–10% (Englyst et al. 1992). RS is fermented by the microflora present in colon (Björck et al. 1987) and release compounds like acetic, propionic, and butyric acid during fermentation. These compounds produced by microflora constitute the main energy source for colonocytes which has a role in the prevention of colonic disease (Åkerberg et al. 1998).

3.2.2 Cellulose

Cellulose is insoluble type of DF which is widely present in fruits and vegetables as one of the main constituents (Raigond et al. 2015). Cellulose is an essential morphogenic polysaccharide which is mainly present in the cell wall. It is a long, unbranched straight chain of glucose having several thousand glucose units with β-1,4-glycosidic linkages (Lampugnani et al. 2019). It is characterized by mechanical strength, resistance to enzymatic digestion, low aqueous solubility, and resistance to acid hydrolysis, attributed mainly due to the hydrogen bonding within microfibrils. It is not digested by the hydrolytic enzymes present in the human gastrointestinal system (Dhingra et al. 2012).

3.2.3 Hemicellulose

Hemicellulose is a hydrophilic and non-cellulosic component of cell wall which is mainly present in vegetative and storage tissues of annual and perennial plants. Being hydrophilic, it binds strongly with cellulose microfibril by hydrogen bonds. It consists of various substituted sugar units arranged in different proportions (Jarvis 1992). These polysaccharides are solubilized by aqueous alkali after the removal of water-soluble and polysaccharide-rich in pectin. Hemicellulose can be classified into four general groups, namely xylans, mannans, mixed linkage β-glucans, and xyloglucans (Rowley and Black 2011). The backbone of hemicellulose is composed of glucose unit with β-1,4-glycosidic linkages and it differs from cellulose. The size is smaller and contains sugar like xylose, galactose, mannose, arabinose, and other branched sugars (Antia 1973).

3.2.4 Gum

Gum is the water-soluble exudate from the plants which contain long-chain polysaccharides. It is unbranched, branched or covalently cross-linked and interacts with water (Walker 1984). Gum may be used as a supplement to make it denser without adding extra calories. It is the type of plant fibre that is not part of the cell wall but is secreted by modified plant cells (Singh et al. 2018). The DF along with gum is generally present on the peel of potatoes (Toma et al. 1979). Its incorporation in the food products is a feasible method for enhancing fibre component. Fibre component with gel-forming properties helps to reduce the level of low-density lipoprotein (LDL)-cholesterol without hampering the high-density lipoprotein

(HDL)-cholesterol fraction (Chawla and Patil 2010). It is soluble and can help to lower the cholesterol level which may have an impact on coronary heart disease. Gum is a high molecular weight, low viscosity, and impartial taste compound present in plants and plant parts. Similar to other DF, the gum is also a prebiotic fibre which enhances the fermentation of lactate bacteria and *Bifidobacteria* (Mariod 2018). Gum also reduces the starch digestibility and enhancement of RS content in potato, this further decreases the release of glucose from food and lowers the glucose level in blood (Sasaki et al. 2015).

3.2.5 Pectin

Pectin is a complex polysaccharide found in the primary cell wall and middle lamella. The composition of pectin includes homogalacturonan (HG), rhamnogalacturonan (RG-I), and rhamnogalacturonan II (RG-II) (Keijbets and Pilnik 1974; Sasaki et al. 2015). It is a highly water-soluble compound which provides firmness and coherence to the tissues. Pectin is composed of a complex group of polysaccharides and principal component is D-galacturonic acid (Dhingra et al. 2012; Yang et al. 2018).

3.2.6 Beta-glucan

Beta-glucan is a water-soluble DF found in oats and barley, and it is often used as a functional fibre when added to foods. Beta-glucans have been shown to reduce cholesterol levels and help control blood sugar levels.

3.2.7 Lignin

Lignin is a branched polymer of phenylpropanoid compounds and is mainly composed of the cell wall. It is a complex polymer resulted from degenerative polymerization of about 40 oxygenated phenylpropane units such as coniferyl, sinapyl, and p-coumaryl alcohols. Lignin is the insoluble DF that has the property of high water holding capacity and helps to increase the faecal bulk in the human intestine and peristaltic movement (Higuchi 1990). It is generally present in the woody plants which are rigid and stout (Dhingra et al. 2012).

3.3 Most Preferred Methods for Estimation of DF

DF can be estimated by various methods which mainly include enzymic, gravimetric, and enzymic-chemical methods. DF is generally estimated using the American Association of Cereal Chemists (AACC) proposed method 32–25. Estimation of the DF method includes the removal of starch by an enzymic process using thermostable alpha-amylase and amyloglucosidase. This is followed by precipitation of soluble polysaccharide with 80% ethanol and finally hydrolysis of amylase resistant polysaccharide (precipitated and insoluble) with sulphuric acid. The final quantification is done by injecting the released neutral sugar as alditol acetates in gas-liquid chromatography (GLC). It is assumed that in uncooked potatoes, RS type III is not present and it is developed only after cooking. This RS type III is calculated by subtracting the average content of glucose residues present in the uncooked sample from the glucose content in the cooked sample. So any increase in glucose residue corresponds to RS type III (Zhao et al. 2018).

DF like uronic acid can be determined by colourimetric method and Klason lignin can be determined gravimetrically. The given method is good for the estimation of DF for vegetables and cereals. The final total DF is determined as the sum of neutral sugar residues, uronic acid residues, and Klason lignin residue in food (Theander et al. 1995). The earlier methods for estimation of crude fibre show the fibre content in food and food products. However, the estimation of crude fibre does not express the real amount of food unavailable to human, this is due to the loss of fibre content, which takes place during chemical treatment. Gradually in vitro method for estimation of DF was proposed where digestive enzymes like pepsin, pancreatin, and lipase were used. The addition of pepsin and pancreatin gave maximum digestion of protein and starch which gave the minimal residue. Cellulose and lignin were determined by using acid detergent fibre (ADF) method (Holst et al. 1976).

3.4 Effect of Growing Conditions, Storage as Well as Cooking on DF

3.4.1 Growing Conditions

The chemical composition of potato tubers may be influenced by various factors like variety, soil, climate, production area, the agricultural package of practices, storage, and processing conditions (Arvanitoyannis et al. 2008). DF content is negatively correlated with starch, ash, and phosphorus content, whereas positively correlated with total sugar content (Leonel et al. 2017). When compared with other vegetables the fibre content in potato is low ranging from 0.34 to 0.80 g 100 g^{-1} fresh weight basis (Leonel et al. 2017). It is also reported that the nutrient level in the soil affects the DF content in potato. When potatoes are cultivated in the phosphorus sufficient soil, it leads to a decrease in the fibre content in tubers (Leonel et al. 2017).

3.4.2 Storage

The bulk production of potato in India and other parts of the world and their post-harvest storage leads to a change in the intrinsic property of potato. Misra and Kulshrestha (2003) reported that there was not much significant change in, protein, mineral, vitamins, dietary fibre, and total starch except moisture and ascorbic acid when potato flour is stored for three and six months in refrigerated condition.

3.4.3 Cooking Method

For human consumption potatoes are cooked, however, the raw form of potato can be given as feed for animals (Onwubuemeli et al. 1985). There are a variety of culinary techniques that differentially affect the nutritional quality, taste, texture, and bioavailability of nutritional constituents of potato and its products (Liu et al. 2007). The enhancement of nutrient bioavailability of macro and micronutrients can be done by modification in a culinary method like cooking at low temperatures, bringing down the cooking time, and cooking under vacuum (Jayanty et al. 2019).

Cooking brings out many physical and chemical changes in starch which increase not only digestibility and palatability but also the bioavailability of many vital nutrients. The cooking of potato improves food safety as well as sensory qualities like texture, flavour, and taste. The cooking method like boiling, microwaving, baking, and frying (Jayanty et al. 2019) change the DF content. However, no significant change in the DF and starch content is reported in the extruded samples (Dhingra et al. 2012).

The molecular structure of DF changes after cooking the potato tubers. Frying leads to gelatinization of starch on the surface of the potato flesh which is amylose-lipid complex and acts as RS (Tahvonen et al. 2006). The processing of potato tuber changes the texture, physical, chemical, and molecular properties which ultimately affect the DF content. Cooking methods such as boiling, microwaving, and deep-frying increase RS up to 2.9%, 7.3%, 6.2%, and 9.1%, respectively (Thed and Phillips 1995). Cellulose content is also affected by the processing of potatoes. When par-fried French fries were kept for different intervals of time, viz., 24 h, 48 h, 15 day, 30 day, 60 day, 120 day, and 180 day at −20 °C (freezing), its cellulose content increased and overall DF also increased along with RS in finish fried French fries (Raigond et al. 2015).

Deep fat frying of potato chips and potato fries reduces in vitro digestible starch along with the increase in RS and water-insoluble fibres which include cellulose, hemicellulose, pullulan, pectin, and lignin (Kumar et al. 2012; Raigond et al. 2015; Chen et al. 2017; Zhao et al. 2018).

Another method of cooking is the extrusion method, which is a process of cooking food and food product at a high temperature for a short time under high pressure. This affects the food's physical and chemical properties which are present in the matrix. The in vitro experiments have been conducted where the French fried processing was evaluated and found that it can bind to bile acid. Cooking by the extrusion method enhances the binding of potato peel with cholic, deoxycholic, and glycolic acids. However, the binding of deoxycholic acid is highly correlated with DF content (Camire et al. 1997; Camire 2016). The cooking of potato by the extrusion process reduced the overall starch content and increase total DF, however,

DF in the abrasion peels was not affected (Camire et al. 1997). This process significantly increases the uronic acid in the soluble fibre portion.

3.5 Health Benefits

The changing lifestyle including sedentary work and less physical exercise leads to various chronic diseases like obesity, type-II diabetes, colonic cancer, damage to internal organs like kidney, liver, etc. Overconsumption of junk food and packaged food leads to an increase in blood sugar, lipid, and cholesterol. Diabetes mellitus case is rising very fast and became a serious problem for the human population. There are many medications and supplements available in the market which acts as DF. For instance, digestion-resistant dextrins and cross-linked starches are available which is created by chemical modifications (Oki et al. 2019). In potato, the main DF constituent is derived from cell walls. The cell walls of potato peel (periderm) make about 1–2% of total potato weight. These fibres are non-lignified and play an important role in decreasing the level of blood cholesterol (Lazarov and Werman 1996).

The role of RS is very important in the gut; it is a part of DF which has lots of nutritional properties and serves as prebiotics in the large intestine. It provides a fermentation substrate of *Ruminoccocus bromii* (probiotic species), which enhances the growth of other probiotic bacteria (Ze et al. 2012). Properties of RS include lowering of glucose and insulin responses, increasing faecal output, whereas reducing faecal transit time in the gut, lowing calorific values in foods, and promoting the growth of beneficial gut bacteria. These bacteria after fermentation produce short-chain fatty acids (SCFA) like acetic acid, propionic acid, and butyric acid which results in beneficial effects on colonocytes and maintain colonic health (Emenaker and Basson 1998; Brown et al. 2001). The combination of DF with other compounds like polyphenol content may be beneficial for diabetics. Studies suggest that DF along with many natural antioxidants such as caffeic acid and chlorogenic acids are abundant in potato extract, which may help in the prevention of type 2 diabetes (Liang and Kitts 2015; Santana-Gálvez et al. 2017).

3.6 Genetic Modifications to Improve DF Content

Various attempts have been made to modify the DF content of the potato, initially by physical and chemical changes and recently genomic and biotechnological tools are also being utilized. The physical addition or fortification of DF (in the form of hydrocolloids) in food products enhances the quality of food and food products. It was reported in rice and potato that the addition of pullulan suppresses the process of gelatinization in rice and potato and increases the fraction of SDSt and RS (Chen et al. 2017, 2019). In potato, Chen et al. (2019) studied the effect of pullulan and pectin addition before frying on DF content. There was a significant increase in SDSt and RS along with the change in the structure of starch granules. The transgenic lines of potato and rice having attributes of RS such as high amylose content, the long chain of amylopectin, high gelatinization temperature, and other properties were developed (Zhu et al. 2011; Zhao et al. 2018).

There are various mechanisms for enhancing RS and DF in potato and other starchy foods. By the addition of compounds like pullulan, guar-gum, pectin, agar, xanthan, carboxymethyl cellulose, and glucomannan the digestibility can be reduced. The mechanism for the reduction of digestibility and enhancement of DF is the coating of starch granules with mentioned compounds, compete with starch for available water or interfere once the digestion process. Pectin being the anionic polysaccharide is also used to enhance the DF content in fried potato (Chen et al. 2019).

When the transgenic potato line T-2012 (uncooked), with the high amylose line, was observed to contain 30% RS and the parental line Dinamo has 56% RS content. After cooking (boiling) the RS content of T-2012 was 13–20% of dry matter which is around three-fold higher than the parental line (Dinamo) (Zhao et al. 2018). This shows that the RS starch content was not affected by cooking in the transgenic line which is already a high amylose line. Although the RS content decreased after cooking in both transgenic and parental line, amylose rich transgenic line contained more RS even after cooking compared to the parental line.

3.7 Conclusion/Future Direction

DF has many health benefits and potatoes are a good source of DF. However, DF is affected by various factors including cultivation practices, storage and most importantly cooking methods. The cooking of potato with peel is the most suitable method to retain the maximum concentration of DF in potato. The consumption of food rich in RS and SDSt may aid in reducing the risk of diseases associated with metabolic syndromes like obesity, type-II diabetes, cancer, and heart diseases. Potato varieties rich in SDSt and RS can be developed using various genomics and biotechnological tools that can help in popularizing potato as healthy food and development of healthy products from potato.

Along with the health benefits of DF, it can also be used in the production of various drugs and delivery system. More research is needed in this area for including potato starch and DF in the production of an efficient drug delivery system. Our understanding of DF, its role in potato and potato products has greatly increased over the past few decades. The application of advanced technology in the field of food science and technology helped scientists to understand the insight of the structure and function of food and food products.

References

Åkerberg AKE, Liljeberg HGM, Granfeldt YE et al (1998) An in vitro method, based on chewing, to predict resistant starch content in foods allows parallel determination of potentially available starch and dietary fiber. J Nutr 128:651–660. https://doi.org/10.1093/jn/128.3.651

Antia FP (1973) Clinical dietetics and nutrition. Oxford University Press, London

Araya H, Contreras P, Alviña M et al (2002) A comparison between an in vitro method to determine carbohydrate digestion rate and the glycemic response in young men. Eur J Clin Nutr 56:735–739. https://doi.org/10.1038/sj.ejcn.1601386

Arvanitoyannis IS, Vaitsi O, Mavromatis A (2008) Potato: a comparative study of the effect of cultivars and cultivation conditions and genetic modification on the physico-chemical properties of potato tubers in conjunction with multivariate analysis towards authenticity. Crit Rev Food Sci Nutr 48:799–823. https://doi.org/10.1080/10408390701691059

Birch PRJ, Bryan G, Fenton B et al (2012) Crops that feed the world 8: potato: are the trends of increased global production sustainable? Food Secur 4:477–508. https://doi.org/10.1007/s12571-012-0220-1

Björck I, Nyman M, Pedersen B et al (1987) Formation of enzyme resistant starch during autoclaving of wheat starch: studies in vitro and in vivo. J Cereal Sci 6:159–172. https://doi.org/10.1016/S0733-5210(87)80052-8

Blennow A, Bay-Smidt AM, Leonhardt P et al (2003) Starch paste stickiness is a relevant native starch selection criterion for wet-end paper manufacturing. Starch 55:381–389. https://doi.org/10.1002/star.200300169

Brown RA, Walsh MF, Ren J (2001) Influence of gender and diabetes on vascular and myocardial contractile function. Endocr Res 27:399–408. https://doi.org/10.1081/ERC-100107864

Camire ME (2016) Potatoes and human health. In: Advances in potato chemistry and technology, 2nd edn. Taylor & Francis Group, Milton Park, pp 685–704

Camire ME, Violette D, Dougherty MP, McLaughlin MA (1997) Potato peel dietary fiber composition: effects of peeling and extrusion cooking processes. J Agric Food Chem 45:1404–1408. https://doi.org/10.1021/jf9604293

Camire ME, Kubow S, Donnelly DJ (2009) Potatoes and human health. Crit Rev Food Sci Nutr 49:823–840. https://doi.org/10.1080/10408390903041996

Chawla R, Patil GR (2010) Soluble dietary fiber. Compr Rev Food Sci Food Saf 9:178–196. https://doi.org/10.1111/j.1541-4337.2009.00099.x

Chen L, Tian Y, Zhang Z et al (2017) Effect of pullulan on the digestible, crystalline and morphological characteristics of rice starch. Food Hydrocoll 63:383–390. https://doi.org/10.1016/j.foodhyd.2016.09.021

Chen L, Zhang H, McClements DJ et al (2019) Effect of dietary fibers on the structure and digestibility of fried potato starch: a comparison of pullulan and pectin. Carbohydr Polym 215:47–57. https://doi.org/10.1016/j.carbpol.2019.03.046

Dhingra D, Michael M, Rajput H, Patil RT (2012) Dietary fibre in foods: a review. J Food Sci Technol 49:255–266. https://doi.org/10.1007/s13197-011-0365-5

Dhital S, Shrestha AK, Hasjim J, Gidley MJ (2011) Physicochemical and structural properties of maize and potato starches as a function of granule size. J Agric Food Chem 59:10151–10161. https://doi.org/10.1021/jf202293s

Dupuis JH, Liu Q (2019) Potato Starch: a review of physicochemical, functional and nutritional properties. Am J Potato Res 96:127–138

Dupuis JH, Liu Q, Yada RY (2014) Methodologies for increasing the resistant starch content of food starches: a review. Compr Rev Food Sci Food Saf 13:1219–1234. https://doi.org/10.1111/1541-4337.12104

Dupuis JH, Lu ZH, Yada RY, Liu Q (2016) The effect of thermal processing and storage on the physicochemical properties and in vitro digestibility of potatoes. Int J Food Sci Technol 51:2233–2241. https://doi.org/10.1111/ijfs.13184

Emenaker NJ, Basson MD (1998) Short chain fatty acids inhibit human (SW1116) colon cancer cell invasion by reducing urokinase plasminogen activator activity and stimulating TIMP-1 and TIMP-2 activities, rather than via MMP modulation. J Surg Res 76:41–46. https://doi.org/10.1006/JSRE.1998.5279

Englyst HN, Kingman SM, Cummings JH (1992) Classification and measurement of nutritionally important starch fractions. Eur J Clin Nutr 46(Suppl 2):S33–S50

Fuentes-Zaragoza E, Riquelme-Navarrete MJ, Sánchez-Zapata E, Pérez-Álvarez JA (2010) Resistant starch as functional ingredient: a review. Food Res Int 43:931–942

García-Segovia P, Andrés-Bello A, Martínez-Monzó J (2007) Effect of cooking method on mechanical properties, color and structure of beef muscle (M. pectoralis). J Food Eng 80:813–821. https://doi.org/10.1016/J.JFOODENG.2006.07.010

Higuchi T (1990) Lignin biochemistry: biosynthesis and biodegradation. Wood Sci Technol 24:23–63

Holst A, Kostrzewa M, Lask H (1976) Process for the production of water adsorbing but water-insoluble cellulose ethers. US3936441A

Jansen G, Flamme W, Schüler K, Vandrey M (2001) Tuber and starch quality of wild and cultivated potato species and cultivars. Potato Res 44:137–146. https://doi.org/10.1007/BF02410100

Jarvis MC (1992) Self-assembly of plant cell walls. Plant Cell Environ 15:1–5

Jayanty SS, Diganta K, Raven B (2019) Effects of cooking methods on nutritional content in potato tubers. Am J Potato Res 96:183–194

Keijbets MJH, Pilnik W (1974) Some problems in the analysis of pectin in potato tuber tissue. Potato Res 17:169–177. https://doi.org/10.1007/BF02360384

Kumar V, Sinha AK, Makkar HPS et al (2012) Dietary roles of non-starch polysachharides in human nutrition: a review. Crit Rev Food Sci Nutr 52:899–935

Lampugnani ER, Flores-Sandoval E, Tan QW et al (2019) Cellulose synthesis – central components and their evolutionary relationships. Trends Plant Sci 24:402–412

Lazarov K, Werman MJ (1996) Hypocholesterolaemic effect of potato peels as a dietary fibre source. Med Sci Res 24:581–582

Leeman M, Östman E, Björck I (2008) Glycaemic and satiating properties of potato products. Eur J Clin Nutr 62:87–95. https://doi.org/10.1038/sj.ejcn.1602677

Leonel M, do Carmo EL, Fernandes AM et al (2017) Chemical composition of potato tubers: the effect of cultivars and growth conditions. J Food Sci Technol 54:2372–2378. https://doi.org/10.1007/s13197-017-2677-6

Li H, Yang J, Gao M et al (2019) Washing rice before cooking has no large effect on the texture of cooked rice. Food Chem 271:388–392. https://doi.org/10.1016/j.foodchem.2018.07.172

Liang N, Kitts DD (2015) Role of chlorogenic acids in controlling oxidative and inflammatory stress conditions. Nutrients 8:16

Lin Ek K, Wang S, Brand-Miller J, Copeland L (2014) Properties of starch from potatoes differing in glycemic index. Food Funct 5:2509–2515. https://doi.org/10.1039/c4fo00354c

Liu Q, Weber E, Currie V, Yada R (2003) Physicochemical properties of starches during potato growth. Carbohydr Polym 51:213–221. https://doi.org/10.1016/S0144-8617(02)00138-8

Liu Q, Tarn R, Lynch D, Skjodt NM (2007) Physicochemical properties of dry matter and starch from potatoes grown in Canada. Food Chem 105:897–907. https://doi.org/10.1016/j.foodchem.2007.04.034

Lovedeep K, Jaspreet S (2016) Microstructure, starch digestion, and glycemic index of potatoes. In: Advances in potato chemistry and technology, 2nd edn. Academic, New York, pp 369–402

Mariod AA (2018) Gum arabic dietary fiber. In: Gum Arabic: structure, properties, application and economics. Elsevier, London, pp 237–243

Mikulíová D, Masá Š, Kraic J (2008) Biodiversity of legume health-promoting starch. Starch 60:426–432. https://doi.org/10.1002/star.200700693

Misra A, Kulshrestha K (2003) Effect of storage on nutritional value of potato flour made from three potato varieties. Plant Foods Hum Nutr 58:1–10. https://doi.org/10.1023/B:QUAL.0000040316.18576.31

Nayak B, De Berrios J, Tang J (2014) Impact of food processing on the glycemic index (GI) of potato products. Food Res Int 56:35–46. https://doi.org/10.1016/j.foodres.2013.12.020
Oki S, Sato C, Kubota N et al (2019) Changes in the molecular structure of potato starch conferred digestion resistance by fatty acid addition and heat treatment. Jpn J Food Eng 20:15–20. https://doi.org/10.11301/jsfe.18534
Onwubuemeli C, Huber JT, King KJ, Johnson C (1985) Nutritive value of potato processing wastes in total mixed rations for dairy cattle. J Dairy Sci 68:1207–1214. https://doi.org/10.3168/JDS.S0022-0302(85)80948-6
Priestley H (2006) How to think like consumers... and win! In: Potato dev chang Eur. Springer, New York, pp 189–198
Raigond P, Ezekiel R, Raigond B (2014) Resistant starch in food: a review. J Sci Food Agric 95 (10):1968–1978
Raigond P, Singh B, Kaundal B (2015) Changes in the starch and fiber fractions of French fries: response of freezing. Potato J 42:8–17
Razdan MK (2005) Genetic improvement of solanaceous crops, volume 1: potato. CRC Press, Boca Raton
Rowley J, Black S (2011) Determination of an efficient extraction method of xylan from biomass. p 1
Santana-Gálvez J, Cisneros-Zevallos L, Jacobo-Velázquez DA (2017) Chlorogenic acid: recent advances on its dual role as a food additive and a nutraceutical against metabolic syndrome. Molecules 22:358
Sanz T, Salvador A, Fiszman SM (2008) Resistant starch (RS) in battered fried products: functionality and high-fibre benefit. Food Hydrocoll 22:543–549. https://doi.org/10.1016/j.foodhyd.2007.01.018
Sasaki T, Sotome I, Okadome H (2015) In vitro starch digestibility and in vivo glucose response of gelatinized potato starch in the presence of non-starch polysaccharides. Starch 67:415–423
Singh J, Dartois A, Kaur L (2010) Starch digestibility in food matrix: a review. Trends Food Sci Technol 21:168–180
Singh A, Kaur V, Kaler RSS (2018) A review on dietary fiber in cereals and its characterization. J Appl Nat Sci 10:1216–1225. https://doi.org/10.31018/jans.v10i4.1894
Slavin JL (2005) Dietary fiber and body weight. Nutrition 21:411–418
Tahvonen R, Hietanen RM, Sihvonen J, Salminen E (2006) Influence of different processing methods on the glycemic index of potato (Nicola). J Food Compos Anal 19:372–378. https://doi.org/10.1016/j.jfca.2005.10.008
Theander O, Aman P, Westerlund E et al (1995) Total dietary fiber determined as neutral sugar residues, uronic acid residues, and Klason lignin (the Uppsala method): collaborative study. J AOAC Int 78:1030–1044
Thed ST, Phillips RD (1995) Changes of dietary fiber and starch composition of processed potato products during domestic cooking. Food Chem 52:301–304. https://doi.org/10.1016/0308-8146(95)92828-8
Tian J, Chen J, Ye X, Chen S (2016) Health benefits of the potato affected by domestic cooking: a review. Food Chem 202:165–175. https://doi.org/10.1016/j.foodchem.2016.01.120
Toma RB, Orr PH, D'Appolonia B et al (1979) Physical and chemical properties of potato peel as a source of dietary fiber in bread. J Food Sci 44:1403–1407. https://doi.org/10.1111/j.1365-2621.1979.tb06448.x
Trowell H, Burkitt D, Heaton K (1985) Dietary fibre, fibre-depleted foods and disease
Walker B (1984) Gums and stabilisers in food formulations. Gums Stab Food Ind 2:137–161
Wang S, Li C, Copeland L et al (2015) Starch retrogradation: a comprehensive review. Compr Rev Food Sci Food Saf 14:568–585. https://doi.org/10.1111/1541-4337.12143
Wani AA, Singh P, Shah MA et al (2013) Physico-chemical, thermal and rheological properties of starches isolated from newly released rice cultivars grown in Indian temperate climates. LWT-Food Sci Technol 53:176–183. https://doi.org/10.1016/J.LWT.2013.02.020

Wiesenborn DP, Orr PH, Casper HH, Tacke BK (1994) Potato starch paste behavior as related to some physical/chemical properties. J Food Sci 59:644–648. https://doi.org/10.1111/j.1365-2621.1994.tb05583.x

Xie F, Wang Y, Wu J, Wang Z (2016) Functional properties and morphological characters of soluble dietary fibers in different edible parts of angelica keiskei. J Food Sci 81:C2189–C2198. https://doi.org/10.1111/1750-3841.13399

Xie F, Li M, Lan X et al (2017) Modification of dietary fibers from purple-fleshed potatoes (Heimeiren) with high hydrostatic pressure and high pressure homogenization processing: a comparative study. Innov Food Sci Emerg Technol 42:157–164. https://doi.org/10.1016/j.ifset.2017.05.012

Yang JS, Mu TH, Ma MM (2018) Extraction, structure, and emulsifying properties of pectin from potato pulp. Food Chem 244:197–205. https://doi.org/10.1016/j.foodchem.2017.10.059

Ze X, Duncan SH, Louis P, Flint HJ (2012) Ruminococcus bromii is a keystone species for the degradation of resistant starch in the human colon. ISME J 6:1535–1543. https://doi.org/10.1038/ismej.2012.4

Zhang G, Ao Z, Hamaker BR (2006) Slow digestion property of native cereal starches. Biomacromolecules 7:3252–3258. https://doi.org/10.1021/bm060342i

Zhao X, Andersson M, Andersson R (2018) Resistant starch and other dietary fiber components in tubers from a high-amylose potato. Food Chem 251:58–63. https://doi.org/10.1016/j.foodchem.2018.01.028

Zhu L-J, Liu Q-Q, Wilson JD et al (2011) Digestibility and physicochemical properties of rice (Oryza sativa L.) flours and starches differing in amylose content. Carbohydr Polym 86:1751–1759. https://doi.org/10.1016/J.CARBPOL.2011.07.017

Potato Proteins

4

Som Dutt, Pinky Raigond, Brajesh Singh, Anshul Sharma Manjul, and Swarup Kumar Chakrabarti

Abstract

Proteins play various functional as well as structural roles in all living cells and are one of the main components of *human nutrition*. Of the various sources of proteins, potato alone contributes about 3% of the recommended dietary allowance of protein globally. With the protein content ranging from 1 to 3% of tuber fresh weight potato proteins have excellent biological value of 90–100. Potato proteins are mainly classified into three groups: patatins, protease inhibitors and other proteins. Attempts have been made to improve the protein in potato in terms of quantity and quality. Potato protein, products and hydrolysates are considered the "Generally Recognized as Safe" and non-allergenic and are being incorporated in various vegetarian/vegan products. Potato proteins have also been found to be industrially useful due to their emulsifying, foaming and gelling properties. Processes for extraction of proteins from potato/potato juices at large industrial scale have been developed keeping in view the ample scope and potential global market available for potato proteins.

Keywords

Patatin · Protease inhibitor · Genetic engineering · Amino acids · Protein quality

S. Dutt (✉) · P. Raigond · B. Singh · A. S. Manjul
Crop Physiology, Biochemistry and Post-Harvest Technology Division, ICAR - Central Potato Research Institute, Shimla, India
e-mail: sd_bio@yahoo.com

S. K. Chakrabarti
ICAR-Central Potato Research Institute, Shimla, Himachal Pradesh, India

P. Raigond et al. (eds.), *Potato*, https://doi.org/10.1007/978-981-15-7662-1_4

4.1 Introduction

Ever increasing world population, industrialization, urbanization and global climatic change have overburdened the existing agriculture lands and food resources. There are several crops which are being grown by human being to meet the ever rising global food demand. One such crop is potato. With an annual production ranging from 329 to 374 million tonnes for the last 10 years (year 2011–2018; Fig. 4.1), potato (*Solanum tuberosum* L.) is the fourth most important agricultural crop grown for human consumption after rice, wheat and corn (FAO 2020). Well known for high agricultural yields and significant nutritive value, potatoes are a source of several nutrients like carbohydrate, protein, vitamin C, potassium, magnesium and potentially health-promoting metabolites (Dutt et al. 2019; Camire et al. 2009). Recently, metabolites having medicinally active compounds have also been reported in potato (Raigond et al. 2018). Among major food crops, potato produces highest nutrition and dry matter on per unit area and time basis. In 2008, FAO declared potato "the crop to address future global food security and poverty alleviation". Apart from consumption as vegetable and staple food, potato is an industrial crop and is used in various processed products, Potato is also used in the production of starch, alcohol, biofuel, animal feed, etc. (Scott and Suarez 2012; Liang and McDonald 2014). Seven major classes of nutrients, *viz* carbohydrates, proteins, vitamins, lipids, fibre, minerals and water, form central core of human nutrition. Of these, protein performs various structural and function roles and thus is one of the essential components of diets of human beings for which we depend on various plant and animal based food sources. There is a rapid expansion in plant-based protein market because of certain advantages associated with the plant proteins in comparison to the animal proteins; these include ease of digestibility, affordability, relative higher nutrition value, etc. (Zhang et al. 2016). When compared with sources of proteins potato has a low total protein content on a fresh weight (FW) basis, and thus potatoes are not usually considered as a rich source of protein compared with other foods. However, with high per capita consumption of total potato, it can make a significant nutritional contribution to the diet because of its high nutritional value. Compared to proteins from other vegetable and cereal sources, potato proteins are of

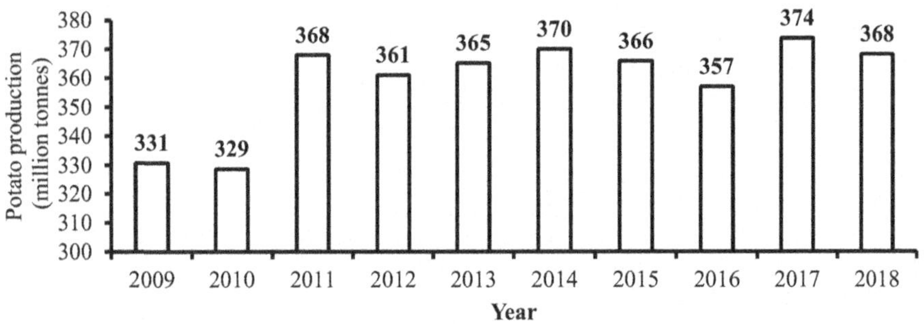

Fig. 4.1 Yearly world potato production during 2009–2018 (data source: FAO 2020). http://www.fao.org/faostat/

great potential as food ingredients because of their higher nutritional quality, ability to regulate serum cholesterol levels and to reduce food intake (Ralet and Guéguen 2001; Liyanage et al. 2008; Komarnytsky et al. 2011). Potato proteins are considered "the Generally Recognized as Safe" (GRAS) and non-allergenic and can be incorporated in various vegetarian/vegan products (David and Livney 2016). Although potato proteins have also been found to be industrially useful due to their emulsifying, foaming and gelling properties (Gambuti et al. 2016; Spelbrink et al. 2015; Baier and Knorr 2015), however, utilizing potato for extraction of proteins at industrial level in sustainable manner will require producing more potatoes, which in turn will require growing potato in unconventional areas (such as tropical) and/or improving productivity of potato by breaking critical yield barriers such as environmental controlled tuberization mechanisms (Dutt et al. 2017). Here, in this chapter a general overview of potato proteins including consumption, different classes of potato proteins, their concentration, and works on increasing the protein content/quality is presented.

4.2 Content, Consumption and Quality of Potato Protein

Protein content of potato tubers ranges from 1 to 3% of tuber fresh weight (Ortiz-Medina and Donnelly 2003; Sato et al. 2017; Nouri-Ellouz et al. 2015; Vaitkevičienė 2019) and is considered as one of the richest protein sources of any root or tuber crop (Shewry 2003) (Table 4.1). Global annual average potato consumption per capita is 32.84 kg (Table 4.2). In a comparison of 157 countries in 2015, Belarus topped the list in potato consumption per capita with 171 kg followed and was flowed by Ukraine (116 kg) and Latvia (115 kg) (Table 4.2). We have estimated the percentage of recommended dietary allowance (RDA) of protein which is supplied by the potato

Table 4.1 Protein contents of major tuber crops (adapted from Shewry 2003)

Crop	Species (family)	Origin of tubers	Approx. protein content of tubers (% dry weight)
Potato	*Solanum tuberosum* (Solanaceae)	Stem	6–12
Sweet potato	*Ipomoea batatas* (Convolvulaceae)	Root	1–10
Cassava	*Manihot esculenta* (Euphorbiaceae)	Root	1–2
Taro	*Colocasia esculenta* (Araceae)	Corm	2.0
	Cytosperma chamissonis (Araceae)	Corm	0.8
	Xanthosoma sagittifolium (Araceae)	Corm	2.0
	Alocasia macrorrhiza (Araceae)	Stem	0.6
Yam	*Dioscorea* spp. (Dioscoreaceae)	Hypocotyl	1–3

Table 4.2 World-wide percent recommended dietary allowance (RDA) of protein supplied by potato (*per capita per year consumption data year: 2015; calculations are based on RDA of protein as 60 g/day, average protein content of potato tubers is taken as 2% on fresh weight basis. In highlighted countries protein supply through potato consumption is >5% RDA for protein*)

Country	Potato consumption / per capita/per year	Protein supply (% RDA)	Country	Potato consumption / per capita / per year	Protein supply (% RDA)	Country	Potato consumption / per capita / per year	Protein supply (% RDA)
Afghanistan	6.29	0.6	Ghana	0.14	0.0	Nicaragua	5.19	0.5
Albania	44.93	4.1	Greece	69.27	6.3	Niger	3.01	0.3
Algeria	65.65	6.0	Guatemala	11.29	1.0	Nigeria	4.53	0.4
Angola	13.78	1.3	Guinea	3.88	0.4	Norway	51.86	4.7
Argentina	36.94	3.4	Guinea Bissau	0.36	0.0	Oman	22.76	2.1
Armenia	43.6	4.0	Guyana	13.15	1.2	Pakistan	14.46	1.3
Australia	49.81	4.5	Haiti	2.3	0.2	Panama	11.19	1.0
Austria	55.42	5.1	Honduras	3.79	0.3	Paraguay	2.94	0.3
Azerbaijan	71.04	6.5	Hungary	48.03	4.4	Peru	83.26	7.6
Bahamas	5.32	0.5	Iceland	39.03	3.6	Philippines	2.36	0.2
Bangladesh	45.27	4.1	India	24.68	2.3	Poland	96.26	8.8
Belarus	170.64	15.6	Indonesia	4.46	0.4	Portugal	64.6	5.9
Belgium	90.68	8.3	Iraq	11.53	1.1	Romania	100.2	9.2
Belize	10.18	0.9	Iran	53.13	4.9	Russia	112.54	10.3
Benin	0.15	0.0	Ireland	72.22	6.6	Rwanda	93.99	8.6
Bolivia	65.99	6.0	Israel	34.66	3.2	Saudi Arabia	13.23	1.2
Bosnia and Herzegovina	78.32	7.2	Italy	36.34	3.3	Senegal	7.36	0.7
Botswana	11.79	1.1	Ivory Coast	0.83	0.1	Serbia	42.22	3.9
Brazil	15.56	1.4	Jamaica	8.33	0.8	Sierra Leone	0.28	0.0
Bulgaria	26.04	2.4	Japan	18.3	1.7	Slovakia	50.34	4.6
Burkina Faso	0.29	0.0	Jordan	24.21	2.2	Slovenia	45.13	4.1
Cambodia	0.07	0.0	Kazakhstan	108.1	9.9	South Africa	31.36	2.9
Cameroon	6.25	0.6	Kenya	36.91	3.4	South Korea	10.1	0.9
Canada	66.62	6.1	Kuwait	50.03	4.6	Spain	57.31	5.2
Central African Republic	0.16	0.0	Kyrgyzstan	95.36	8.7	Sri Lanka	9.47	0.9
Chad	2.17	0.2	Laos	4.24	0.4	Sudan	10.11	0.9
Chile	54.21	5.0	Latvia	115.48	10.5	Suriname	16.69	1.5
China	42.27	3.9	Lebanon	41.66	3.8	Swaziland	12.52	1.1
Colombia	33.36	3.0	Lesotho	56.94	5.2	Sweden	51.64	4.7
Congo	1.61	0.1	Liberia	0.36	0.0	Switzerland	37.82	3.5
Costa Rica	14.92	1.4	Lithuania	76.19	7.0	Tajikistan	36.46	3.3
Croatia	38.95	3.6	Luxembourg	35.62	3.3	Tanzania	18.41	1.7
Cuba	11.41	1.0	Macedonia	63.06	5.8	Thailand	3.28	0.3
Cyprus	23	2.1	Madagascar	5.73	0.5	Togo	0.39	0.0
Czech Republic	49.21	4.5	Malawi	48.13	4.4	Trinidad and Tobago	29.24	2.7
Dem. People's Republic of Korea	48.15	4.4	Malaysia	12.37	1.1	Tunisia	30.52	2.8
Denmark	57.79	5.3	Maldives	12.13	1.1	Turkey	47.62	4.3
Djibouti	16.16	1.5	Mali	7.34	0.7	Turkmenistan	30.86	2.8
Dominican Republic	8.1	0.7	Malta	42.42	3.9	Uganda	3.66	0.3
East Timor	1.56	0.1	Mauritania	8.35	0.8	Ukraine	116.15	10.6
Ecuador	23.14	2.1	Mauritius	21.92	2.0	United Arab Emirates	10.96	1.0
Egypt	34.62	3.2	Mexico	15.3	1.4	United Kingdom	86.14	7.9
El Salvador	13.15	1.2	Moldova	46.76	4.3	USA	50.22	4.6
Estonia	68.5	6.3	Mongolia	39	3.6	Uruguay	28.47	2.6
Ethiopia	6.41	0.6	Montenegro	26.07	2.4	Uzbekistan	52.29	4.8
Fiji	28.47	2.6	Morocco	46.3	4.2	Venezuela	15.77	1.4
Finland	55.97	5.1	Mozambique	8.35	0.8	Vietnam	3.6	0.3
France	51.07	4.7	Myanmar	7.53	0.7	Yemen	7.74	0.7
Gabon	1.31	0.1	Namibia	17.42	1.6	Zambia	2.24	0.2
Gambia	1.08	0.1	Nepal	74.24	6.8	Zimbabwe	4.1	0.4
Georgia	49.3	4.5	Netherlands	83.48	7.6	World	32.84	3.0
Germany	64.61	5.9	New Zealand	46.34	4.2			

Table 4.3 Protein content in flesh and peels of coloured potatoes (Reference used: Vaitkevičienė 2019)

Potato cultivar	Potato tissue	Protein (g kg^{-1} dry weight)	Dry matter (g kg^{-1} fresh weight)	% dry matter	% protein on dry weight basis	% protein on fresh weight basis
Red emmalie	Flesh	103.3	172.2	17.22	10.33	1.78
	Peel	147.2	159.6	15.96	14.72	2.35
Rosalinde	Flesh	121.1	195.5	19.55	12.11	2.37
	Peel	161.1	169.2	16.92	16.11	2.73
Highland burgundy	Flesh	107.1	228.3	22.83	10.71	2.45
	Peel	165.3	192.2	19.22	16.53	3.18
Violetta	Flesh	119.3	165.9	16.59	11.93	1.98
	Peel	148.9	151.8	15.18	14.89	2.26
Valfi	Flesh	89.4	208.1	20.81	8.94	1.86
	Peel	139	174.9	17.49	13.9	2.43
Salad blue	Flesh	119.4	188.7	18.87	11.94	2.25
	Peel	141.1	185.1	18.51	14.11	2.61

consumption (Table 4.2), by taking into account the average protein content of potato tubers as 2% (on fresh weight basis) of potatoes and RDA of protein as 60 g/day. From the derived data it is amazingly surprising to see that in more than 25 countries potato consumption supplies more than 5% of the RDA for protein, highest being in Belarus with 15.6% (Table 4.2). This is a significant contribution of potato as a protein source with global average of 3% of the RDA for protein. Further, peel of potato tubers also contains significantly higher concentration of protein (1.2–1.6 times higher as compared to that of the flesh) (Table 4.3), though the potato peel is yet to be exploited as a source for extraction of protein at industrial level.

In order to evaluate various nutritional characteristics of proteins, various nutritional parameters have been developed. The main parameters include; "Biological Value", "Protein Efficiency Ratio", "True Digestibility", "Amino Acid Score" and "Protein Digestibility Corrected Amino Acid Score". Various nutritional parameters of potato protein isolates are high in comparison to proteins from other sources. Potato proteins have a True Digestibility of more than 0.98. Also, potato protein isolate exhibits the highest Protein Digestibility Corrected Amino Acid Score and Protein Efficiency Ratio, of the vegetable proteins. The quality of the protein is directly correlated and reflects the quantity of different amino acids it is composed of. Biological value of protein is a method to estimate the quality of protein. Biological value of the proteins of potatoes is 90–100, and is categorized as excellent (Table 4.4) (Camire et al. 2009). Apart from these high molecular weight proteins, potato also contains some small peptides and free amino acids. Asparagine, glutamate and aspartate represent major part of the free amino acids content of potato. The amino acids composition of potato is given in Table 4.5. This table shows that potato protein isolates have a balanced composition of majority of the essential amino acids.

Table 4.4 Biological value of different sources of protein (adapted from Storey 2007)

Protein source	Biological value
Rice	83
Maize	72–86
Beans	73
Cow milk	84–88
Whole chicken egg	100
Fish	83
Potato	90–100
Whole egg and potato (35%: 65%)	137
Whole egg and milk (71%: 29%)	122

Table 4.5 Amino acid composition of potato tuber protein (adapted from Lisinska and Leszczynski 1989)

Amino acid	Range (%)	Amino acid	Range (%)
Alanine	4.62–5.32	Lysine	6.70–10.1
Arginine	4.74–5.70	Methionine	1.20–2.15
Aspartic acid	11.9–13.9	Phenylalanine	4.80–6.53
Cysteine	0.20–1.25	Proline	4.70–4.83
Glutamic acid	10.2–11.8	Serine	4.90–5.92
Glycine	4.30–6.05	Threonine	4.60–6.50
Histidine	2.10–2.50	Tryptophan	0.30–1.85
Isoleucine	3.73–5.80	Tyrosine	4.50–5.68
Leucine	9.70–10.3	Valine	4.88–7.40

4.3 Classification of Potato Proteins

The potato proteins have been named and classified on the basis of various physical and chemical properties including structure, solubility, isoelectric point, molecular weight, enzymatic activity, etc. Work on potato protein has been reported as early as 1896. In 1896, Osborne and Campbell named the globular protein of potato tubers as "tubrin". Subsequently, in 1947, Groot et al. suggested potato tubers contain 70% tubrin and 30% "tubernin" protein fractions. Levitt (1951) reported almost equal quantities of water soluble and insoluble protein fractions. Nakasone et al. (1972) also observed about equal amounts of albumin and globulin. Lindner et al. (1960) were the earliest workers to classify the total tuber proteins into albumin, globulin, and prolamin and glutelin fractions. Kapoor et al. (1975) confirmed these percentages, but contradicted by Seibles (1979) who reported albumin and globulin as 75% and 25%, respectively. Snyder and Desborough (1980) studied the relationship of the albumin, globulin, prolamin, and glutelin fractions to the total protein in different potato germplasm groups and observed that the range of individual albumin fractions was from 46 to 74% in both the Andigena and Phureja-Tuberosum populations. For globulin content the range was 10 to 17% and 7 to 14% for Phureja-Tuberosum and Andigena, respectively.

4.3.1 Major Types of Potato Proteins

Based on all the above data, and the works of Ahldén and Trägårdh (1992), Ralet and Guéguen (2000), and Peksa et al. (2009) summarized that approximately 75% of the potato proteins are water soluble; which includes albumin (50–60%), globulin (25–25%), glutelin (9%), prolamin (2–4%) and residual (9%). On functional basis, soluble proteins present in potato tubers have been divided into three groups: patatins (30–40%), protease inhibitors (40–50%) and other proteins (10–15%) (it includes starch metabolism enzymes, kinases, phosphorylases, etc.) (Fig. 4.2) (Racusen and Foote 1980; Pouvreau et al. 2001; Man et al. 1997). These three major types of protein of potato tubers are discussed below.

4.3.1.1 Patatin

Patatin is the predominant potato tuber protein and represents a group of homologous glycoprotein isoforms as storage proteins with molecular weights of 40–45 kDa and have 4.5–5.2 isoelectric point (Alting et al. 2011; Fu et al. 2016). Patatin was given its common name by Racusen and Foote (1980). Potato patatins have potential cancer chemo-preventive activities and regulate serum cholesterol levels (Sun et al. 2013; Liyanage et al. 2008). Patatin also has antioxidant activity and prevents DNA from free radical damage (Nieto et al. 2009; Udenigwe et al. 2016). Functionally, patatin has enzymatic activity (lipid acyl hydrolase). Based on structural properties, surface charge and molecular weight patatins have been further divided into four isoforms, namely A, B, C and D, respectively (Pots et al. 1999). At neutral pH, patatin exists as a dimer with MW of 88 kDa held together by non-covalent hydrophobic forces. Its denaturation in the presence of SDS results into release of monomeric subunits of 40–45 kDa. Patatin is comprised of about 365 amino acids. These amino acid residues with both negative and positive charges are randomly distributed over the sequence. Therefore, patatin does not exhibit specific hydrophilic or hydrophobic properties. Patatin is a highly structured and stabilized protein at secondary as well as tertiary levels (Fig. 4.3). The a-helical amount of patatin is stable up to 45 °C, containing about 33% α helix and 46% β-sheet (Pots et al. 1998).

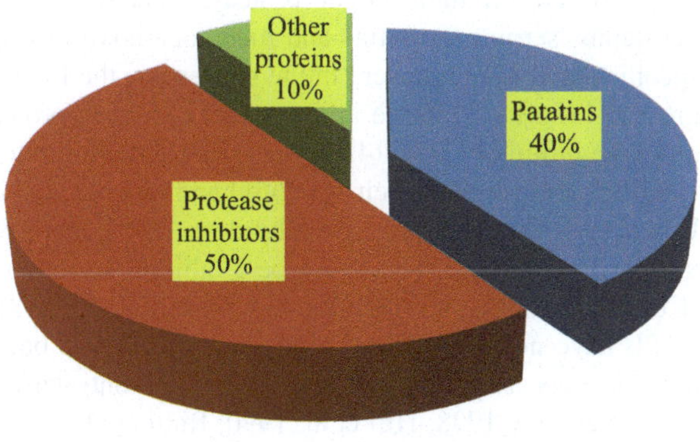

Fig. 4.2 Major types of potato proteins: (1) Patatin range 30–40%; (2) Protease inhibitors range 40–50% and (3) Other proteins range 10–15%

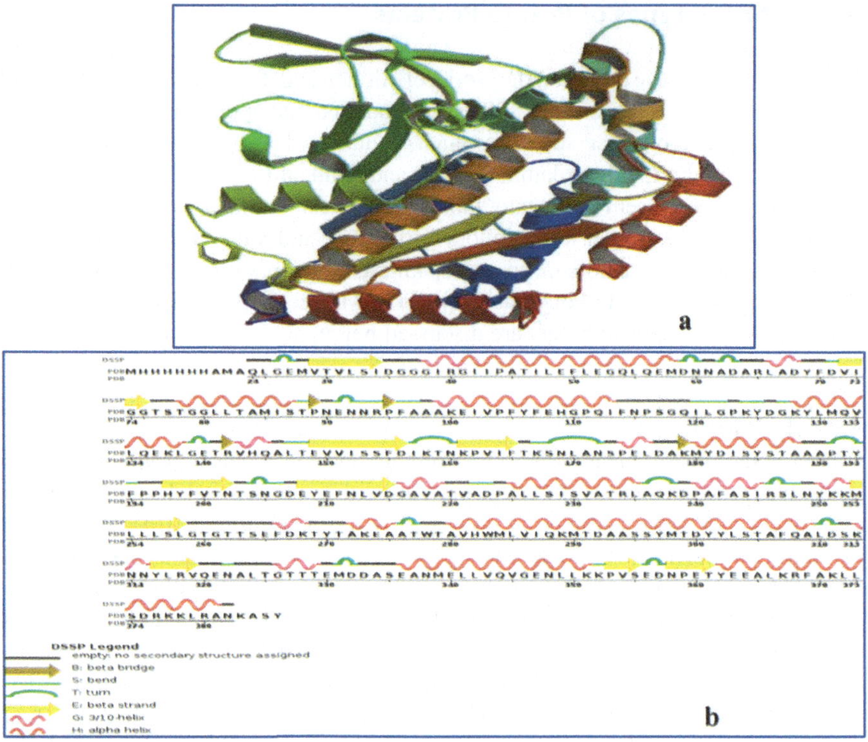

Fig. 4.3 Structure of patatin from potato. (**a**) Crystal structure, (**b**) Secondary structure (Source: RCBS Protein Data Bank https://www.rcsb.org/structure/4PK9)

4.3.1.2 Protease Inhibitors

Protease inhibitors (PIs) represent peptides with low-molecular mass (5 and 25 kDa) are more heterogeneous group of proteins than patatin. In plants, PIs also participate in many physiochemical processes including plant development, interaction of plants with biotic factors pathogens and insect pests, regulation of programmed cell death, storage proteins mobilization, etc. PIs have been classified in various ways. Based on their target protease specificity, PIs have been classified as pepstatins, serpins, cystatins and metallocarboxy protease inhibitors. Based on the specific reactive site present in the sequences, the PIs have been classified into four groups (1) serine protease inhibitors, (2) cysteine protease inhibitors, (3) metalloid protease inhibitors and (4) aspartic protease inhibitors, In plants, PIs are also classified according to their structural and biochemical characteristics (for example, Bowman–Birk serine protease inhibitors, metallocarboxypeptidase inhibitors, mustard trypsin inhibitors, potato type I inhibitors, potato type II protease inhibitors, (Kunitz) inhibitors, etc.) (Birk 2003; Rustgi et al. 2018).

PIs have not possess anti-nutritional properties, but also have anti-carcinogenic activities, act as positive dietary agents and anti-dermatitis agent (Pouvreau et al. 2001; Kennedy 1998; Hill et al. 1990; Rafii et al. 1999; Ruseler-van Embden et al.

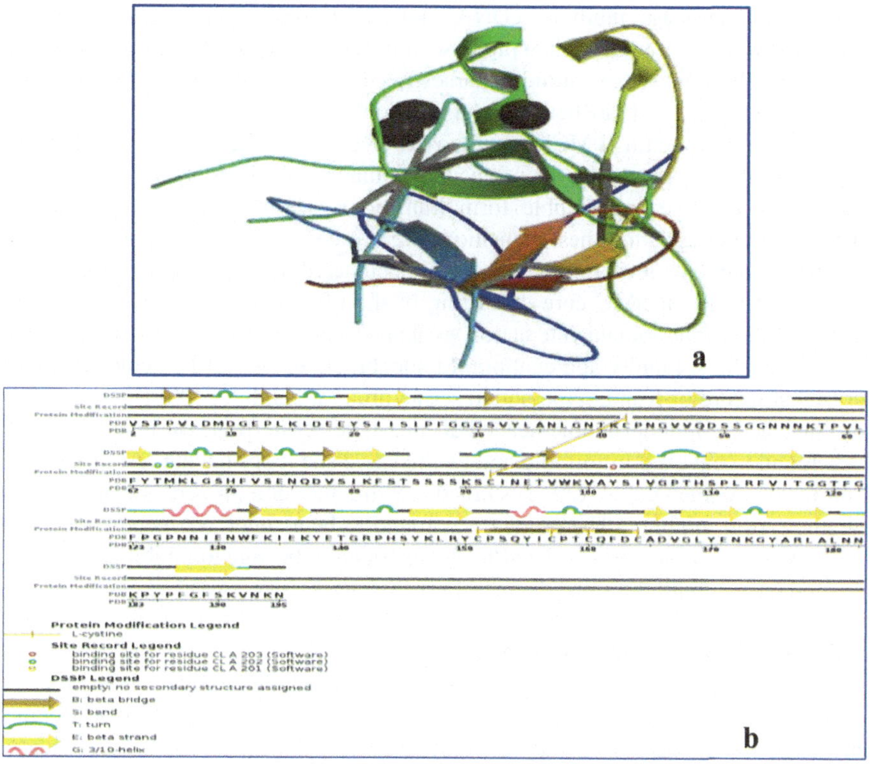

Fig. 4.4 Structure of Kunitz-type cysteine protease inhibitor from potato. (**a**) Crystal structure, (**b**) Secondary structure (Source: RCBS Protein Data Bank https://www.rcsb.org/structure/5DVH)

2004). Potato protease inhibitors generally inhibit chymotrypsin, trypsin and human leukocyte elastase (Pouvreau et al. 2001). In potato several PIs have been isolated and reported by various researchers. Structures of PIs from potato have been deciphered at secondary as well as at tertiary levels (Fig. 4.4). Shah et al. (2016) isolated and characterized a 20 kDa Kunitz-type serine PI from potato tubers. The isolated PI possessed lectin and hemagglutination activities with the retention at temperature as high as 80 °C, in a pH range of 4–9 and had dissociation constants of 2.8×10^{-3}M and 1.5×10^{-3}M respectively, for two interacting sugars galactose and mannose.

Guo et al. (2015) reported the first X-ray structure of another glycoprotein having 188 amino acids, from potato, named cathepsin D inhibitor contains. The cathepsin D inhibitor inhibited the serine protease trypsin as well as aspartic protease cathepsin D Guo et al. X-ray structure at a resolution of 2.1 Å showed that the cathepsin D inhibitor adopted a β-trefoil fold. β-trefoil fold is specific feature of the Kunitz-family protease inhibitors. Fischer et al. (2014) reported novel in vitro inhibitory activities of potato tuber protease inhibitors. They randomly sequenced more than 9000 cDNA clones derived from mature tubers of 10 potato cultivars and identified

120 unique protease inhibitor cDNA clones. Of those, 88 represented novel sequence variants of known plant protease inhibitor families. Those clones encoding protease inhibitors were further characterized functionally using heterologous expression system and biochemical and kinetic studies.

Potato multicystatin (PMC) is composed of eight cysteine proteases inhibiting repeating units, has several trypsin-sensitive domains linked cystatins and undergoes between crystalline and soluble form transitions. Though, significance and the mechanism(s) regulating these transitions are not completely understood. Keeping this background in view Green et al. (2013) deciphered the 2.2-Å crystal structure of the trypsin-resistant PMC core consisting of the fifth, sixth and seventh domains. The observed interdomain interaction explained PMC's resistance to trypsin and pH-dependent solubility/aggregation. Meulenbroek et al. (2012) elucidated the crystal structure of Kunitz-type potato serine protease inhibitor (PSPI) representing the first heterodimeric double-headed Kunitz-type serine protease inhibitor. PSPI was found to have a β-trefoil fold and two reactive site loops having residues Phe75 and Lys95. Valueva et al. (2012) studied the structure and properties of the potato chymotrypsin inhibitor. The protein was found to be a single polypeptide chain of molecular, weighing 23 kDa. The PKCI was found to be able to inhibit chymotrypsin and trypsin with the same degree of effectiveness.

4.3.1.3 Other Proteins in Potato

Proteins from potato tubers that do not belong to the patatin family or show protease inhibiting activities are grouped as other proteins. They have molecular weights higher than 40 kDa and they include lectins polyphenol oxidases, oxidase, lipooxygenases, enzymes involved in starch, phosphorylase isozymes, etc. These have been studied by various researchers (Hasan et al. 2014). As there are several proteins belonging to this group, however, they represent a relative smaller fraction of the total protein of potato tubers. Hence, these proteins have not been described here individually in this chapter.

4.4 Free Amino Acid in Potato

The free amino acid pool in potato accounts for 40–60% of the total nitrogen. This free amino acid pool does not proportionately contribute to the nutritional value of the potato as more than 90% of the free amino acid pool is constituted by nonessential amino acids consisting mainly of the amides, glutamine and asparagines. The free amino acid pool is influenced by environmental cues, soil type, fertilizer applications and cultivars and crop maturity. In general, it has been observed that increased nitrogen fertilizer application results in increase of free amino acid pools, particularly the amides. For example, 18% and 43% increase in asparagines and glutamine, respectively, as a result of high nitrogen supply, has been reported by Mulder and Bakema (1956). Further, Hoff et al. (1971), Coutrez-Geerinck (1975), and Eppendorfer and Bille (1996), have also reported variability in the free amino acid pool of potato tubers due to differential nitrogen fertilizer application. During storage of potatoes, the amides and free amino acids content have been found to be constant (Burton 1978).

4.5 Extraction and Estimation of Potato Proteins

Various methods have been employed for fractionating/isolating/purifying proteins from potato tubers. These methods include ultrafiltration, ion exchange, gel permeation, affinity, mixed-mode chromatography, carboxymethyl cellulose complexation, acid and heat coagulation, ethanol and salt-precipitation methods. Schoenbeck et al. (2013) developed direct capture membrane adsorption chromatography based method to purify proteins from potato fruit juice. They tested modules with fleece polymer spacers and extruded polymer spacers, as well as different spacer channel sizes for their binding capacities and long-term stability. They observed that modules with extruded polymer spacers channel size 250 μm had the highest binding capacities, while the modules with extruded polymer spacers channel size 480 μm showed the best long-term stability. Waglay et al. (2014) explored various techniques to isolate proteins from potato fruit juice. The employed extraction techniques included thermal/acidic combination, acidic, $FeCl_3$, $MnCl_2$, ethanol and $(NH4)_2SO_4$ precipitations, and carboxymethyl cellulose complexation. $(NH4)_2SO_4$ precipitation led to the highest yield (98.6%) and to the recovery of protein isolates enriched in patatin. $FeCl_3$ and $MnCl_2$ proved to be the best precipitating agents for >15 kDa protease inhibitors enrichment. Ralla et al. (2012) separated patatins and protease inhibitors from potato fruit juice using natural clay minerals as cation exchangers. The developed method also reduced the glycoalkaloids significantly in addition to separating protease inhibitors and patatins. Waglay and Karboune (2017) demonstrated a novel enzymatic approach based on the use of multi-enzymatic systems for the recovery of enriched protein extracts from potato pulp. They assessed ten commercially available multi-enzymatic systems for the recovery of patatin and protease inhibitors from potato pulp. Of those assessed, Depol 670L (DEP) and Ceremix 2XL (CER) were found to be most efficient with enrichment of up to 60% for patatin and up to 72% for protease inhibitors. Utilizing the thermotolerance property of lectins, Feng et al. (2018) developed a rapid method for purification of a heat-resistant lectin from potato. The developed method involved only one anion exchange chromatographic step beyond the ammonium sulphate precipitation and the heating treatment. With this method, the potato lectin was purified to homogeneity with 9513.3 u/mg of specific hemagglutination activity in 76.8% yield. Kowalczewski et al. (2019) used cross-flow ultrafiltration for concentrating protein fraction of potato juice from fresh potato juice obtained from starch processing industry. The concentrated fraction of potato juice was evaluated for biological activity, the amino acid composition, mineral content and antioxidant properties.

Various methods have been employed for estimation of proteins in potato tubers. Veilleux et al. (1981) have reviewed the methods employed to determine potato protein. These methods have been used to measure crude protein, all nitrogen-containing compounds, soluble or true protein. Among those listed for crude protein are Kjeldahl, amino acid analyses and alkali-phenol extraction. There are five methods used invariably for estimation of soluble protein in potato. These include (1) Lowry, (2) Potty, (3) Orange G, (4) Bromophenol blue, and (5) Coomassie blue. Non-destructive methods have also been employed for estimation of protein content in potato tubers. López et al. (2013) have demonstrated the use of near infrared spectroscopy (NIRS) to estimate the crude protein in potato. The estimation of crude protein content of lyophilized samples of potato by NIRS exhibited correlation coefficients of 0.86, 0.95 and 0.88 was obtained for cross validation, calibration and external validation, respectively, with low standard errors.

4.6 Effect of Growing Conditions, Storage as Well as Cooking on Protein and Free Amino Acids Content

Growing condition, storage environment, storage duration and various cooking methods have been envisaged to have effect on protein content, quality (in terms of amino acid composition of potato tubers) and/or free amino acid content and composition. In these aspects various studies have been conducted by various researchers across the globe. Generally it has been observed that when adequate nitrogen is supplied to the growing potato plants, the tubers contain proteins with the same amino acid composition. Sato et al. (2017) estimated the protein content in potato tubers of four cultivars (Toyoshiro, Kitahime, Snowden and Poroshiri) from three different locations in Japan (Tokachi, Kamikawa and Abashiri areas of Hokkaido). The protein content ranged from 1.37 to 1.92% (on fresh weight basis) and did not show any location specific significant variations. Leonel et al. (2017) studied the effect of cultivars and growth conditions on protein content. The study involved evaluating the chemical composition of tubers of five potato cultivars that were grown under the same cultural practices in soils with low, medium and high availability of phosphorus. They reported that increased availability of phosphorus in soil allowed the production of tubers with higher dry matter content, higher percentage of protein. One attribute of immature tubers is higher amounts of protein compared to mature tubers of the same genotype.

The protein pattern of the mature tubers has been reported to be stable during 9 months of cold storage (Snyder and Desborough 1978). In 1974, Desborough and Weiser reported that some tubers stored in cold storage for 6 months; a small loss of 3% in tuber protein was detected. A significant loss (63%) of isoleucine was noted, accompanied by a gain of tyrosine. They further suggested that because isoleucine may be a limiting amino acid in certain clones, the content of this amino acid should be checked after long periods of cold storage. Pęksa et al. (2018) studied the effect of storage conditions on amino acid composition of flesh-coloured potatoes. They observed decline in total protein content and amino acid content with increased storage time. Decrease of 6–19% and 21–38% was recorded for majority of amino acids after 3 and 6 months storage period, respectively. Twenty-five percent increase was observed for the coagulable protein content at 3 months' storage. Storage temperature was found to have no effect on the coagulable protein content, serine, glycine, cysteine, tyrosine and phenylalanine. However, potatoes stored at 2 °C had slightly more amino acids than tubers stored at 5 °C. Independent of the storage conditions, potatoes of yellow-fleshed and red-fleshed varieties were found to have comparatively high nutritive value, limited by leucine, methionine plus cysteine and leucine, respectively.

Klaassen et al. (2019) identified allele-specific quantitative trait loci (QTLs) for tuber protein content in cultivated potato at the tetraploid level. In a three year field trials they analysed 496 full-sib F1 clones. The analysis identified potential additive QTLs on different chromosomes, explained 12–17% of the phenotypic variance and the alleles of the QTLs provided both positive and negative effects on the phenotype. The results have indicated the complex genetic architecture of protein content trait which can be utilized for developing potential breeding strategies for improving protein content in potato. Bárta et al. (2012) studied cultivar variability of patatin in table versus processing potatoes. They compared the biochemical characteristics in tubers of 20 potato cultivars to evaluate their genotype differences with respect to utility groups, table potato cultivars and processing potato cultivars and found no signification variation in protein content, dry matter and relative abundance of patatin.

4.7 Improving Protein/Essential Amino Acids in Potato Through Genetic Engineering

Genetic engineering exploits the metabolic pathways genes or the associated transcription factors for manipulating intermediary or end product of the pathway. Three main genetic engineering based strategies being employed for improving protein quantity/quality are: (a) genetic engineering of essential amino acids, (b) genetic engineering to enhance the levels of natural high quality proteins in potato tubers and (c) improving the nutritional quality of potato tuber proteins through protein engineering and/or design (Dutt et al. 2019). Various attempts have been/are being made for increasing content of various essential amino acids in potato by targeting the genes involved in metabolic pathway (Fig. 4.5). Works in potato on the improvement of various quality attributes including proteins have been compiled recently (Dutt et al. 2019). For the first time, Altenbach et al. (1989) demonstrated the significant enhancement of methionine content in tobacco seed proteins through expressing transgene encoding a methionine-rich protein from Brazil nut. Chakraborty et al. (2000) developed transgenic potato over-expressing the sunflower albumin or an amaranth seed albumin (*AmA1*), which resulted in five to seven folds increase in total methionine level in tubers. Further analysis of transgenic potato lines with enhanced methionine amino acid via tuber-specific expression of a seed protein, *AmA1* (*Amaranth albumin* 1) revealed an increase in total protein contents up to 60% in comparison to the transformed potato (Chakraborty et al. 2010). Subsequently in 2013, Agrawal et al. performed comparative proteomics to elucidate the mechanism of action of Ama1 mediated increase in protein and higher tuber yield

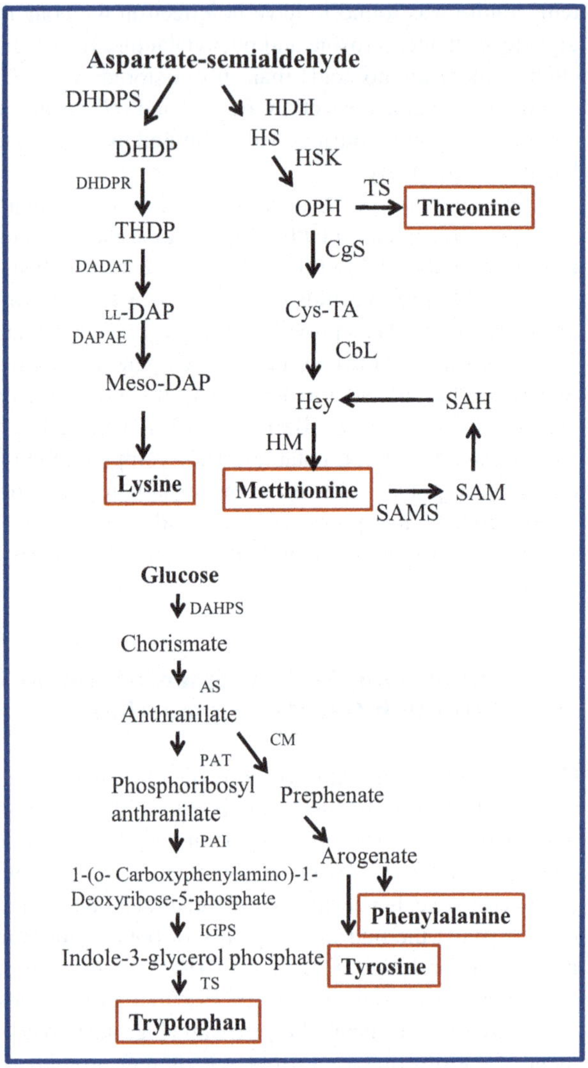

Fig. 4.5 Metabolic pathways involved in biosynthesis of essential amino acids. *ASD* aspartate semialdehyde, *DHDP* dihydrodipicolinate, *DHDPS* dihydrodipicolinate synthase, *HS* homoserine, *HDH* homoserine dehydrogenase, *OPH* o-phospho-homoserine, *HSK* homoserine kinase, *TS* Thr synthase, *CysTA* cystathionine, *CgS* cystathionine-synthase, *HcY* homocysteine, *CbL* cystathionine-lyase, *SAH* S-adenosyl-homocysteine, *HM* homocysteine methyltransferase, *SAM* S-adenosyl-methionine, *SAMS* S-adenosyl-methionine synthetase, *THDP* tetrahydro-dipicolinate, *DAPAE* DAP epimerase, *DAP* diaminopimelate, *DAPAT* DAP-aminotransferase. *AS* Anthranilate synthase, *PAT* phosphoribosylanthranilate transferase; *PAI* phosphoribosyl anthranilate isomerase, *IGPS* indole-3-glycerol phosphate synthase, *TS* Trp synthase, *AH* arogenate dehydrogenase, *DAHPS* DAHP synthase (partly adapted from Dutt et al. 2019)

in potato. Mass spectrometric analysis of AmA1-responsive protein spots (ARPs) led to the identification of 80 ARPs presumably associated with cell differentiation, regulating diverse functions, viz. protein biogenesis and storage, bioenergy and metabolism, and cell signaling. Metabolome study indicated up-regulation of amino acids paralleling the proteomics analysis. They suggested that these ARPs might work in coordinated fashion and result in increased protein synthesis and storage reserve accumulation in AmA1 tubers on one hand and organ development on the other. Goo et al. (2013) developed the potato transgenic having approximately 3.5-folds more methionine. They over-expressed a gene from *Perilla,* encoding *PrLeg* polypeptide using tuber-specific patatin promoter. It was also reported that higher isoleucine accumulation in transgenic tubers enhanced the methionine accumulation via methionine gamma-lyase (MGL) catabolism pathway (Huang et al. 2014). Kumar and Jander (2017) over-expressed cystathionine γ-synthase gene (isolated from Arabidopsis) in potato and observed increased methionine levels in tubers. Further, silencing methionine γ-lyase gene of potato has been shown to increase methionine content in potato tubers. Concomitant overexpression of cystathionine γ-synthase gene and silencing of methionine γ-lyase gene have been reported to accumulate higher free methionine levels than either single (either over-expressed or silenced) transgenic line. Zeh et al. (2001) reported up to 30-fold increase in methionine levels in potato upon silencing threonine synthase gene (involved in synthesis of threonine).

4.8 Other Specific Activity of Potato Proteins and Peptides

In addition to patatins, protease inhibitors and other proteins, potato tubers contain various peptides. Also, the digestion of potato proteins with various enzymes (proteases) results in potato protein hydrolysate having a range of peptides. Since peptides have lower molecular weights and less complex structures than proteins, their solubility, digestibility and absorbability are higher than those of proteins. Peptides isolated from potato protein have been reported to exhibit antioxidant, anticancer, antiobesity, antihyperlipidemic and antifungal/antibiotic activities and were also reported to exhibit angiotensin-converting enzyme (ACE) inhibition in vitro. Alcalase-generated potato protein hydrolysate (APPH) has been reported to be potential bioactive peptide against diabetes mellitus. Hu et al. (2015) reported that the heart protection effect of APPH is through IGF1R-PI3K-Akt compensatory reactivation in aging rats on high fat diets. It was observed that APPH treatments resulted in reduction of total cholesterol, low density lipoprotein and serum triacylglycerol levels to the normal levels. Asokan et al. (2019) studied the antidiabetic effects of a short peptide of potato protein hydrolysate in STZ-induced diabetic mice and the results demonstrated that the short synthetic peptide-DF might be effective in imparting protection against diabetes mellitus-associated damages.

Asokan et al. (2019) investigated the effect of APPH on high fat diet associated liver damages and recommended that APPH could be an efficient therapeutic agent to ameliorate high fat diet related liver damages. In 2018, Marthandam Asokan et al. reported anti-hypertrophic and anti-apoptotic effects of short peptides of potato protein hydrolysate against hyperglycaemic condition in cardiomyoblast cells. They showed that the small peptides, DF and IF of potato protein hydrolysate provided effective protection against high glucose-induced cardiomyocyte damages by ameliorating the apoptotic and hypertrophic effects. Bártová et al. (2018) investigated antifungal activity of potato tuber proteins isolated from starch production waste under different temperature regimes. They reported that the protein samples (consisting predominantly low-molecular proteins of 7–17 kDa) of 80 °C exhibited the highest antifungal activity when tested against *Fusarium* strains. Hu et al. (2019) reported that oral administration of APPH reduced aging and high fat diet induced hypertrophy and fibrosis in rats. Cotabarren et al. (2018) isolated a novel cystine-knot metallocarboxypeptidase inhibitor (chuPCI) from tubers of *Solanum tuberosum*, subsp. andigenum cv. Churqueña. These cystine-knot metallocarboxypeptidase have been suggested to have interesting pharmacological properties. Hasan et al. (2014) purified a new chitin-binding lectin from a potato cultivar of Bangladesh having bactericidal as well as antifungal activity.

4.9 Regulatory Status and Safety Issues Related to Potato Proteins

Potato protein based products have a regulatory status depending on the type of processing used. Because of their long history of safe use all over the globe potato protein products and hydrolysates derived from the potato proteins have obtained the GRAS status in the USA. Also, European Union has approved the potato protein based products as Novel Food in the EU. For above two categories the safe limits for glycoalkaloids have been set as <150 mg/kg product, and for sulphite the set limit is 100 ppm. These regulations highlight the essentiality and importance of keeping the glycoalkaloids content in potato cultivars to be used for extraction of protein, as low as possible. Similarly, while considering the peel of potato as a source for extraction of proteins this becomes even more important to have a plan for tackling (excluding/removing) glycoalkaloids during protein extraction, as peel contains comparatively higher amount of glycoalkaloids as compared to that of potato flesh (Singh et al. 2016; Dale et al. 1998). Of both, animal protein and vegetable sources, potato protein consumption exhibits the lowest incidence of allergenicity due to their faster and higher digestibility (Majamaa et al. 2001).

4.10 Conclusion

Potato is an important food crop consumed globally and contributes significantly in terms of quantity as well as in nutrition terms. It is a significant source of protein, yet improvement of its protein quantity and quality will be very useful for ameliorating various serious malnutrition problems (such as Kwashiorkor) associated with protein deficiency. Recent outbreak of COVID-19 pandemic has resulted in drastic reduction of animal based food consumption and thus intake of protein has come down significantly. This reduction in protein intake may have serious health related repercussions in the coming times. Therefore, it is speculated that the plant-based sources of proteins might play even more vital role in near future. Recent refinements in gene and genome editing technologies may be of great help in developing potato cultivars with high protein content and high quality protein.

References

Alting AC, Pouvreau L, Giuseppin MLF, van Nieuwenhuijzen NH (2011) Potato proteins. In: Phillips, G. O. & Williams, P. A. eds. Handbook of Food Proteins, Woodhead Publishing, 316-334

Agrawal L, Narula K, Basu S et al (2013) Comparative proteomics reveals a role for seed storage protein AmA1 in cellular growth, development, and nutrient accumulation. J Proteome Res. https://doi.org/10.1021/pr4007987

Ahldén I, Trägårdh G (1992) A study of soluble proteins from four potato varieties used in the Swedish starch industry. Food Chem. https://doi.org/10.1016/0308-8146(92)90321-R

Altenbach SB, Pearson KW, Meeker G et al (1989) Enhancement of the methionine content of seed proteins by the expression of a chimeric gene encoding a methionine-rich protein in transgenic plants. Plant Mol Biol. https://doi.org/10.1007/BF00027311

Asokan SM, Hung TH, Li ZY et al (2019) Protein hydrolysate from potato confers hepatic-protection in hamsters against high fat diet induced apoptosis and fibrosis by suppressing Caspase-3 and MMP2/9 and by enhancing Akt-survival pathway. BMC Complement Altern Med 19:283. https://doi.org/10.1186/s12906-019-2700-8

Baier AK, Knorr D (2015) Influence of high isostatic pressure on structural and functional characteristics of potato protein. Food Res Int. https://doi.org/10.1016/j.foodres.2015.05.053

Bárta J, Bártová V, Zdráhal Z, Šedo O (2012) Cultivar variability of patatin biochemical characteristics: table versus processing potatoes (Solanum tuberosum L.). J Agric Food Chem. https://doi.org/10.1021/jf3003448

Bártová V, Bárta J, Vlačihová A et al (2018) Proteomic characterization and antifungal activity of potato tuber proteins isolated from starch production waste under different temperature regimes. Appl Microbiol Biotechnol. https://doi.org/10.1007/s00253-018-9373-y

Birk Y (2003) In Plant protease inhibitors: significance in nutrition, plant protection, cancer prevention, and genetic engineering. Springer: Berlin/Heidelberg, Germany, pp 1–170

Burton WG (1978) Post-harvest behaviour and storage of potatoes. In: Applied biology. Academic, London

Camire ME, Kubow S, Donnelly DJ (2009) Potatoes and human health. Crit Rev Food Sci Nutr. https://doi.org/10.1080/10408390903041996

Chakraborty S, Chakraborty N, Datta A (2000) Increased nutritive value of transgenic potato by expressing a nonallergenic seed albumin gene from Amaranthus hypochondriacus. Proc Natl Acad Sci U S A. https://doi.org/10.1073/pnas.97.7.3724

Chakraborty S, Chakraborty N, Agrawal L et al (2010) Next-generation protein-rich potato expressing the seed protein gene AmA1 is a result of proteome rebalancing in transgenic tuber. Proc Natl Acad Sci U S A. https://doi.org/10.1073/pnas.1006265107

Cotabarren J, Tellechea ME, Tanco SM et al (2018) Biochemical and MALDI-TOF mass spectrometric characterization of a novel native and recombinant cystine knot miniprotein from Solanum tuberosum subsp. andigenum cv. Churqueña. Int J Mol Sci. https://doi.org/10.3390/ijms19030678

Coutrez-Geerinck D (1975) The effect of mineral nutrition on the distribution of free amino acids in tubers of cv. Gari Potato Res. https://doi.org/10.1007/BF02361772

Dale MFB, Griffiths DW, Bain H (1998) Effect of bruising on the total glycoalkaloid and chlorogenic acid content of potato (Solanum tuberosum) tubers of five cultivars. J Sci Food Agric. https://doi.org/10.1002/(sici)1097-0010(199808)77:4<499::aid-jsfa72>3.3.co;2-7

David S, Livney YD (2016) Potato protein based nanovehicles for health promoting hydrophobic bioactives in clear beverages. Food Hydrocoll. https://doi.org/10.1016/j.foodhyd.2016.01.027

Desborough SL, Weiser CJ (1974) Improving potato protein I. Evaluation of selection techniques. Am Potato J. https://doi.org/10.1007/BF02851409

Dutt S, Manjul AS, Raigond P et al (2017) Key players associated with tuberization in potato: potential candidates for genetic engineering. Crit Rev Biotechnol 37(7):942–957

Dutt S, Manjul AS, Chauhan M et al (2019) Biotechnology for nutritional and associated processing quality improvement in potato. In: Nutritional quality improvement in plants. Springer, New York, pp 429–483

Eppendorfer WH, Bille SW (1996) Free and total amino acid composition of edible parts of beans, kale, spinach, cauliflower and potatoes as influenced by nitrogen fertilisation and phosphorus and potassium deficiency. J Sci Food Agric. https://doi.org/10.1002/(SICI)1097-0010(199608)71:4<449::AID-JSFA601>3.0.CO;2-N

FAO (2020) FAO statistical databases FAOSTAT. http://www.fao.org/faostat/en/#home

Feng Y, Song J, Zhao Z et al (2018) A rapid and effective method for purification of a heat-resistant lectin from potato (Solanum tuberosum) tubers. Glycoconj J. https://doi.org/10.1007/s10719-018-9836-5

Fischer M, Kuckenberg M, Kastilan R et al (2014) Novel in vitro inhibitory functions of potato tuber proteinaceous inhibitors. Mol Gen Genomics. https://doi.org/10.1007/s00438-014-0906-5

Fu Y, Wu W, Zhu M, Xiao Z (2016) In silico assessment of the potential of patatin as a precursor of bioactive peptides. J Food Biochem. https://doi.org/10.1111/jfbc.12213

Gambuti A, Rinaldi A, Romano R et al (2016) Performance of a protein extracted from potatoes for fining of white musts. Food Chem. https://doi.org/10.1016/j.foodchem.2015.05.067

Goo YM, Kim TW, Lee MK, Lee SW (2013) Accumulation of PrLeg, a perillalegumin protein in potato tuber results in enhanced level of sulphur-containing amino acids. Comptes Rendus. https://doi.org/10.1016/j.crvi.2013.09.002

Green AR, Nissen MS, Mohan Kumar GN et al (2013) Characterization of Solanum tuberosum multicystatin and the significance of core domains. Plant Cell. https://doi.org/10.1105/tpc.113.121004

Groot EH, Janssen LW, Kentie A et al (1947) A new protein in potatoes. Biochim Biophys Acta 1:410–414

Guo J, Erskine PT, Coker AR et al (2015) Structure of a Kunitz-type potato cathepsin D inhibitor. J Struct Biol. https://doi.org/10.1016/j.jsb.2015.10.020

Hasan I, Ozeki Y, Kabir SR (2014) Purification of a novel chitin-binding lectin with antimicrobial and antibiofilm activities from a Bangladeshi cultivar of potato (Solanum tuberosum). Indian J Biochem Biophys 51(2):142–148

Hill AJ, Peikin SR, Ryan CA, Blundell JE (1990) Oral administration of proteinase inhibitor II from potatoes reduces energy intake in man. Physiol Behav. https://doi.org/10.1016/0031-9384(90)90307-P

Hoff JE, Jones CM, Wilcox GE, Castro MD (1971) The effect of nitrogen fertilization on the composition of the free amino acid pool of poato tubers. Am Potato J. https://doi.org/10.1007/BF02861758

Hu WS, Ting WJ, Chiang WD et al (2015) The heart protection effect of alcalase potato protein hydrolysate is through IGF1R-PI3K-akt compensatory reactivation in aging rats on high fat diets. Int J Mol Sci. https://doi.org/10.3390/ijms160510158

Hu WS, Ting WJ, Tamilselvi S et al (2019) Oral administration of alcalase potato protein hydrolysate-APPH attenuates high fat diet-induced cardiac complications via TGF-$β$/GSN axis in aging rats. Environ Toxicol 34:5–12

Huang T, Joshi V, Jander G (2014) The catabolic enzyme methionine gamma-lyase limits methionine accumulation in potato tubers. Plant Biotechnol J. https://doi.org/10.1111/pbi.12191

Kapoor AC, Desborough SL, Li PH (1975) Potato tuber proteins and their nutritional quality. Potato Res 18:469–478

Kennedy AR (1998) Chemopreventive agents: protease inhibitors. Pharmacol Ther 78:167–209

Klaassen MT, Bourke PM, Maliepaard C, Trindade LM (2019) Multi-allelic QTL analysis of protein content in a bi-parental population of cultivated tetraploid potato. Euphytica. https://doi.org/10.1007/s10681-018-2331-z

Komarnytsky S, Cook A, Raskin I (2011) Potato protease inhibitors inhibit food intake and increase circulating cholecystokinin levels by a trypsin-dependent mechanism. Int J Obes. https://doi.org/10.1038/ijo.2010.192

Kowalczewski PŁ, Olejnik A, Białas W et al (2019) The nutritional value and biological activity of concentrated protein fraction of potato juice. Nutrients. https://doi.org/10.3390/nu11071523
Kumar P, Jander G (2017) Concurrent overexpression of Arabidopsis thaliana cystathionine γ-synthase and silencing of endogenous methionine γ-lyase enhance tuber methionine content in Solanum tuberosum. J Agric Food Chem. https://doi.org/10.1021/acs.jafc.7b00272
Leonel M, de Carmo EL, Fernandes AM et al (2017) Chemical composition of potato tubers: the effect of cultivars and growth conditions. J Food Sci Technol 54:2372–2378. https://doi.org/10.1007/s13197-017-2677-6
Levitt J (1951) The isolation and preliminary fractionation of proteins from dormant and growing potato tubers. Plant Physiol. https://doi.org/10.1104/pp.26.1.59
Liang S, McDonald AG (2014) Chemical and thermal characterization of potato peel waste and its fermentation residue as potential resources for biofuel and bioproducts production. J Agric Food Chem. https://doi.org/10.1021/jf5019406
Lindner K, Jaschik S, KORPACZY I, others (1960) Ammo acid composition and biological value of potato protein fractions. Qual Plant Mater Veg 7:289–294
Lisinska G, Leszczynski W (1989) Potato science and technology. Springer, New York
Liyanage R, Han KH, Watanabe S et al (2008) Potato and soy peptide diets modulate lipid metabolism in rats. Biosci Biotechnol Biochem. https://doi.org/10.1271/bbb.70593
López A, Arazuri S, Jarén C et al (2013) Crude protein content determination of potatoes by NIRS technology. Proc Technol. https://doi.org/10.1016/j.protcy.2013.11.064
Majamaa H, Seppälä U, Palosuo T et al (2001) Positive skin and oral challenge responses to potato and occurrence of immunoglobulin E antibodies to patatin (Sol t 1) in infants with atopic dermatitis. Pediatr Allergy Immunol. https://doi.org/10.1034/j.1399-3038.2001.00096.x
Man AL, Purcell PC, Hannappel U, Halford NG (1997) Potato SNF1-related protein kinase: molecular cloning, expression analysis and peptide kinase activity measurements. Plant Mol Biol. https://doi.org/10.1023/A:1005765719873
Marthandam Asokan S, Yang JY, Lin WT (2018) Anti-hypertrophic and anti-apoptotic effects of short peptides of potato protein hydrolysate against hyperglycemic condition in cardiomyoblast cells. Biomed Pharmacother. https://doi.org/10.1016/j.biopha.2018.08.070
Meulenbroek EM, Thomassen EAJ, Pouvreau L et al (2012) Structure of a post-translationally processed heterodimeric double-headed Kunitz-type serine protease inhibitor from potato. Acta Crystallogr Sect D Biol Crystallogr. https://doi.org/10.1107/S090744491201222X
Mulder EG, Bakema K (1956) Effect of the nitrogen, phosphorus, potassium and magnesium nutrition of potato plants on the content of free amino-acids and on the amino-acid composition of the protein of the tubers. Plant Soil. https://doi.org/10.1007/BF01343724
Nakasone K, Hayashi R, Hata T (1972) Composition of potato protein. Agric Chem Soc Jpn J 46:45
Nieto G, Castillo M, Xiong YL et al (2009) Antioxidant and emulsifying properties of alcalase-hydrolyzed potato proteins in meat emulsions with different fat concentrations. Meat Sci. https://doi.org/10.1016/j.meatsci.2009.03.005
Nouri-Ellouz O, Zeghal N, Makni S et al (2015) New food from a potato somatic hybrid: nutritional equivalence and safety assessment by a feeding study on rats. J Sci Food Agric. https://doi.org/10.1002/jsfa.6898
Ortiz-Medina E, Donnelly DJ (2003) Concentration and distribution of total soluble protein in fresh and stored potato tubers. In: Acta horticulturae. Springer, New York
Osborne TB, Campbell GF (1896) The proteids of the potato. J Am Chem Soc 18:575–582
Peksa A, Rytel E, Kita A et al (2009) The properties of potato proteins. Food 3:79–87
Pęksa A, Miedzianka J, Nemś A (2018) Amino acid composition of flesh-coloured potatoes as affected by storage conditions. Food Chem. https://doi.org/10.1016/j.foodchem.2018.06.026
Pots AM, De Jongh HHJ, Gruppen H et al (1998) Heat-induced conformational changes of patatin, the major potato tuber protein. Eur J Biochem. https://doi.org/10.1046/j.1432-1327.1998.2520066.x
Pots AM, Gruppen H, Hessing M et al (1999) Isolation and characterization of patatin isoforms. J Agric Food Chem. https://doi.org/10.1021/jf981180n

Pouvreau L, Gruppen H, Piersma SR et al (2001) Relative abundance and inhibitory distribution of protease inhibitors in potato juice from cv. Elkana. J Agric Food Chem. https://doi.org/10.1021/jf010126v

Racusen D, Foote M (1980) A major soluble glycoprotein of potato tubers. J Food Biochem. https://doi.org/10.1111/j.1745-4514.1980.tb00876.x

Rafii F, Ruseler-van Embden JGH, Van Lieshout LMC (1999) Changes in bacterial enzymes and PCR profiles of fecal bacteria from a patient with ulcerative colitis before and after antimicrobial treatments. Dig Dis Sci. https://doi.org/10.1023/A:1026634229934

Raigond P, Kaundal B, Sood A et al (2018) Quantification of biguanide and related compounds (anti-diabetic) in vegetables and fruits. J Food Compos Anal. https://doi.org/10.1016/j.jfca.2018.09.001

Ralet MC, Guéguen J (2000) Fractionation of potato proteins: solubility, thermal coagulation and emulsifying properties. LWT- Food Sci Technol. https://doi.org/10.1006/fstl.2000.0672

Ralet MC, Guéguen J (2001) Foaming properties of potato raw proteins and isolated fractions. LWT- Food Sci Technol. https://doi.org/10.1006/fstl.2000.0684

Ralla K, Sohling U, Suck K et al (2012) Separation of patatins and protease inhibitors from potato fruit juice with clay minerals as cation exchangers. J Sep Sci. https://doi.org/10.1002/jssc.201200030

Ruseler-van Embden JGH, Van Lieshout LMC, Smits SA et al (2004) Potato tuber proteins efficiently inhibit human faecal proteolytic activity: implications for treatment of peri-anal dermatitis. Eur J Clin Investig. https://doi.org/10.1111/j.1365-2362.2004.01330.x

Rustgi S, Boex-Fontvieille E, Reinbothe C et al (2018) The complex world of plant protease inhibitors: insights into a Kunitz-type cysteine protease inhibitor of Arabidopsis thaliana. Commun Integr Biol 11:e1368599

Sato H, Koizumi R, Nakazawa Y et al (2017) Data on the weights, specific gravities and chemical compositions of potato (Solanum tuberosum) tubers for food processing from different areas of Hokkaido. Jpn Data Br. https://doi.org/10.1016/j.dib.2017.03.009

Schoenbeck I, Graf AM, Leuthold M et al (2013) Purification of high value proteins from particle containing potato fruit juice via direct capture membrane adsorption chromatography. J Biotechnol. https://doi.org/10.1016/j.jbiotec.2013.09.018

Scott GJ, Suarez V (2012) The rise of Asia as the centre of global potato production and some implications for industry. Potato J 39:1–22

Seibles TS (1979) Studies on potato proteins. Am Potato J. https://doi.org/10.1007/BF02854059

Shah KR, Patel DK, Pappachan A et al (2016) Characterization of a Kunitz-type serine protease inhibitor from Solanum tuberosum having lectin activity. Int J Biol Macromol. https://doi.org/10.1016/j.ijbiomac.2015.11.068

Shewry PR (2003) Tuber storage proteins. Ann Bot 91(7):755–769

Singh B, Raigond P, Barwal S et al (2016) Glycoalkaloids in peels of Indian potatoes. Potato J 43:86–92

Snyder JC, Desborough SL (1978) Rapid estimation of potato tuber total protein content with coomassie brilliant blue G-250. Theor Appl Genet. https://doi.org/10.1007/BF00264747

Snyder JC, Desborough SL (1980) Total protein and protein fractions in tubers of Group Andigena and Phureja-Tuberosum hybrids. Qual Plant Plant Foods Hum Nutr. https://doi.org/10.1007/BF01099050

Spelbrink REJ, Lensing H, Egmond MR, Giuseppin MLF (2015) Potato patatin generates short-chain fatty acids from milk fat that contribute to flavour development in cheese ripening. Appl Biochem Biotechnol. https://doi.org/10.1007/s12010-015-1569-3

Storey M (2007) The harvested crop. In: Vreugdenhil D, Bradshaw J, Gebhardt C, Govers F, Mackerron DKL (eds) Potato biology and biotechnology. Elsevier, London, p 2007

Sun Y, Jiang L, Wei D (2013) Partial characterization, in vitro antioxidant and antiproliferative activities of patatin purified from potato fruit juice. Food Funct. https://doi.org/10.1039/c3fo60248f

Udenigwe CC, Udechukwu MC, Yiridoe C et al (2016) Antioxidant mechanism of potato protein hydrolysates against in vitro oxidation of reduced glutathione. J Funct Foods. https://doi.org/10.1016/j.jff.2015.11.004
Vaitkevičienė N (2019) A comparative study on proximate and mineral composition of coloured potato peel and flesh. J Sci Food Agric. https://doi.org/10.1002/jsfa.9895
Valueva TA, Parfenov IA, Revina TA et al (2012) Structure and properties of the potato chymotrypsin inhibitor. Plant Physiol Biochem. https://doi.org/10.1016/j.plaphy.2011.10.021
Veilleux RE, Lauer FI, Desborough SL (1981) Breeding behavior for tuber protein in Solanum tuberosum and Tuberosum-phureja hybrids. Euphytica. https://doi.org/10.1007/BF00038782
Waglay A, Karboune S (2017) A novel enzymatic approach based on the use of multi-enzymatic systems for the recovery of enriched protein extracts from potato pulp. Food Chem. https://doi.org/10.1016/j.foodchem.2016.09.147
Waglay A, Karboune S, Alli I (2014) Potato protein isolates: recovery and characterization of their properties. Food Chem. https://doi.org/10.1016/j.foodchem.2013.07.060
Zeh M, Casazza AP, Kreft O et al (2001) Antisense inhibition of threonine synthase leads to high methionine content in transgenic potato plants. Plant Physiol. https://doi.org/10.1104/pp.010438
Zhang D, Mu T, Sun H (2016) Domestic and abroad research progress of potato tuber-specific storage protein patatin. Sci Agric Sin. https://doi.org/10.3864/j.issn.0578-1752.2016.09.011

Lipids in Potato

5

Milan Kumar Lal, Awadhesh Kumar, Rupak Jena, Som Dutt, Nitasha Thakur, Vandana Parmar, Vinod Kumar, and Brajesh Singh

Abstract

Potato tuber is rich in carbohydrates, but it also contains essential lipids. The total lipid content of the potato is approximately 0.15%–0.5% of the fresh weight. The lipids of potato are enriched in linoleic and linolenic acids. It may contain bioactive lipid compounds like glycolipid, phospholipids, sterols and carotenoids which are more desirable for their health-promoting effects. Biosynthesis of fatty acid takes place mainly in two organelles, viz., plastid and endoplasmic reticulum. Presence of lipid, phosphates and low-molecular-weight-proteins in the interior of the starch granules leads to interaction with the starch. These properties affects the starch physicochemical characteristics like gelatinization, retrogradation, swelling and viscosity during the processing of potato. The extraction and estimation of lipid content can be done by using a combination of *n*-propanol and water. In TLC chromatography method chloroform-methanol is used to separate various lipid components and further subjected to gas chromatography. Growing condition of potato crop, storage and cooking of potato may affect the lipid content. Efforts have been made to increase the lipid content in potato through conventional breeding and transgenic approaches. However, high lipid content in potato may affect the starch accumulation and composition which may lead to enhanced sugar content. More research needs to be carried out for the develop-

M. K. Lal (✉) · S. Dutt · N. Thakur · V. Parmar · V. Kumar · B. Singh
ICAR-Central Potato Research Institute, Shimla, Himachal Pradesh, India
e-mail: milan2925@gmail.com

A. Kumar
ICAR-National Rice Research Institute, Cuttack, Odisha, India

R. Jena
ICAR-Directorate of Groundnut Research, Junagadh, Gujarat, India

P. Raigond et al. (eds.), *Potato*, https://doi.org/10.1007/978-981-15-7662-1_5

ment of potato cultivars with an adequate amount of lipid content without affecting starch metabolism.

Keywords

Health benefits · Genetic modification

5.1 Introduction

Potato (*Solanum tuberosum* L.) is the third most important food crop after rice and wheat with increasing popularity in terms of human consumption. Its annual production is 388.19 million tons (MT) and expected to increase in future (FAO 2003). Potato is a well-known staple vegetable for Indians and also used as a staple food in the European countries and is known to add carbohydrates, nutrients and minerals to our diet (Raigond et al. 2017).

Potato is rich in starch content which can go up to 30.4% of its fresh weight (FW), however, protein content is only up to 2% (Jansen et al. 2001). Freshly harvested potatoes are rich in vitamins such as vitamin B_6 (2.5 μg/g FW), vitamin C (0.20 mg/g FW) and niacin (0.164 μg/g FW). Potato is a good source of vitamin C and also have antioxidants and phenolic compounds like phenolic acid and flavonoids (Kostyn et al. 2013). There are diversity in uses of potato as it can be consumed by cooking in several ways like boiling, baking, steaming, frying, microwaving and roasting (Maga 1994).

The total lipid content in potato varies from 0.1 to 0.5% of total fresh weight in potato tuber (Ramadan 2016). Out of this major portion of lipid is present in cell membrane and triacylglycerol (TAG) is accumulated in the flesh of tuber in the form of lipid droplet (LD). The presence of quality lipid, which is mostly unsaturated fatty acid, makes potato a suitable crop for making healthy food and products. The presence of starch, low-molecular weight (LWM) protein and fat in potato and their interaction can influence the processing of potato. Their presence and interaction within starch granules strongly affect the starch behaviour toward retrogradation, gelatinization, viscosity, swelling, glycemic index, resistant starch and leaching of soluble carbohydrate during processing (Kumar et al. 2018a). These above properties may also be influenced by cooking, baking and extrusion (Kitahara et al. 1997; Lin and Czuchajowska 1998).

It is interesting to know that despite storage of starch, potato also accumulate oil in low concentrations. There must be a pathway which is regulating the sucrose import in the tuber and can further convert it into starch or lipid depending upon the demand and regulation at molecular level (Hofvander et al. 2016). Most of North American and European cultivars have been evaluated for lipid content (Cheng and Muneta 1978), but little information is available about lipid content in Indian potato cultivars. Not much research has been carried out on evaluation or modification of potato lipids.

5.2 Biosynthesis and Types of Lipids Present in Potato

Lipids are the group of molecules that are soluble in organic solvent. These can be further classified broadly into three groups, viz., monoglyceride, diglyceride and triglycerides. The complex lipid which is mainly present in membrane system includes phospholipid, sulpholipid, galactolipid and other lipid, like derived lipids such as free fatty acid and long chains of alcohols (Bonnaire et al. 2008).

The different types of lipids present in potato tuber are phospholipids (PL), glycolipids (GL), galactolipids, sulpholipids and neutral lipids (NL). Total lipid content in potato is approximately 0.1–0.5% of fresh weight of potato tuber. Fatty acid composition of potato tuber reflects the composition of cell membranes. Lipid found in potato consists of 47.4% phospholipid, 21.6% galactolipid, 1.3% sulpholipid, and 15.4% triglyceride (Galliard 1973). Most of the lipids present in potato are in the esterified form. Esterified lipids found in potato are 6.4% esterified steryl glucoside and 2.4% cerebroside. Unsaturated fatty acid like omega-3 and omega-6 fatty acids are essential for good health. Polyunsaturated fatty acid (PUFA) and monounsaturated fatty acid (MUFA) are two classes of unsaturated fatty acids which are present in potato and considered as desirable fatty acid for human health. While other fats like trans fatty acid and some unsaturated fats are unhealthy and have adverse effect on health that increases the risk of diseases. The fatty acid content in potato is advantageous as most of it is made up of unsaturated fatty acid like linolenic and linoleic acid (Prescha et al. 2001). Linoleic and linolenic acids are the precursor for a wide range of volatile compounds (Spychalla and Desborough 1990). The tolerance to the heat stress or cold stress is provided by degree of unsaturation of fatty acid in the lipid constituent. More unsaturation of fatty acids provide tolerance to cold and less unsaturation of fatty acids provide tolerance to high temperature stress in plants (Upchurch 2008). It was reported that cold acclimated potato (*Solanum commersonii*) accumulated more linoleic acid (18:2) in the membrane of leaf tissue compared to non-acclimated or sensitive potato (Vega et al. 2004).

Interaction of lipid with other molecules like starch may affect the lipid content. This lipid–saccharide interaction can facilitate or can hamper processing and affect the properties of both starch and lipids. Sometimes the autoxidation of these unsaturated fatty acid can cause the off-flavour development in the dehydrated potato products which make it unsuitable for consumption (Galliard 1973). PUFA present in potato are very much susceptible to non-enzymatic autoxidation. Potato contains 16–19% palmitic acid, 13–24% linolenic acid and 52–60% linoleic acid (Galliard et al. 1975).

Biosynthesis of fatty acid and triacylglycerol synthesis is highly regulated process in plant which requires lots of energy and controlled mechanism. Biosynthesis of fatty acid in plant mainly takes place in plastid with last few steps in endoplasmic reticulum. Plastid is an important organelle where de novo biosynthesis of fatty acid occurs. The biotin containing enzyme acetyl-CoA carboxylase which catalyse the first committed step, results the formation of malonyl-CoA from acetyl-CoA and bicarbonate. Biosynthesis of 16 or 18-carbon fatty acid is performed by

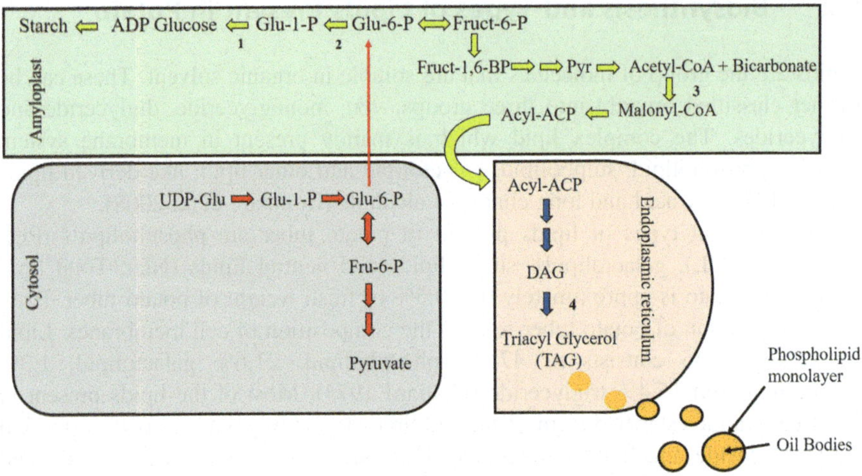

Fig. 5.1 Illustration of lipid and starch pathway in potato (*Solanum tuberosum* L.) tuber. Starch is produced in potato tuber and stored in the form of starch granules. Biosynthesis of lipid occurs in potato tuber by metabolizing stored starch via glycolysis and converted to acetyl-CoA and malonyl-CoA de novo synthesis in starch granules of amyloplast. Further fatty acyl chain which is formed is transported to smooth endoplasmic reticulum for lipid biosynthesis. Depending upon the requirement of plant (either for storage or membrane stability) TAG or DAG (storage lipid) is formed and different categories of membrane lipids are formed. TAG is stored, collected and stored in the protein like membrane structure known as oleosin which is made of phospholipid bilayer. The oil bodies formed from ER is known as lipid. *ADP-glucose* adenosine diphosphate glucose, *Glu-1-P* glucose-1-phosphate, *Glu-6-P* glucose-6-phosphate, *Fruct-6-P* fructose-6-phosphate, *Fruct-1,6-BP* fructose-1,6-bisphosphate, *Pyr* pyruvate, *Acyl-ACP* acyl carrier protein, *DAG* diacyl glycerol, *UDP-Glu* uridine diphosphate glucose, (1) ADP-glucose pyrophosphorylase, (2) Phosphoglycerate mutase, (3) Acetyl-CoA carboxylase, (4) diacylglycerol acyltransferase

multi-subunit complex consisting of monofunctional enzymes, fatty acid synthase (FAS). This enzyme is a large and a complex of seven multi-subunit which can be easily dissociated (Brown et al. 2006). FAS catalyses the transfer of malonyl group from CoA to acyl carrier protein (ACP) (which serves as carrier for growing acyl chain). Subsequent condensation of acetyl-CoA with ACP leads to formation of 16-carbon palmitoyl-ACP to 18-carbon stearol-ACP which was catalysed by 3-ketoacyl-ACP synthases II (KASII) (Shimakata et al. 1982). The fatty acids or the acyl-ACP are synthesized and exported to endoplasmic reticulum (ER), where they become the precursor for production of storage lipids like diacylglycerol (DAG) and triacylglycerol (TAG), membrane lipids like phosphatidic acid (PA), phosphatidylcholine (PC) and phosphatidylethanolamine (PE) (Gidda et al. 2009). Glycerol-3-phosphate is produced by glycerol-3-phosphate dehydrogenase which is then converted to PA. The enzymes present in endoplasmic reticulum, *sn1*-glycerol-3-phosphate acyltransferase (GPAT) (which attack at *sn1* position of glycerol backbone), lysophosphatidic acid acyltransferase (LPAAT) and diacyl-glycerol acyltransferase (DGAT) sequentially add acyl group on a glycerol-3-phosphate backbone chain to produce DAG and subsequently TAG (Fig 5.1) (Weselake et al.

2009). After formation of TAG in endoplasmic reticulum, which is also known as oil bodies is budded off which consists of single layer phospholipid membrane. This is also known as micelles, which contain polar head group (hydrophilic) and is in contact with cytosol and the non-polar tails (hydrophobic), which remain in internal neutral lipids. This is coated with oleosin, which is small protein that coats lipid droplets and thus stabilizing TAG from disintegration by cytosolic enzymes which can withstand strain of both dehydration and rehydration (Siloto et al. 2006). Moreover, it was observed in transgenic potato that there is tight association of LD and endoplasmic reticulum, which implies that LD is formed on the smooth surface of endoplasmic reticulum (Thiam et al. 2013).

The impact of lipid biosynthesis or accumulation in potato was found to have severe effects on other major carbon metabolism like starch synthesis, sucrose synthesis and other metabolites' production. Attempts were made to increase lipid and fat content in potato which were reported to hamper the starch synthesis. Overexpression of WRI1, DGAT1 or OLEOSIN in potato negatively affects the starch accumulation in potato lines with high lipid content (Liu et al. 2017). There was an increase in the accumulation of soluble sugars due to enhancement in lipid biosynthesis. It was also reported that the upregulation of fatty acid biosynthesis and TAG assembly resulted in the reduction of starch accumulation in tobacco plants (Vanhercke et al. 2014).

5.3 Most Preferred Methods for Estimation of Lipids

Lipid extraction in potato may hinder due to formation of amylose-lipid complex. *N*-propanol-water in the ratio of 3:1 v/v is used in most preferred method for extraction of lipid from potato. This combination is used as cold and hot extraction into surface and internal lipid fraction. Potato starch contains high level of surface lipid (0.32 g/100 g), while internal lipid content is 0.21 g/100 g. Hot extraction in potato showed higher lipid content as compared to cold extraction. Moreover, it was found that C20:1 (n-9) could be extracted from potato starch by hot extraction method only (Błaszczak et al. 2003). It was reported in other experiments that the hot extraction (*n*-propanol:water 3:1 v/v at 90–100 °C.) from potato starch leads to increase in the amount of saturated fatty acid.

Potato starch granules were fractionated in very small, small medium, large and very large groups out of which total lipid was found to be the highest in smaller potato starch granules. Fatty acid found in potato starch are in the following order: C16:0>C18:2>C18:3>C18:1. As discussed earlier hot extraction gives more amount of lipid content, but the cold extract of potato starch contained high level of saturated fatty acid (42.1%) as compared to hot extraction (Dhital et al. 2011). Lipids were extracted with chloroform-methanol extraction solution and further subjected to thin layer chromatography (TLC) or column chromatography. With this method neutral lipids, phospholipids and glycolipids were reported to be 6.5%, 45.5% and 38.1%, respectively (Lepage 1968).

The total fat can be extracted and determined in the peeled tuber using the extraction-gravimetrical method (Prescha et al. 2001). This is the modified method of Bligh and Dyer 1959. The chloroform extract can be used to extract lipid content in potato from its various tissues. Then it can be evaporated under lyophilizer or under stream of dry nitrogen. The fat which was extracted is fractionated in non-polar and polar fraction which was later separated using silica gel. The polar lipid was eluted in chloroform and polar lipids in methanol. Fatty acid composition was determined by using gas chromatography method by converting lipid to fatty acid methyl esters (FAMEs) (Prescha et al. 2001). Most of the lipids in potato are located between the peel and vascular ring of the tuber. So, while estimating lipid content in potato this is to be considered.

5.4 Effect of Growing Conditions, Storage as Well as Cooking on Lipids

5.4.1 Growing Condition

Potato tuber is the modification of stem known as tuber, which is produced underground. This tuber is sink tissue in the later part of the plant development. It acts as the storage organ mainly for starch. It also accumulates protein (predominantly the glycoprotein patatin) and triacylglycerol which is accumulated over the growing season. The stored starch is used as energy reserve for germination, sprouting and early plant development (Höfgen and Willmitzer 1990). This helps the seedlings to develop into plant until it starts to photosynthesize its own assimilate. The growing condition of potato affects the nutritional quality of potato. Effect of growing condition on potato lipids content and composition is not studied yet.

5.4.2 Storage

Potato when stored for longer time develops off-flavour due to autoxidation of unsaturated fatty acid present in potato. This development of off-flavour is mainly observed in the dehydrated products of potato (Cheng and Muneta 1978). The storage of potato tuber leads to loss of membrane integrity between 7 and 26 months of storage at 4 °C and 95% relative humidity (RH). The process of aged induced decline in membrane integrity is due to increased peroxidation of damage of membrane lipid of potato (Kumar and Knowles 1993).

5.4.3 Cooking Method

The cooking methods of potato include boiling, microwaving, baking and frying. Cooking of potato affects the nutritionally quality, taste and bioavailability of various bioactive compounds present in the tuber (Liu et al. 2007). The raw potato contains low levels of flavour volatiles and these volatiles enhance during cooking. The thermal processing influences flavour generation in potato due to temperature change. Processing method affects the concentration of flavouring compounds and hence flavour of dish/products (Maga 1994; Raigond et al. 2015). These flavouring compounds are reducing sugar, amino acid and product of fatty acid which are involved in the formation of pyrazines when Maillard reaction takes place. During cooking of potato, amino acids were produced from branched chain carbonyl and alcohol volatiles by Strecker degradation. However, molecules that are straight chain including ketones, aldehyde, alcohol and alkyl furans are derived from oxidation. Alkyl furan are derived from oxidation of unsaturated fatty acids (Dobson et al. 2004). The products of lipid oxidation in potato which give strong flavour on heating are 2,4-decadienal (gives earthy and fatty flavour), c4-heptenal (also provide earthy aroma at low level) and 1-octen-3-ol (gives earthy and mushroom flavour) (Maga 1994). During potato slicing/chopping disruption of tissue takes place that provide opportunities for lipoxygenase enzyme to react with the substrates during boiling (Duckham et al. 2001).

When lipid is added exogenously or present endogenously in a food system, this may improve the processing quality of final food products of wheat and maize (Wang et al. 2016; Sun et al. 2019). The exogenous application of lipids or emulsifier is commonly used in foods to enhance the quality of final food product. The linear amylose fraction in starchy food can interact with the lipid, and can form single-helix complexes. Chao et al. (2018) reported that monoglyceride formed more starch-lipid complex with maize starch as compared to palmitic acid. However, palmitic acid forms more stable structure than that of complex formed with monoglyceride. There is less evidence on these studies in potato and these areas need to be explored (Nemś et al. 2015).

Free fatty acids (FFA) form helical complex structure with amylose (Tufvesson et al. 2003). However, the dissociation temperature of lipid-amylose complexes increases with increasing the length of chain in lipids and increasing the number of unsaturation (double bond) in the hydrocarbon chain. When amylose is complexed with lipid, it shows resistance to hydrolysis by amylase enzyme in small intestine. Thus amylose escape from digestion which ultimately lowers the glycemic effect of that food product (Seneviratne and Biliaderis 1991; Kumar et al. 2018b). Therefore, the increase in amylose content and resistant starch and the combination with lipid and protein can help to control diabetes and related disorders (Jayanty et al. 2019). The amylose fraction of the starch in potato interact with lipid which mainly include fatty acids or monoglycerides to form single-helix inclusion under stable conditions (Sun et al. 2019). It was reported that the starch chemistry, thermal and pasting properties affected severely with more oil content in potato and there was accumulation of highest level of TAG (Mitchell et al. 2017).

The effect of addition of external source of lipid in potato is variable. Potato is fried with different types of oil for consumption purposes. Frying of potato can be done for industrial purposes as well as for at domestic use for preparation of variety of starch-based food products (Chen et al. 2019). Overconsumption of any deep fried food can lead to overweight, obesity and different life style related diseases.

Attempts are being made to decrease the use of oil so as to reduce the calorie content and also maintain the sensory nature of food products. There is limited information on impact of starch granule size on oil absorption. However, most of the previous studies in the area of impact of oil absorption during frying and chips making identified various factors which affects its quality (Sothornvit 2011; Zeng et al. 2016). There is report on the change in the size of potato starch by oil absorption. It was reported that the starch granules having low granular size absorb more oil as compared to the bigger one (Chen et al. 2019).

5.5 Health Benefits

The major component of our diet is carbohydrate which is generally derived from starchy food like rice, wheat, maize and potato. It is utilized in human body for production of energy but where ever it is not utilized, it is stored in internal organ like liver in the form of glycogen. As discussed earlier that fat content in potato is up to 0.5% of total dry matter of tuber. This is high quality fat due to presence of unsaturated fatty acid and it enhances potato palatability. Due to small fat content in potato the daily intake of this quality fat from potato is also low. The quality fat includes MUFA and PUFA. There are three major unsaturated fatty acid present in potato are oleic acid, linoleic acid and linolenic acid. The concentration of linoleic acid is 32.13 mg/100 g and linolenic acid is 22.75 mg/100 mg in potato (Prescha et al. 2001).

Potato products like chips and French fries contribute 8–10% of daily recommended fat. French fries provide 13–15% of total fat of which more than 75% was found to be MUFA and PUFA (McGill et al. 2013).

5.6 Genetic Modifications to Improve Lipid Quality

Potato contains qualitative essential lipids which is beneficial for human consumption. As lipids are also source of calories, their demand is increasing in the recent years due to change in dietary pattern and exploding world population (FAO 2003). The potato crop which has been domesticated produces either lipids or starch. However, recent development in the technology and metabolic engineering has provided the opportunity for generating plants which could be capable of producing both starch and lipids simultaneously (Mitchell et al. 2017). In plants, fatty acid biosynthesis processes are highly regulated by various factors which involves spatial separation enzymes present in the different organelle compartments or involves temporal separation which include timing of production of various enzymes (Weselake et al. 2009). Increase in the lipid content in starchy foods like potato will not only improve the lipid content but will also enhance the nutritional quality of crop which will be beneficial for food, feed, fuel, purposes and industries. Researchers focused on increasing lipid content in potato tuber by increasing TAG content up to 1% dry weight (Klaus et al. 2004; Hofvander et al. 2016).

WRINKLED1 is the major gene which is responsible for oil accumulation in *Arabidopsis thaliana*. It belongs to the class of AP2/EREB domain family and it controls the process of fatty acid synthesis and glycolysis (Cernac and Benning 2004). It is the transcription which is involved in fatty acid synthesis in seed embryo. However, when it was expressed in potato there was change in the tuber metabolism. Hofvander et al. 2016 reported that potato transformed with WRINKLED1 transcription factor accumulates TAG in tuber which was reported up to 12 nmol TAG/mg dry weight. This increase in TAG affected the pattern of starch accumulation and its composition which was correlated with increase in sugar content and polar membrane lipid content. A single gene, acetyl-CoA carboxylase of *Arabidopsis* was expressed in amyloplast of potato tuber which shown fivefold increase in accumulation of TAG (Hofvander et al. 2016). The fat content (oil content) in potato can be correlated with higher phosphate content, increased amylopectin and decreased amylose content with smaller starch granules. The phosphorylation of starch is inversely related with amylose content and highly dependent on amylopectin content. Potatoes with overexpressed WRI1 gene were reported to show higher activity of α-glucan water dikinase (GWD) and phosphoglucan water dikinase (PWD) genes (Hofvander et al. 2016).

It was reported that amylose (hydrophilic molecule) undergoes a conformational change in the presence of lipid molecule (hydrophobic molecule). This results in the change in the inner cavity of the helix which is strongly hydrophobic due to presence of glycosidic linkage of amylose and methylene group. However, outer surface is strongly hydrophilic due to the presence of glycosyl hydroxyl group. The complex of starch and lipid forms V-type polymorph structure (Karademir et al. 2019). Prescha et al. 2001 developed transgenic potatoes with overexpression of 14-3-3-gene to enhance the total lipid content. There was 69% increase in total fat content as compared to the wild type potato. The increase was reported mainly in qualitative lipids like oleic acid, linoleic acid and linolenic acid content in transgenic potatoes as compared to control. In other experiment, the overexpression of acetyl Co-A carboxylase from *Arabidopsis* in the amyloplast of potato showed four- to fivefold increase in TAG biosynthesis as compared to wild type (Klaus et al. 2004).

In previous work on other Solanaceous crops like tobacco, it was reported that the co-overexpression of WRI1, DGAT1 and OLEOSIN gene leads to 15% of total dry weight increase in TAG (Vanhercke et al. 2014). As discussed earlier oleosin protein provides stability to lipid droplet, but this is not highly expressed in the non-seed oil storing tissue like in mesocarp of oil palm, olive and avocado. Recently identified protein that is associated with lipid droplet formation is lipid droplet-associated protein (LDAP) (Horn et al. 2013). So, not only oleosin but other proteins are also involved in providing stability to TAG against lipase and other cytosolic enzymes that may degrade the quality of TAG.

5.7 Conclusion/Future Direction

Though potato is a starchy crop, there is much scope of increasing lipid content in it. Potato is enriched in quality fatty acids like linoleic and linolenic acid which is beneficial for human consumption and preparation of other pharmaceutical products. The fatty acid content in potato has important implication both as the functioning of biological membrane as well as for post-harvest application. Moreover, lipid form complex with amylose (amylose-lipid) in potato which gradually decreases the glycemic response and increases resistant starch which is beneficial for diabetic person. The lipid accumulation also affects the tuber metabolism and increases the concentration of sugars. The digestibility of starch-lipid complex due to enzymatic action is not attempted and needs to be explored. The formation of starch-lipid complexes and their impact on digestibility can be explored for development or formation of new healthy potato products with low GI. The genetic and biotechnological approaches for enhancement of lipid biosynthesis increase the biosynthesis of lipid droplet and thus increase the TAG production which is known to have health benefits. It was reported in transgenic line that TAG accumulation was about 3.3% on dry weight, basis which was comparable to cotton seed-oil production. This indicates that potato can be a cheaper source of oil. Introduction of novel fatty acid in potato will not only have health benefit but, will also increase opportunities for industrial uses. Research needs to be focused in the area of lipid biology of potato where ample scope is there to increase the quality lipid content both for consumption as well as for industrial purposes. Enhancing lipid accumulation mechanism in potato through various approaches will further enhance the potato utilization for product development as it will also increase the industrial uses.

References

Błaszczak W, Fornal J, Amarowicz R, Pegg RB (2003) Lipids of wheat, corn and potato starch. J Food Lipids 10:301–312. https://doi.org/10.1111/j.1745-4522.2003.tb00023.x

Bligh EG, Dyer WJ (1959) A rapid method of total lipid extraction and purification. Can J Biochem Physiol 37:911–917. https://doi.org/10.1139/y59-099

Bonnaire L, Sandra S, Helgason T et al (2008) Influence of lipid physical state on the in vitro digestibility of emulsified lipids. J Agric Food Chem 56:3791–3797. https://doi.org/10.1021/jf800159e

Brown AP, Affleck V, Fawcett T, Slabas AR (2006) Tandem affinity purification tagging of fatty acid biosynthetic enzymes in Synechocystis sp. PCC6803 and Arabidopsis thaliana. J Exp Bot 57:1563–1571. https://doi.org/10.1093/jxb/erj150

Cernac A, Benning C (2004) WRINKLED1 encodes an AP2/EREB domain protein involved in the control of storage compound biosynthesis in Arabidopsis. Plant J 40:575–585. https://doi.org/10.1111/j.1365-313X.2004.02235.x

Chao C, Yu J, Wang SS et al (2018) Mechanisms underlying the formation of complexes between maize starch and lipids. J Agric Food Chem 66:272–278. https://doi.org/10.1021/acs.jafc.7b05025

Chen L, Ma R, McClements DJ et al (2019) Impact of granule size on microstructural changes and oil absorption of potato starch during frying. Food Hydrocoll 94:428–438. https://doi.org/10.1016/j.foodhyd.2019.03.046

Cheng FC, Muneta P (1978) Lipid composition of potatoes as affected by storage and potassium fertilization. Am Potato J 55:441–448. https://doi.org/10.1007/BF02852148

Dhital S, Shrestha AK, Hasjim J, Gidley MJ (2011) Physicochemical and structural properties of maize and potato starches as a function of granule size. J Agric Food Chem 59:10151–10161. https://doi.org/10.1021/jf202293s

Dobson G, Griffiths DW, Davies HV, McNicol JW (2004) Comparison of fatty acid and polar lipid contents of tubers from two potato species, Solanum tuberosum and Solanum phureja. J Agric Food Chem 52:6306–6314. https://doi.org/10.1021/jf049692r

Duckham SC, Dodson AT, Bakker J, Ames JM (2001) Volatile flavour components of baked potato flesh. A comparison of eleven potato cultivars. Nahrung – Food 45:317–323. https://doi.org/10.1002/1521-3803(20011001)45:5<317::AID-FOOD317>3.0.CO;2-4

FAO (2003) World agriculture: towards 2015/2030: an FAO perspective. http://www.fao.org/3/y4252e/y4252e05c.htm. Accessed 3 Aug 2019

Galliard T (1973) Lipids of potato tubers. 1. Lipid and fatty acid composition of tubers from different varieties of potato. J Sci Food Agric 24:617–622. https://doi.org/10.1002/jsfa.2740240515

Galliard T, Berkeley HD, Matthew JA (1975) Lipids of potato tubers. Effects of storage temperature on total, polar and sterol lipid content and fatty acid composition of potato tubers. J Sci Food Agric 26:1163–1170. https://doi.org/10.1002/jsfa.2740260814

Gidda SK, Shockey JM, Rothstein SJ et al (2009) Arabidopsis thaliana GPAT8 and GPAT9 are localized to the ER and possess distinct ER retrieval signals: functional divergence of the dilysine ER retrieval motif in plant cells. Plant Physiol Biochem 47:867–879. https://doi.org/10.1016/j.plaphy.2009.05.008

Höfgen R, Willmitzer L (1990) Biochemical and genetic analysis of different patatin isoforms expressed in various organs of potato (Solanum tuberosum). Plant Sci 66:221–230. https://doi.org/10.1016/0168-9452(90)90207-5

Hofvander P, Ischebeck T, Turesson H et al (2016) Potato tuber expression of Arabidopsis WRINKLED1 increase triacylglycerol and membrane lipids while affecting central carbohydrate metabolism. Plant Biotechnol J 14:1883–1898. https://doi.org/10.1111/pbi.12550

Horn PJ, James CN, Gidda SK et al (2013) Identification of a new class of lipid droplet-associated proteins in plants. Plant Physiol 162:1926–1936. https://doi.org/10.1104/pp.113.222455

Jansen G, Flamme W, Schüler K, Vandrey M (2001) Tuber and starch quality of wild and cultivated potato species and cultivars. Potato Res 44:137–146. https://doi.org/10.1007/BF02410100

Jayanty SS, Diganta K, Raven B (2019) Effects of cooking methods on nutritional content in potato tubers. Am J Potato Res 96:183–194

Karademir Y, Gökmen V, Öztop HM (2019) Investigation of lipid-derived formation of decadien-1-amine, 2-pentylpyridine, and acrylamide in potato chips fried in repeatedly used sunflower oil. Food Res Int 121:919–925. https://doi.org/10.1016/j.foodres.2019.01.020

Kitahara K, Tanaka T, Suganuma T, Nagahama T (1997) Release of bound lipids in cereal starches upon hydrolysis by glucoamylase. Cereal Chem 74:1–6. https://doi.org/10.1094/CCHEM.1997.74.1.1

Klaus D, Ohlrogge JB, Neuhaus HE, Dörmann P (2004) Increased fatty acid production in potato by engineering of acetyl-CoA carboxylase. Planta 219:389–396. https://doi.org/10.1007/s00425-004-1236-3

Kostyn K, Szatkowski M, Kulma A et al (2013) Transgenic potato plants with overexpression of dihydroflavonol reductase can serve as efficient nutrition sources. J Agric Food Chem 61:6743–6753. https://doi.org/10.1021/jf400645s

Kumar G, Knowles NR (1993) Changes in lipid peroxidation and lipolytic and free-radical scavenging enzyme activities during aging and sprouting of potato (Solanum tuberosum) seed-tubers. Plant Physiol 102:115–124. https://doi.org/10.1104/pp.102.1.115

Kumar A, Sahoo S, Sahu S et al (2018a) Rice with pulses or cooking oils can be used to elicit lower glycemic response. J Food Compos Anal 71:1–7. https://doi.org/10.1016/j.jfca.2018.05.003

Kumar A, Sahoo U, Baisakha B et al (2018b) Resistant starch could be decisive in determining the glycemic index of rice cultivars. J Cereal Sci 79:348–353. https://doi.org/10.1016/j.jcs.2017.11.013

Lepage M (1968) The lipid components of white potato tubers (Solanum tuberosum). Lipids 3:477–481. https://doi.org/10.1007/BF02530889

Lin PY, Czuchajowska Z (1998) Role of phosphorus in viscosity, gelatinization, and retrogradation of starch. Cereal Chem 75:705–709. https://doi.org/10.1094/CCHEM.1998.75.5.705

Liu Q, Tarn R, Lynch D, Skjodt NM (2007) Physicochemical properties of dry matter and starch from potatoes grown in Canada. Food Chem 105:897–907. https://doi.org/10.1016/j.foodchem.2007.04.034

Liu Q, Guo Q, Akbar S et al (2017) Genetic enhancement of oil content in potato tuber (Solanum tuberosum L.) through an integrated metabolic engineering strategy. Plant Biotechnol J 15:56–67. https://doi.org/10.1111/pbi.12590

Maga JA (1994) Potato flavor. Food Rev Int 10:1–48. https://doi.org/10.1080/87559129409540984

McGill CR, Kurilich AC, Davignon J (2013) The role of potatoes and potato components in cardiometabolic health: a review. Ann Med 45:467–473

Mitchell M, Pritchard J, Okada S et al (2017) Oil accumulation in transgenic potato tubers alters starch quality and nutritional profile. Front Plant Sci 8:554. https://doi.org/10.3389/fpls.2017.00554

Nemś A, Pęksa A, Kucharska AZ et al (2015) Anthocyanin and antioxidant activity of snacks with coloured potato. Food Chem 172:175–182

Prescha A, Świędrych A, Biernat J, Szopa J (2001) Increase in lipid content in potato tubers modified by 14-3-3 gene overexpression. J Agric Food Chem 49:3638–3643. https://doi.org/10.1021/jf010258o

Raigond P, Singh B, Kaundal B (2015) Changes in the starch and fiber fractions of French fries: response of freezing. Potato J 42:8–17

Raigond P, Singh B, Dutt S et al (2017) Potential of Indian potatoes for the management of hyperglycemia. Indian J Hortic 74:103–108. https://doi.org/10.5958/0974-0112.2017.00023.8

Ramadan MF (2016) Potato lipids, 2nd edn. Elsevier, London

Seneviratne HD, Biliaderis CG (1991) Action of α-amylases on amylose-lipid complex superstructures. J Cereal Sci 13:129–143. https://doi.org/10.1016/S0733-5210(09)80030-1

Shimakata T, Stumpf PK, Heilmann I et al (1982) Isolation and function of spinach leaf beta-ketoacyl-[acyl-carrier-protein] synthases. Proc Natl Acad Sci U S A 79:5808–5812. https://doi.org/10.1073/pnas.79.19.5808

Siloto RMP, Findlay K, Lopez-Villalobos A et al (2006) The accumulation of oleosins determines the size of seed oilbodies in *Arabidopsis*. Plant Cell 18:1961–1974. https://doi.org/10.1105/tpc.106.041269

Sothornvit R (2011) Edible coating and post-frying centrifuge step effect on quality of vacuum-fried banana chips. J Food Eng 107:319–325. https://doi.org/10.1016/J.JFOODENG.2011.07.010

Spychalla JP, Desborough SL (1990) Fatty acids, membrane permeability, and sugars of stored potato tubers. Plant Physiol 94:1207–1213. https://doi.org/10.1104/pp.94.3.1207

Sun S, Hong Y, Gu Z et al (2019) An investigation into the structure and digestibility of starch-oleic acid complexes prepared under various complexing temperatures. Int J Biol Macromol. https://doi.org/10.1016/J.IJBIOMAC.2019.07.166

Thiam AR, Antonny B, Wang J et al (2013) COPI buds 60-nm lipid droplets from reconstituted water–phospholipid–triacylglyceride interfaces, suggesting a tension clamp function. Proc Natl Acad Sci 110:13244–13249. https://doi.org/10.1073/pnas.1307685110

Tufvesson F, Wahlgren M, Eliasson A-C (2003) Formation of amylose-lipid complexes and effects of temperature treatment. Part 2. Fatty acids. Starch 55:138–149. https://doi.org/10.1002/star.200390028

Upchurch RG (2008) Fatty acid unsaturation, mobilization, and regulation in the response of plants to stress. Biotechnol Lett 30:967–977. https://doi.org/10.1007/s10529-008-9639-z

Vanhercke T, El Tahchy A, Liu Q et al (2014) Metabolic engineering of biomass for high energy density: oilseed-like triacylglycerol yields from plant leaves. Plant Biotechnol J 12:231–239. https://doi.org/10.1111/pbi.12131

Vega SE, del Rio AH, Bamberg JB, Palta JP (2004) Evidence for the up-regulation of stearoyl-ACP (Δ9) desaturase gene expression during cold acclimation. Am J Potato Res 81:125–135. https://doi.org/10.1007/BF02853610

Wang S, Wang J, Yu J, Wang S (2016) Effect of fatty acids on functional properties of normal wheat and waxy wheat starches: a structural basis. Food Chem 190:285–292. https://doi.org/10.1016/j.foodchem.2015.05.086

Weselake RJ, Taylor DC, Rahman MH et al (2009) Increasing the flow of carbon into seed oil. Biotechnol Adv 27:866–878

Zeng H, Chen J, Zhai J et al (2016) Reduction of the fat content of battered and breaded fish balls during deep-fat frying using fermented bamboo shoot dietary fiber. LWT 73:425–431. https://doi.org/10.1016/J.LWT.2016.06.052

Minerals in Potato

6

Milan Kumar Lal, Awadhesh Kumar, Ashok Kumar, Rupak Jena, Pinky Raigond, Dharmendra Kumar, Nitasha Thakur, and Brajesh Singh

Abstract

In the last few decades, the research is focused on sustainable development towards food security. Present research scenario is emphasized on nutritional security in the developing countries. In terms of consumption, potato is the fourth in number after rice, wheat and maize. It is considered a complete food and used as a staple food in various European countries. Potato is rich in carbohydrate, quality protein, antioxidant and micronutrients. It is also a rich source for minerals such as potassium, calcium, phosphorus, magnesium, selenium, iodine, iron and zinc. There is wide variation in the minerals content of germplasm available for potato which provides ample scope for plant breeder and genetic engineers to explore the mechanism of mineral uptake in potato and enhance the bioavailability of minerals. So, biofortification of potato for mineral content through various approaches, viz., agronomic approach, conventional breeding and biotechnology is the best way to deal with the problem of malnutrition.

Keywords

Minerals · Estimation · Storage · Cooking · Genetic modifications

M. K. Lal (✉) · P. Raigond · D. Kumar · N. Thakur · B. Singh
ICAR–Central Potato Research Institute, Shimla, Himachal Pradesh, India
e-mail: milan2925@gmail.com

A. Kumar
ICAR–National Rice Research Institute, Cuttack, Odisha, India

A. Kumar
ICAR–Directorate of Onion and Garlic Research, Pune, Maharashtra, India

R. Jena
ICAR - Directorate of Groundnut Research, Junagadh, Gujarat, India

P. Raigond et al. (eds.), *Potato*, https://doi.org/10.1007/978-981-15-7662-1_6

6.1 Introduction

The potato (*Solanum tuberosum* L.) is an important staple food crop after rice, maize and wheat. It was originated in Andes Mountain and used as a food 10,000 years ago and domesticated during the pre-Columbian period over 8000 years ago (Staiger 2008). These cultivated potatoes have spread form Andes Mountain to 160 countries around the world. The initial form of consumption of potato during pre-Columbian times was the production of chuño, which was used in soups and stews (Woolfe and Poats 1987). Along with the energy requirement provided by potato consumption, it also supplements the requirement of vitamins, minerals and high-quality proteins. It can be categorized as a healthy and versatile component of nutrition and balanced diet food. Micronutrient deficiency affects about two billion people worldwide, which is a major concern for developing countries. Potato also contributes to the micronutrient deficiency which is also referred to as hidden hunger and is the major health problem of the developing nation (Bailey et al. 2015).

The potato is rich in health-promoting components like dietary fibre, vitamin C, vitamin B6, vitamin B3, phenolic compounds, high-quality proteins (comparable to cereals grains), potassium, phosphorus, iron, zinc and elemental micronutrients (Burlingame et al. 2009). It was reported that the potato and its products were found to have favourable impact on human health. Many of the compound present in potato have a positive effect on human health and highly demanded in the vegetarian diet (Katan and De Roos 2004). It helps to lower the blood pressure of human, improves cardiometabolic activity, improves lipid profiles and decreases inflammation. However, there is a need for controlled clinical trials to define the impact of potatoes on cardiometabolic health (McGill et al. 2013). According to USDA (United States Department of Agriculture) nutrient database, 100 grams of potatoes contains 4% of RDA (Recommended Daily Allowance) calorie intake, 12% of the RDA for potassium and 33% of the RDA of vitamin C (Navarre et al. 2009). The nutrient-rich potato was referred to as a "phytochemical jewel box" at the Potato Association of American conference (Camire et al. 2009).

Potatoes are genetically diverse than other crops which indicate that potato can be grown divergently in a different environment. The genetically variable potato can be a valuable resource for improving tuber nutritional content in the breeding programme (Navarre et al. 2009). It is very much important for developing countries to prevent their children from malnutrition due to micronutrient deficiency. The relationship between agricultural production and the nutritional status of young children in the rural area of Peru was deciphered by Creed-Kanashiro et al. (2015). They reported that potato consumption from farm and percentage of recommended dietary intake for micronutrient iron (Fe) and zinc (Zn) intakes by children showed a positive correlation.

Micronutrients are essential for the growth and development of human, to sustain life and for optimal physiological function. It was estimated that six billions people of the world are suffering from micronutrient deficiency out of which 60–80% are Fe deficient, 30% are Zn deficient, 30% are I deficient and 15% are Se deficient (Kennedy et al. 2003). Pregnant women and their children suffer from micronutrient deficiency and are at high risk of micronutrient deficiency. Iron, iodine, folate, vitamin A and zinc are the major micronutrients which are required for sustaining life and performing optimal physiological functions (Bailey et al. 2015). Zinc plays an important and vital role in the human diet. It as the cofactor for the various enzymes in the cell. The developing countries face a severe problem of Zn deficiency (Brown et al. 2011). Genetic diversity of potato will facilitate the developing countries to adopt potato to combat micronutrient deficiencies.

Biofortification of potato with Fe, Zn and other elemental micronutrients will help developing countries to overcome this deficiency. The potential of biofortification of Fe and Zn in staple food like rice, wheat and potato must be assessed in the context of prospect for genetic improvement (Brown et al. 2010). On the other hand, another elemental nutrient is potassium, which plays an important role in human diet. Its requirement for an adult is 4700 mg/day for adults (National Research Council 1989). Potassium content in human blood serum varies between 3.7 and 5.2 mEq/L which is necessary for healthy electrolyte balance.

There is some common misconception about the potato that most of its nutrients are found on the skin. Though potato skin contains more than 50% of dietary fibre, more than half of the total nutrients are found in the flesh only. The bioavailability of the nutrients also depends upon the cooking method of potato. Due to various cooking methods, there is a chance of nutrient loss particularly water-soluble vitamins and minerals. Highest nutrient loss takes place when there is involvement of water or extended period of cooking is done under high temperatures (baking) (Bethke and Jansky 2008).

Approximately three billion people are affected by micronutrient deficiencies. These micronutrient deficiencies affect the human potential and hinder the socio-economic development particularly in the developing nation. Micronutrient deficiency includes deficiency of iron, zinc and vitamin A which are the most devastating amongst the economically weaker section in developing countries. The developing countries are fighting with hidden hunger with the help of the World Health Organization (WHO) and the Consultative Group on International Agricultural Research (CGIAR) under high priority category. For nutrient supplementation, WHO emphasizes on food fortification and biofortification which can address the malnutrition issues. Moreover, CGIAR plays an important role in biofortification programme through HarvestPlus programme which has an emphasis on staple foods like rice, wheat maize, beans, cassava, potato, millets and sweet potato. This biofortification programme has been undertaken through conventional breeding and biotechnological approaches (Khush et al. 2012).

Many of the projects on biofortification virtually ignored potato which is rich in carbohydrate, quality protein and will be composed of essential amino acids which are necessary for human diet (Haynes et al. 2012).

6.2 Minerals Present in Potato

A wide range of mineral elements occur in fruits and vegetables, which act as primary dietary source. Minerals can generally be classified as major minerals such as calcium (Ca), potassium (K), magnesium (Mg), sodium (Na), phosphorus (P), cobalt (Co), manganese (Mn), nitrogen (N), chlorine (Cl), and nutritionally essential minor and trace minerals such as iron (Fe), zinc (Zn), copper (Cu), selenium (Se), Nickel (Ni), lead (Pb), sulphur (S), boron (B), iodine (I), silicon (Si) and bromine (Br). The importance of optimal mineral intake to maintain good health is widely recognized (Avioli 1988). The maximum concentration of different minerals present in raw potato are potassium (564 mg/g FW), phosphorus (30–60 mg/g FW) and calcium (6–18 mg/g FW). The %RDA of potassium, phosphorus and calcium in potato is 22, 6 and 6, respectively (Burton 1989). Mainly the skin of potato is considered to be a good source of potassium. Baking potato with skin is a good cooking method to retain the minerals (True et al. 1979). The flesh of potato contains little phosphorus which is present in the form of phytic acid. However, the majority of phytate forms an unabsorbable complex with important micronutrients such as zinc and iron. High phytic acid is directly correlated with the lesser bioavailability of micronutrients. It is also considered as a natural antioxidant and suggested to have property of reducing lipid oxidation and act as a preservative in food (Zhou and Erdman 1995). The importance of micronutrient deficiencies has been identified and recognized in human, where zinc, iron, iodine and beta-carotene deficiencies lead to major risk for the global and regional burden of diseases (Ezzati et al. 2002).

6.2.1 Potassium

Potassium uptake takes place in the form of univalent cation which is highly related to metabolic activity. Potassium is one of the essential elements which is responsible for plant growth and development (Marschner 2011). It is highly mobile in both soil solution and plant system. It can be transported to long distance inside the plant through xylem and phloem. The important function where potassium is involved in the plant is carbohydrate metabolism, protein synthesis and cell elongation (Epstein and Bloom 2004). Potassium accumulates in the cytosol of the plant cell. It plays an important role as a cofactor in the enzyme. Potassium affects potato plant by influencing tuber yield, tuber quality, susceptibility to black spot bruise, the sugar level in tubers and enhancing the storage of tuber (Rowe 1993). The concentration of potassium in potato is found to be abundant varying from 150 to 1386 mg/100 g FW (Nassar et al. 2012). In human, it plays an important role in the development and functioning of the nervous system. The boiled potato (100 g) provides about 16% of the total potassium recommended for adults, which is around 4700 mg/day. It was also reported that the deficiency of potassium can lead to impairment of phloem loading and thus inhibiting the sucrose transport in the phloem. Thus, there can be an accumulation of sucrose in the leaves of potato (Koch et al. 2019).

6.2.2 Calcium

For the development of bone and tooth structure in human, calcium plays the most important role. It is also involved in blood clotting and nerve transmission. Its deficiency can lead to malformation of skeletal tissue and blood pressure abnormalities. The RDA for calcium in a young adult is 1300 mg which helps to reduce osteoporosis. This also varies for different age groups and gender. Calcium deficiency is rare in the soil but may show symptoms under acidic soil condition. Deficiency of calcium in the vegetable crop may reduce the quality of fruit (Bangerth 1979). The deficiency of calcium can be observed in the whole plant or a particular organ of the plant. The symptoms of deficiency mainly include reduction in growth, browning phenomena of developing fruit and necrosis in the plant tissue. The major location of calcium in the plant cell is in the apoplast region. When the cell needs calcium it comes from the apoplast region inside the cell through various transporters. It also maintains the integrity of the plasma membrane and stabilizes the pectin both intra- and inter-molecularly (White and Broadley 2005).

Calcium content in potato tuber ranges from 2 to 20 mg/100 g FW, contributing no more than 2% of the Estimated Average Requirement (EAR) of calcium for adult (800–1100 mg/day). Calcium level in potato has shown resistance against the internal brown spot, heat necrosis and infection by soft rot pathogen *Pectobacterium* spp., heat tolerance and freezing tolerance (Brown et al. 2012). Andre et al. 2007 reported calcium content in 74 Andean landraces ranging from 271 to 1093 μg/g DW. They also reported that the calcium content has been found to be the maximum in wild Solanum species. Higher calcium level is associated with higher resistance to the pathogen and abiotic stress (Andre et al. 2007).

6.2.3 Phosphorus

Phosphorus also plays a significant role in the human body. It helps in the formation and maintenance of healthy cell, teeth and bones. The daily requirement of phosphorus in human is around 800–1000 mg. The deficiency of phosphorus in human can result in abnormal low serum phosphate level, bone pain, lowering of bone density, ricketosteomalacia, susceptibility to infection, anaemia, numbness and difficulty in walking (Fuller et al. 1976). The RDA for phosphorus ranges from 800 to 1200 mg for adults (National Research Council 1989).

The role of phosphorus in the plant is very much essential where it is required in the energy transfer by dephosphorylation of adenosine triphosphate (ATP) to adenosine diphosphate (ADP). It is the important constituents of many co-enzymes, phospholipid in biomembrane and ribonucleic acid (Marschner 1995; Raghothama 1999). The phosphorus content in potato ranges from 1300 to 6000 g/g DW (Sanchez-Castillo et al. 1998).

Phosphorus is also highly mobile in the plant but immobile or less mobile in soil solution. It is present in the plant in an oxidized form. The uptake of phosphorus takes place in the form of $H_2PO_4^-$, where it can remain in the plant as an inorganic form, can be esterified into sugar or can be found in energy-rich pyrophosphate. In the plant, phosphorus has a significant role in maintaining the structure of nucleic acid. The tuber of potato contains 0.25% dry wt of phosphorus. The role of phosphorus is not only limited to the setting of potato tubers in early stages but at later stages, it also enhances the tuber maturity (Rosen et al. 2014; Hopkins et al. 2014). Potato can tolerate moderate phosphorus stress without any severe symptoms until the inner mechanism like photosynthesis and respiration is affected which finally leads to a reduction in carbohydrate accumulation. This leads to the changing of leaf colour to dark green to purple leaf discolouration (Hoppo et al. 1999).

6.2.4 Magnesium

Magnesium is a cofactor for more than 300 enzymes which are involved in the energy metabolism, carbohydrate, protein and nucleic acid synthesis. It is the second most intracellular cation which in ionized is the physiologically active element (Elin 1994). It is used in the treatment of acute myocardial infarction (heart disease) and also in atherosclerosis. About more than half of the magnesium in the body is present intracellularly in soft tissues and another half of it is present in bone tissues. Less than 1% of total magnesium is found to be present in the blood plasma. Most of the magnesium present in plasma is bound with albumin, globulin and protein. Deficiency of serum magnesium concentration can lead to the development of various diseases (Ahsan 1998). The instruments used to measure the concentration of magnesium in serum are an ion-selective electrode and in soft tissue is nuclear magnetic resonance spectroscopy. However, the present clinical status of laboratory examination is the evaluation of total magnesium in serum and 24-h urinary excretion (Elin 1994).

Magnesium is also referred to as 'the forgotten element in crop production' as its supply and needs to the plant is underestimated (Cakmak and Yazici 2010). The concentration of magnesium in potato ranged from 16 to 40 mg/100 g FW. It was estimated that 100 gram of boiled potato contributes around 11% of the EAR of magnesium for adults (265–340 mg/day, respectively). It is involved in various roles in the plant which includes photosynthesis, protein synthesis, enzyme regulation and DNA synthesis. Its deficiency can lead to impaired growth and tuber yield (Senbayram et al. 2015).

6.2.5 Copper

Copper is an essential trace element for human health, where it acts as the metalloenzymes in oxidase which help in the reduction of molecular oxygen. It is involved in the neurological function of human (Desai and Kaler 2008). Copper is required for various physiological functions in human such as respiration, peptide bond formation, neurotransmitter biosynthesis, pigment formation (melanin) and providing strength to the connective tissue. A small amount of copper is stored in the human body. The average adult can store only about 50–120 mg copper in the body and most of the copper is excreted out from bile and a small amount through urine (Erdman et al. 2012). It acts as the cofactor for various enzymes which are involved in the development of the central nervous system (CNS) and deficiency of this element can result in the incomplete development of CNS. It is also involved in brain development and is located in the basal ganglia, hippocampus, cerebellum, numerous synaptic membrane and cerebellar granular neuron (Madsen and Gitlin 2007). The RDA of copper in an adult ranges from 1.2 to 3 mg/day (Ikeda and Yamashita 1994).

Copper is a micronutrient in the plant where it is present in the normal range of 3–10 ppm. The ideal range of copper is 20 times lesser than iron in the plant tissue (Ed Bloodnick 2018). The visual symptoms of copper deficiency appear on the tip of the young leaves and further, it develops into necrotic lesions. The photosynthesis mechanism is also impaired along with the non-photochemical quenching which is the result of lacking plastocyanin pigment. Copper is the important cofactor of plastocyanin pigment in photosystem I which is present in the thylakoid membrane (Abdel-Ghany and Pilon 2008). The physiologically active form of copper is the reduced form. The redox-active free copper can lead to the formation of reactive oxygen species (ROS) which further leads to the damage of DNA, protein and other bioactive molecules (Hänsch and Mendel 2009).

6.2.6 Selenium

Selenium is an important essential element required for both animal and human metabolism as Selenium incorporates into the selenoprotein. The enzyme such as glutathione peroxidase contains selenium as a cofactor which is involved in its activation (Rayman 2012). Deficiency of selenium in human is associated with the increased in the risk of mortality, poor immune function, and cognitive decline in the health. The deficiency of selenium affects around hundreds of millions of people worldwide. Its deficiency also affects the development of the reproductive organs both in male and female, autoimmune thyroid diseases and increase the risk of prostate cancer. The supplementation of selenium can overcome these deformities and disorder (White and Brown 2010).

For plants, selenium is not essential but rather it is one of the beneficial elements and required in low concentrations. It is beneficial for some group of plant where it is utilized to increase the activity of various enzymes (Ramos et al. 2011). Selenium is also found to activate the enzymes like superoxide dismutase, catalase and ascorbate peroxidase. It also delays the ripening of tomato by inhibiting ethylene biosynthesis and enhancing the antioxidant system (Zhu et al. 2017). Selenium is applied in potato for its biofortification in the dose of 0.75 mg kg^{-1}, which result in enhanced production of tubers (de Oliveira et al. 2019). Addition of selenium in the form of fertilizer or foliar spray improves the physiological and agronomic characteristics of the potato crop (Ježek et al. 2011).

6.2.7 Iodine

Iodine is the essential micronutrient for the animals as well as for human health. Its deficiency is a widespread problem and has attracted a lot of attention in the past. Generally, its deficiency can be overcome by addition of iodine in the table salt which is the common prophylaxis tool (Van Der Straeten et al. 2017). The RDA for iodine ranges from 40 to 200 μg day^{-1}. Iodine is mainly required for development and maintenance of thyroid gland (Fuge and Johnson 2015). The deficiency of iodine results in the insufficient secretion of thyroid hormone and ultimately leads to goitre, the enlargement of the thyroid gland (Zimmermann et al. 2008).

Most of the cereals, vegetable, fruits and other foods lack iodine leading to iodine deficiency disorders. The absorption of iodine in plant takes place in two chemical forms such as potassium iodide (I^-) and potassium iodate (IO^{3-}). Plants can tolerate high levels of iodine as $IO^-{}_3$ better than I^- in the root environment (Caffagni et al. 2011). Biofortification of potato with iodine can be a better strategy for enhancing iodine uptake in plants and thus finally the consumption of these tubers which can provide an adequate amount of iodine in food.

6.2.8 Iron

Iron is the important constituent of haemoglobin and its deficiency has affected about 1.7 billion people worldwide. Due to severe iron deficiency, more than 60,000 women die during pregnancy. About 500 million women who attended that age of childbearing were affected by anaemia. However, the dietary iron requirement depends on age, gender and diet. This problem is one of the most widespread in world as reported by Vo et al. 2020. Iron content in potato has the potential to ameliorate the anaemic nutritional deficiency and stunting growth of children (Woolfe and Poats 1987). About 40% of the world's poor people suffer from iron deficiency and anaemia. The countries like Eastern and western Africa, Caribbean region and Mesoamerica have a high rate of anaemic patients (Black et al. 2008).

Iron is an essential micronutrient for the plants as well as animals and iron plays a significant role in various physiological and biochemical pathways. It plays a major role in the synthesis of DNA, photosynthesis and respiration. It acts as the prosthetic group of many enzymes, where the presence of iron can activate these enzymes. As iron is mobile in the plant, its deficiency can lead to chlorosis in the older leaves. In the process of respiration, iron can act as a vital component in enzymes like cytochromes. It is highly soluble in water and primarily found in the ferric form at neutral pH level. It is also involved in the synthesis of chlorophyll, maintenance of chloroplast structure and function (Rout and Sahoo 2015). Iron content in potato tuber has been reported to be 0.3–2.3 mg in 100 g tuber (True et al. 1978). The potato cultivars with red-skin were found to have more iron content (Brown 2008). Brown 2008 reported that the iron content in potato tuber of various genotype and varieties varied from 18 to 65 μg/g DW. In 74 Andean landraces, Andre et al. 2007 reported the iron content ranges from 29.87 to 157.96 μg/g DW. The RDA of iron in 150 g serving of potato is about 6% (O'Neill 2005).

6.2.9 Zinc

Zinc is one of the most essential elements whose deficiency can have a significant effect on the human health. Zinc plays an important role in increasing the resistance against the disease. There are three major biological roles of zinc, viz., catalyst, structural and as a regulatory ion. Many zinc-binding motifs are found on the proteins which are encoded by the human genome. It also plays an important role in the development of cognitive skill in children and its deficiency can lead to the impairment of brain development. The developing countries lack in providing zinc supplements, thus the brain development in these populations have the issue of poor brain development (Nestel et al. 2006).

Zinc also plays an important role in maintaining the homeostasis, immune function, apoptosis and lowering the rate of the ageing process. Its deficiency can also lead to the development of several chronic diseases like atherosclerosis, neurological disorder, autoimmune diseases and ageing-related disorder. Zinc helps to alleviate the oxidative stress and thus helps to lower the chronic inflammation in human (Chasapis et al. 2012). The children under the age of 5 years with diarrhoea are given zinc supplementation. Moreover, under the case of pneumonia, zinc supplementation reduced the time of sickness and increased recovery rates. About one million deaths in the world yearly are attributed to zinc deficiency (Lopez et al. 2006). The RDA for zinc is 15–20 mg. Zinc is involved in maintaining the immunity in the body and also involved in the cell growth, cell division and wound healing.

Zinc is also an essential element for the growth and development of the plant. It is also toxic when it is absorbed in excess amount. The uptake of zinc takes place form soil solution in the form of Zn^{2+} and it is highly mobile in the plant (Marschner 1995). Micronutrient stored in the tuber is affected by anti-nutritional factors like phytic acid which binds to micronutrients like iron and zinc, thus inhibiting their bioavailability in the human gut. The plant deficient in zinc shows retarded growth as its deficiency hinders auxin production and auxin is involved in the growth of the plant and development of the organ in a plant (Taiz and Zeiger 2015). It also acts as the cofactor of many enzymes that tightly bound and is very much essential for their function. Its deficiency leads to inhibition of protein synthesis, energy production and growth regulation in the plant (Salisbury and Ross 1992). Zinc content in potato tuber has been reported in Andean landraces ranging from 12.6 to 28.83 $\mu g g^{-1}$ of DW (Andre et al. 2007). Dugo et al. 2004 reported zinc content in the yellow-fleshed potato from different cultivars ranging from 0.5 to 4.6 μg/g FW.

6.3 Most Preferred Methods for Estimation of Minerals

6.3.1 Potassium and Magnesium

Determination of potassium and magnesium from plant tissue involves drying of tissue and further digestion of tissue with mineral acids. The tuber of potato can be dried in the hot-air oven or can be freeze-dried using freeze-dryer under subminus temperature. Further, the dried samples are digested with 65% (v/v) nitric acid and 30% (v/v) hydrogen peroxide for 75 min at 200 °C in the oven. The extract after filtration and volume makeup can be analysed with the help of plasma optical emission spectrometry (Koch et al. 2019).

6.3.2 Calcium

The calcium in potato tuber can be extracted by using water or by using HCL. These procedures (wet ashing) may show higher calcium content as compared to the dry-ashing technique of extraction. Extraction can be done by using 1 N HCL at room temperature and further it is incubated at 70 °C in a hot-air oven. Mainly the cortex and pith of potato are used for wet ashing. The final extract is then used in atomic absorption spectrophotometer for detail analysis of calcium content (Bretzloff 1971).

6.3.3 Phosphorus

The estimation of phosphorus also requires wet acid digestion. Initially, the tissue of potato is dried in hot-air oven or freeze-dried with the help of lyophilizer. The wet digestion can be done by using the method developed by Murphy and Riley 1962, where diacid digestion of HNO_3 and $HClO_4$ in 9:4 ratio is performed. The principle behind the development of blue colour by the addition of the digested sample in an acidified solution of ammonium molybdate containing ascorbic acid and a small amount of antimony. This reagent reacts rapidly with phosphate ion yielding a blue-purple compound which contains antimony and phosphorus in a 1:1 atomic ratio.

6.3.4 Selenium

For analysis of Selenium content in the tuber, it is dried in a hot-air oven to remove all the moisture inside the tuber. The analysis is performed with dried ground sample of potato tuber. The dried powder of tuber is digested with nitric acid (Element 2007). The extracts are analysed using atomic absorption spectrometry with electro-thermal atomization in a graphite furnace. The contents of Selenium can be expressed in the dry mass of plant tissues (de Oliveira et al. 2019).

6.3.5 Copper

The estimation of copper requires the same process as that of selenium where the plant tissue such as tuber is dried either in the hot-air oven (65 °C for 48 h) or freeze-dried. Then the sample is digested under monoacid digestion. The dried sample is digested in nitric acid and heated for 2 h at 60 °C and then 6 h at 130 °C. The final digest can be analysed using inductively coupled plasma-atomic emission spectrometry (Abdel-Ghany and Pilon 2008).

6.3.6 Iodine

Iodine in potato tuber can be measured using the standard protocol developed by the U.S. Environmental Protection Agency method 3052 [nitric acid (HNO_3)–hydrogen peroxide (H_2O_2), microwave digestion] with the combination of the use of plasma mass spectrometry (ICP-MS) analysis procedure (Dai et al. 2004).

6.3.7 Iron and Zinc

For estimating iron and zinc, the combination of acids has been used for digestion of dry potato sample. The samples are oven-dried at 60 °C, and using 0.25 g of dry powdered sample it was digested at 200 °C in 5 mL of a solution (5:1) HNO_3 (65%, W/W) to $HClO_4$ (70%, W/W) ratio. These methods are adopted for digestion by Zasoski and Burau 1977. Then the extract can be analysed using atomic absorption spectrophotometer.

Estimation of iron and zinc from potato can be done from the same sample. The wet ashing is done by using 5 mL of 12 N nitric acid in the combustion chamber. The final aliquot is injected into an Inductively Coupled Argon Plasma Emission Spectrophotometer (ICAPES). The final iron and zinc content is expressed as μg/gram dry weight (Brown et al. 2010). The new technology such as X-ray fluorescence spectrometry can be used for estimation of iron and zinc in freeze-dried and milled potato tuber sample. The calibration and the external and independent validations in using X-ray fluorescence show high coefficients of determination, 0.93–0.96 for iron and 0.92–0.97 for zinc, indicating that iron and zinc can be estimated by X-ray fluorescence spectrometry with high precision (Sosa et al. 2018).

6.4 Effect of Growing Conditions, Storage As Well As Cooking on Minerals

6.4.1 Growing Condition

The minerals which are present in potato tuber such as potassium, phosphorus, calcium and magnesium may change during irrigation and fertilization. These applications also affect the physiological maturity of potato tubers (Ilin et al. 2002). Moreover, the concentration of micronutrients such as iron and zinc increases with the application of fertilizer along with the increase in calcium content. Whereas, the content of phosphorus and molybdenum shows a significant reduction with the application of fertilizer. Along with the fertilizer application in potato, environment and genotype also affect the mineral content (Frossard et al. 2000).

In many types of soil, micronutrients are deficient resulting in their low uptake by crop and food. Merely the application of conventional NPK fertilizers is not sufficient for production and productivity enhancement. For growth and development of crop, micronutrients are also essential. It has been reported that micronutrients enhance nutritional quality, crop yield, biomass production and resiliency to biotic and abiotic stresses. The application of micronutrients in the form of nanoparticles has a profound effect on the crop response to other fertilizers and helps to increase the uptake of other nutrients (Dimkpa and Bindraban 2016). Agronomic approaches are successful in optimizing the mineral fertilizer and improving the content of micronutrients in the edible part. This approach is implemented by the mobilization of micronutrient. Moreover, breeding for increasing of promoter substances such as ascorbate, β-carotene and cysteine-rich polypeptides which stimulate the absorption of micronutrients by the gut and reduction of antinutrients such as oxalate, polyphenols and phytate that interfere with the absorption of micronutrient have been suggested (White and Broadley 2009).

Fortification also plays an important role in enhancing the micronutrients in potato and tomato. When fed with 0.05% of iodine salt in both potato and tomato plants there was an increase in iodine content showing of iodate (IO^{-3}) as 272 and 527 μg/100 g fresh weight (FW), respectively. Similarly, iodite (I^{-}) concentration upon supllementation in potato and tomato was reported as 1875 and 3900 μg/100 g FW, respectively. The important thing to note is that the uptake levels in both the crop is more than RDA, which is about 150 μg day^{-1} for adults (Caffagni et al. 2011).

The concentration of iron and zinc are significantly affected by the growing environment. The composition of soil also affects the mineral content and its concentration in crop. The soil which is rich in sandy texture favours oxidation process and thus reducing the uptake of available iron (Lombardo et al. 2013). Hajŝlová et al. 2005 reported the difference in the micronutrient content of the potato grown organically and conventionally. They reported four-year study data on various aspects such as micronutrients, metal, secondary metabolite, enzymic browning and organoleptic properties. The organically grown potatoes were reported to have more micronutrients (iron, zinc, manganese, copper and selenium) content (Hajšlová et al. 2005). The application of micronutrients also affects the tuberization process in potato. Zinc-sulphate when sprayed on crop, changes the hormonal balance increasing the cytokinin content. Moreover, the auxin/cytokinin ratio is decreased and an increase in cytokinin/ABA ratio is found. The zinc-sulphate application was also found to inhibit the apical dominance and enhance the tuber weight, as there was an increase in the number of cork cell layers (Puzina 2004).

6.4.2 Cooking

The heat treatment affects the bioavailability of micronutrients in potato. It was reported by Burgos et al. 2007 that in cooked potatoes the concentration of iron and zinc ranged from 0.29 to 0.69 mg/100 g FW and from 0.29 to 0.48 mg/100 g FW, respectively. The bioavailability of iron and zinc in potato is more than that of cereals and legumes, this may be due to the presence of high level of ascorbic acid in the tuber. Ascorbic acid facilitates the absorption of iron and zinc in the gut. The cooking of potato also lowers the phytic acid in the tuber which is beneficial in the absorption of micronutrients like iron and zinc. Low level of phytic acid is favourable for iron and zinc absorption (Phillippy et al. 2004; Furrer et al. 2018). Moreover, it is also reported that the bioaccessibility of iron in potato is higher than crop such as wheat and beans. In an in vitro intestinal study, it was reported that about 63–79% of bioavailable iron was released from the food matrix (Andre et al. 2015).

The main cause for the loss of mineral nutrient takes place during the cooking is the leaching process. Boiling along with peeling of potato may result in the loss of dietary fibre. Mishra et al. 2020 reported that the washing after peeling can lead to a significant reduction in potassium by about 50% and other soluble vitamins (Fig. 6.1). Along with potassium, other minerals such as iron, zinc, phosphorus and manganese are also lost. The processing of potato such as shredding and cubing can lead to loss of mineral compound up to 50–75% of total mineral content (Bethke and Jansky 2008). Narwojsz et al. 2020 reported that the highest losses of mineral such as manganese, potassium and zinc occurred during boiling of potato tuber and lowest during the steaming of potato. The mineral-like copper, iron, calcium and phosphorus are retained in the maximum amount during these treatments. This phenomenon happened due to the degree of migration of soluble component to water (Narwojsz et al. 2020). The longer is the duration of cooking and higher is the temperature of cooking, the more is the loss of mineral nutrients. There are many food items which are prepared using potato. Iodine was evaluated in the various Italian dishes such as dumplings, vegetable pie and focaccia bread. The stability of iodine was more in focaccia bread. The significant losses of iodine have been observed in the process of boiling of dumpling and baking of vegetable pie (Cerretani et al. 2014).

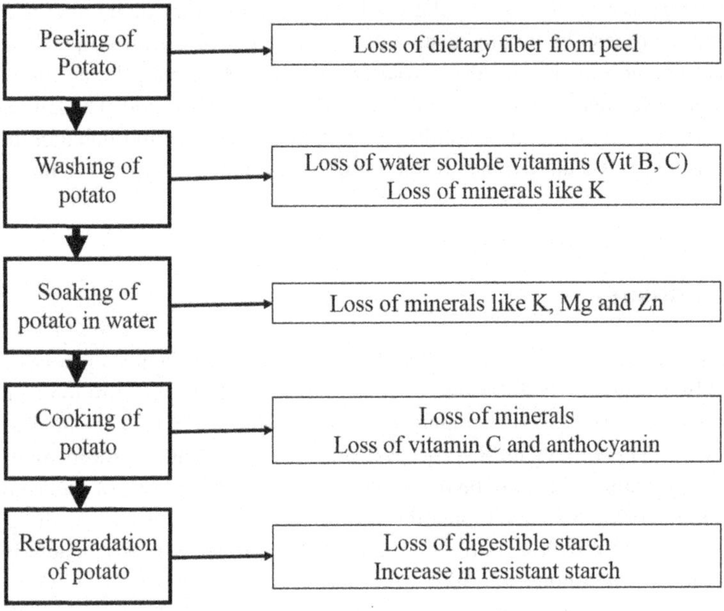

Fig. 6.1 Nutrient lost during various steps of processing of potato

6.5 Health Benefits of Elemental Micronutrient and Role in Addressing Malnutrition

In the last few decades, agricultural research and practices in developing countries are focusing on the production of cereals. The recent research and developments are now focused to reduce hunger as well as on nutrient-rich foods. One in three people were found suffering from hidden hunger which may be caused due to lack of mineral and vitamins (Kennedy et al. 2003).

6.5.1 Promoter and Antinutrients

From the health point of view, the target is not only to enhance the micronutrients but there is a target to increase the ability to acquire the mineral element or to increase the bioavailability of micronutrients. It is also targeted to increase the promoter compound (which help to increase the micronutrients) such as cysteine-rich polypeptides and β-carotene that stimulate the absorption of an essential mineral element from the gut. The other method is to reduce antinutrients contents. Antinutrients are the compounds which inhibit the uptake of micronutrients in the human gut by interfering with micronutrients. Antinutrients include oxalate, phytic acid and polyphenolics are substances which interfere and make the micronutrient unavailable for absorption. These approaches are addressing mineral malnutrition in humans globally (White and Broadley 2009).

6.5.2 Potassium

The consumption of potassium is high in the diet when it is obtained from food. But the consumption of highly processed food leads to removal of potassium combined with low consumption of fruits and vegetables. It was reported that the high potassium diet has the property of lowering blood pressure. This help to reduce cardiovascular disease and mortality. Increase in potassium intake leads to lowering of urinary calcium excretion which helps in the management of hypercalciuria and kidney stones. Intake of potassium is also related to a decrease in the case of osteoporosis. The best way to of potassium intake is the consumption of fruits and vegetable such as potato (He and MacGregor 2008).

6.5.3 Calcium

Calcium is an essential element which is responsible for the formation and maintenance of bones and teeth in human. More than 99% of the total calcium of the body is found in teeth and bones. The homeostasis of calcium in human is maintained by parathyroid hormone (Schulze 2012).

6.5.4 Phosphorus

Phosphorus is the essential nutrient which is required for carrying out the physiological processes. Deficiency of phosphorus leads to deformity in the bone and teeth and development of cardiovascular diseases. It is required for the process of glycolysis, gluconeogenesis, energy metabolism and cellular signal transduction in every cell of the body. About 80% of total phosphorus of the body is stored in the bones and teeth, whereas the intracellular phosphorus exists in the form of the organic compound. It helps to maintain the energy status of cell and the overall human body (Takeda et al. 2012).

6.5.5 Magnesium

Magnesium is another most important micronutrient which is often neglected in the human body. It plays an important role in the activation of more than 300 enzymes present in the human body. Moreover, it is also involved in over 600 enzymatic reactions including energy metabolism and protein synthesis (de Baaij et al. 2015). The deficiency of magnesium in the human body can cause tiredness, stiffness in muscle and difficult to concentrate. It is involved in protein synthesis, DNA replication and also acts as a cofactor for various enzymes. It maintains the function of the nerve cell and muscle tissue. It also provides resiliency against the pathological condition. It provides health benefits against bone strength, heart, stroke, muscular system, dental health, diabetes, alleviation of depression and anxiety (Faryadi 2012).

6.5.6 Copper

The deficiency of copper is characterized by loss in appetite, anaemia, oedema and arthritis. Copper plays an important role in the development and maintenance of the central nervous system. It also acts as the component of metalloenzymes which participates in the redox reaction (Bremner 1998). Copper in diet has been reported to be associated with recovery rate in neurological disorder including Alzheimer's disease and prion diseases (Llanos and Mercer 2002). The other beneficial role of copper in human includes the development of foetal and their growth, development and function of the brain, maintaining bone strength, glucose and cholesterol metabolism and formation of pigments (Stern et al. 2007).

6.5.7 Selenium

The deficiency of selenium is also associated with increased in the mortality rate, reduction in immune function and cognitive decline. The supplementation of selenium is associated with enhancement of the function of the reproductive organ in male and female, reduction in the risk of autoimmune thyroid disease. Biofortification of potato with selenium will enhance the physiological function in the human, which is more beneficial for developing nation (White and Brown 2010).

6.5.8 Iodine

About 30% of the world's population is affected by iodine deficiency. This deficiency of iodine is ameliorated by the help of iodized salt and biofortification of fruits and vegetable with iodine. It helps to prevent reproductive failures, mental retardation, brain damage and most importantly goitre disease (Blasco et al. 2008).

6.5.9 Iron

Deficiency of iron causes anaemia and more than 30% of the world's population is suffering from this condition. It is a problem that affects both rich and poor countries. Moreover, anaemic condition coexists with other health issues such as malaria, parasitic infection and other nutritional deficiencies (WHO 2015). The deficiency also plays a negative effect on immune function, cognitive development, regulation of body temperature, energy metabolism and maintenance of work performance (Dallman 1986). Warman and Havard 1998 reported the iron content ranging from 2.8 to 158 mg kg^{-1} in varieties grown in South America.

6.5.10 Zinc

The essentiality of zinc was first recognized in the 1960s where the deficiency caused stunted growth and delayed sexual maturity in young Egyptian men (Prasad et al. 1963). Zinc plays an important role in human body development and has health benefits. It is needed for normal growth, DNA synthesis, maintaining the neurosensory function, and act as a cofactor for various enzymes (Chasapis et al. 2012). It was estimated that the deficiency of zinc attributed 4.4% of total death among children under the age of 5 years in developing countries of Africa and Asia (Walker et al. 2009).

6.6 Genetic Modifications to Improve Mineral Content

The feeding of an ever-rising population with maintaining nutritional status is a challenging job. To overcome this, biofortification is considered to be the most effective way to increase the elemental nutrient and vitamins in the staple food to maintain nutritional quality. It is low cost and non-recurring investment to increase the nutritional quality without a change in the eating habit of people (Khush 2002). Enhancement of micronutrient level can be done by various approach, viz., agronomic approach, conventional breeding and genetic engineering approach (Fig. 6.2).

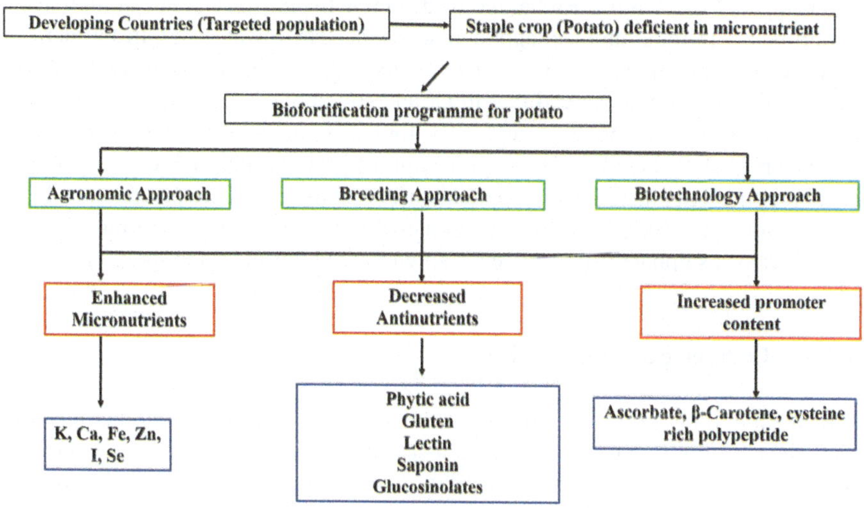

Fig. 6.2 Breeding criteria for biofortification of vegetables

These processes are cost-effective, long term, inexpensive and sustainable source (Meenakshi et al. 2010). There are mainly three strategies for improving the micronutrient content, viz., the addition of supplements, food fortification and development of staple crop which is enriched in micronutrient by genetic modification or breeding programmes. But out of these strategies addition of supplements and fortification of food requires more infrastructure and are economically not viable. Thus, the policymaker and scientist are working for enhancement through breeding, which is an economically feasible method (Bouis 2000, 2002).

6.6.1 Agronomic Approach

Agronomic approach for biofortification is the process of biofortification of staple food with the help of application and management of crop with the spraying and application of fertilizers. This can be achieved without sacrificing on the yield of the crop. The foliar application of crop with iron, zinc, iodine and selenium solution helps to enhance the mineral concentration. The applied micronutrient fertilizer should be in chelated form (Prasad et al. 2014). Spraying of water-soluble sources of zinc and iron is recommended whereas the soil application is not recommended for these micronutrients. Agronomic practices such as tillage of the soil, water management and nutrient interaction also play a vital role in biofortification. Moreover, the root exudates from plant such as phytosiderophores (iron), organic acid (Singh and Pandey 2003) and acid phosphatase enhance the uptake of nutrients (mainly phosphorus) (Pandey et al. 2018).

Agronomic biofortification is the strategy which can provide a short term increase in the micronutrient level but not in the long term or the character of uptake of nutrient which can be transferred to another generation (Saltzman et al. 2013). Kromann et al. 2017 reported that the Andean potato cultivar can be agronomically biofortified with foliar and soil application of zinc. There was a significant increase in zinc concentration in Andean potato cultivars both by foliar and soil application of zinc fertilizers. It was observed that foliar and soil application of zinc can increase zinc concentration up to 2.51 and 1.91 fold zinc in potato tuber, respectively (Kromann et al. 2017). The foliar application of iron is highly effective compared to soil application since the soil application of results in quick conversion to Fe^{3+}, which is an unavailable form (Backhausen et al. 2014). However, the foliar application of ferrous sulphate is considered highly effective fertilizer for spraying.

6.6.2 Conventional Breeding

The breeding programme for biofortification of potato with elemental micronutrient gives the possibility to produce a mineral-rich potato. The breeding for low phytate potato can also enhance the bioavailability of micronutrients such as iron and zinc (Woolfe and Poats 1987). However, this method requires a long period and adequate fund to biofortify the crop. Due to the lack of genetic variation in potato cultivars, the most common practice of biofortification is conventional breeding cross-breeding method (La Frano et al. 2014). The use of tools such as genetic variation in the potato breeding programme for micronutrient is very much essential for biofortification. Genetic parameter such Bayesian model was used to determine micronutrients, iron and zinc (Paget et al. 2014). The other breeding methods which can be utilized are novel marker system that includes single nucleotide polymorphism (SNP) genotyping, genotyping by sequencing (GBS), marker-assisted selection (MAS). The methods like next-generation population such as backcross population, multiparent advanced generation intercross (MAGIC), recombinant inbred line (RIL), targeting induced local lesion in genomes (TILLING) and various QTL (quantitative trait locus) have been established for various important traits (Welch and Graham 2004).

Antinutrients such as phytate and polyphenols apart from inhibiting the nutrient intake, also act as anticarcinogenic compound and decrease the risk of heart disease and diabetes. Thus, it will be an advantage to the breeders to design plant ideotype which should be known for the pleiotropic nature of iron and zinc association (Das et al. 2017). The criteria have to be set for the breeding of potato for micronutrient and vitamins which will meet the demand of farmer as well as the targeted people. Firstly, the potato crop production and productivity should be maintained so that the farmer must get proper remuneration. Secondly, the micronutrient concentration in the potato crop should be achieved significantly so that it will have an impact on human health. Third, the particular trait for biofortification of potato crop should be stable between generations. This should apply to the various ecological and climatic zones. Fourthly, the micronutrient bioavailability should be maintained while cooking and its level should not get altered after cooking. The breeding programmes should be designed keeping in mind the retention of micronutrient when screening and the bioavailability of food product from that crop. Lastly, the taste of biofortified product should be acceptable to the consumer. More publicity and importance of crop should be disseminated to the farmer so that they will know the importance of growing of biofortified vegetables (Welch and Graham 2004).

6.6.3 Genetic Engineering

For accumulation of micronutrients in the edible tissues, there are several physiological barriers which need to be eliminated. The barriers are the result of the tightly controlled mechanism inside the plant cell which regulate metal absorption, translocation and redistribution of micronutrient. There must be an appropriate concentration of minerals so that they should not be toxic to plant tissues (Welch 1995). The transgenic approaches are only feasible to increase the concentration of micronutrient by increasing the expression of mineral transporters such as zinc-regulated transporter (ZRT) for zinc, iron-regulated transporter (IRT)-like protein (ZIP) family for iron and phytosiderophore genes (Vert et al. 2002; White and Broadley 2009). These proteins are involved in the uptake and transport of cations such as iron and zinc in cell and overexpression of these proteins in tuber can be helpful to enhance these minerals. The advantage of the transgenic approach is that it can enhance the genetic system to increase micronutrient concentration, increase promoter for bioavailability and decrease of antinutrients (Welch and Graham 2004).

6.7 Conclusion/Future Direction

Potato is generally consumed after cooking. Potato is the rich source for various nutrients, vitamins and antioxidant molecules which are beneficial for human health. Cooking leads to an increase in the digestibility and bioavailability of macro and micronutrients. The antinutrients which are present in potato get degraded after cooking leading to the bioavailability of micronutrients. There are various means to enhance mineral content in potato. Agronomic practices, conventional breeding, genetic engineering and gene-editing technology can be used to enhance the uptake of mineral from the soil. The variation in genetic diversity of potato which is now being explored is also helpful to provide the target based need for mineral content. The sufficiency in micronutrient will help the developing nations to have nutritional security for their population. The problem of hidden hunger can also be addressed by the process of biofortification to enhance the micronutrient level in the potato tuber. The other challenge of biofortified potato is the acceptance by the consumer. Accepting the biofortified food will solve the problem of increasing malnutrition cases.

The breeding programmes and transgenic approached were made to develop nutrient-rich potato. The breeding strategies must be focused on the target germplasm of potato and also the identification of QTLs. Gene-editing is another promising technique to enhance the mineral content in potato. However, the rate of acceptance by the policymaker and public is very less. Thus, potato can be considered as a hero crop for those human population which rely on cheap staple food for energy. This will not only provide the required energy but also the appropriate vitamins, minerals and other constituents which are essential for living.

References

Abdel-Ghany SE, Pilon M (2008) MicroRNA-mediated systemic down-regulation of copper protein expression in response to low copper availability in Arabidopsis. J Biol Chem 283:15932–15945. https://doi.org/10.1074/jbc.M801406200

Ahsan SK (1998) Magnesium in health and disease. JOURNAL-PAKISTAN Med Assoc 48:246–249

Andre CM, Ghislain M, Bertin P et al (2007) Andean potato cultivars (Solarium tuberosum L.) as a source of antioxidant and mineral micronutrients. J Agric Food Chem 55:366–378. https://doi.org/10.1021/jf062740i

Andre CM, Evers D, Ziebel J et al (2015) In vitro bioaccessibility and bioavailability of iron from potatoes with varying vitamin C, carotenoid, and phenolic concentrations. J Agric Food Chem 63:9012–9021. https://doi.org/10.1021/acs.jafc.5b02904

Avioli LV (1988) Calcium and phosphorus. In: Shils ME, Young E (eds) Modern nutrition in health and disease, 7th edn. Lea & Febiger, Philadelphia, pp 142–158

Backhausen JE, Scheibe R, Ahmad I et al (2014) Physiological, biochemical, and genome-wide transcriptional analysis reveals that elevated CO2 mitigates the impact of combined heat wave and drought stress in Arabidopsis thaliana at multiple organizational levels. Plant Physiol Biochem 30:963–967. https://doi.org/10.1007/s11738-009-0415-z

Bailey RL, West KP Jr, Black RE (2015) The epidemiology of global micronutrient deficiencies. Ann Nutr Metab 66:22–33. https://doi.org/10.1159/000371618

Bangerth F (1979) Calcium-related physiological disorders of plants. Annu Rev Phytopathol 17:97–122

Bethke PC, Jansky SH (2008) The effects of boiling and leaching on the content of potassium and other minerals in potatoes. J Food Sci 73:H80–H85. https://doi.org/10.1111/j.1750-3841.2008.00782.x

Black RRE, Allen LHL, Bhutta ZZA et al (2008) Maternal and child undernutrition: global and regional exposures and health consequences. Lancet 371:243–260

Blasco B, Rios JJ, Cervilla LM et al (2008) Iodine biofortification and antioxidant capacity of lettuce: potential benefits for cultivation and human health. Ann Appl Biol 152:289–299. https://doi.org/10.1111/j.1744-7348.2008.00217.x

Bouis HE (2000) Enrichment of food staples through plant breeding: a new strategy for fighting micronutrient malnutrition. Nutrition 16(7–8):701–704

Bouis HE (2002) Plant breeding: a new tool for fighting micronutrient malnutrition. J Nutr 132:491S–494S

Bremner I (1998) Manifestations of copper excess. Am J Clin Nutr 67:1069S–1073S

Bretzloff CW (1971) Calcium and magnesium distribution in potato tubers. Am Potato J 48:97–104. https://doi.org/10.1007/BF02861720

Brown CR (2008) Breeding for phytonutrient enhancement of potato. Am J Potato Res 85:298. https://doi.org/10.1007/s12230-008-9028-0

Brown CR, Haynes KG, Moore M et al (2010) Stability and broad-sense heritability of mineral content in potato: iron. Am J Potato Res 87:390–396. https://doi.org/10.1007/s12230-010-9145-4

Brown CR, Haynes KG, Moore M et al (2011) Stability and broad-sense heritability of mineral content in potato: zinc. Am J Potato Res 88:238–244. https://doi.org/10.1007/s12230-011-9188-1

Brown CR, Haynes KG, Moore M et al (2012) Stability and broad-sense heritability of mineral content in potato: calcium and magnesium. Am J Potato Res 89:255–261. https://doi.org/10.1007/s12230-012-9240-9

Burgos G, Amoros W, Morote M et al (2007) Iron and zinc concentration of native Andean potato cultivars from a human nutrition perspective. J Sci Food Agric 87:668–675. https://doi.org/10.1002/jsfa.2765

Burlingame B, Mouillé B, Charrondiere R (2009) Nutrients, bioactive non-nutrients and anti-nutrients in potatoes. J Food Compos Anal 22:494–502

Burton WG (1989) The potato, 3rd edn. Longman Scientific & Technical, Harlow

Caffagni A, Arru L, Meriggi P et al (2011) Iodine fortification plant screening process and accumulation in tomato fruits and potato tubers. Commun Soil Sci Plant Anal 42:706–718. https://doi.org/10.1080/00103624.2011.550372

Cakmak I, Yazici AM (2010) Magnesium: a forgotten element in crop production. Better Crop 94:23–25

Camire ME, Kubow S, Donnelly DJ (2009) Potatoes and human health. Crit Rev Food Sci Nutr 49:823–840. https://doi.org/10.1080/10408390903041996

Cerretani L, Comandini P, Fumanelli D et al (2014) Evaluation of iodine content and stability in recipes prepared with biofortified potatoes. Int J Food Sci Nutr 65:797–802. https://doi.org/10.3109/09637486.2014.917155

Chasapis CT, Spiliopoulou CA, Loutsidou AC, Stefanidou ME (2012) Zinc and human health: an update. Arch Toxicol 86:521–534

Creed-Kanashiro H, Bartolini R, Goya C, Bentley M (2015) Changing the Consistency of Animal Source and Micronutrient-Rich Complementary Foods to Improve Acceptability and Consumption among Infants and Toddlers: Formative Research (FR) in Rural Peru. The FASEB Journal.1: 29

Dai JL, Zhu YG, Zhang M, Huang YZ (2004) Selecting iodine-enriched vegetables and the residual effect of iodate application to soil. Biol Trace Elem Res 101:265–276. https://doi.org/10.1385/BTER:101:3:265

Dallman PR (1986) Biochemical basis for the manifestation of iron deficiency. Annu Rev Nutr 6:13–40. https://doi.org/10.1146/annurev.nu.06.070186.000305

Das A, Laha S, Mandal S et al (2017) Preharvest biofortification of horticultural crops. Elsevier, Amsterdam

de Baaij JHF, Hoenderop JGJ, Bindels RJM (2015) Magnesium in man: implications for health and disease. Physiol Rev 95:1–46. https://doi.org/10.1152/physrev.00012.2014

de Oliveira VC, Faquin V, Andrade FR et al (2019) Physiological and physicochemical responses of potato to selenium biofortification in tropical soil. Potato Res 62:315–331. https://doi.org/10.1007/s11540-019-9413-8

Desai V, Kaler SG (2008) Role of copper in human neurological disorders. Am J Clin Nutr 88:855S–858S

Dimkpa CO, Bindraban PS (2016) Fortification of micronutrients for efficient agronomic production: a review. Agron Sustain Dev 36:1–26

Dugo G, La Pera L, Lo Turco V et al (2004) Determination of copper, zinc, selenium, lead and cadmium in potatoes (Solanum tuberosum L.) using potentiometric stripping methods. Food Addit Contam 21:649–657. https://doi.org/10.1080/02652030410001698698

Ed Bloodnick (5 Oct 2018) Role of copper in plant culture. PRO-MIX Greenhouse Growing, Quebec

Element CAS (2007) Method 3051A microwave assisted acid digestion of sediments, sludges, soils, and oils. Z Für Anal Chem 111:362–366

Elin RJ (1994) Magnesium: the fifth but forgotten electrolyte. Am J Clin Pathol 102:616–622. https://doi.org/10.1093/ajcp/102.5.616

Epstein E, Bloom AJ (2004) Mineral nutrition of plants: principles and perspectives. University of California, Davis, CA

Erdman JW, MacDonald I, Zeisel SH, International Life Sciences Institute (2012) Present knowledge in nutrition. International Life Sciences Institute, Washington, DC

Ezzati M, Lopez AD, Rodgers A et al (2002) Selected major risk factors and global and regional burden of disease. Lancet 360:1347–1360. https://doi.org/10.1016/S0140-6736(02)11403-6

Faryadi Q (2012) The magnificent effect of magnesium to human health: a critical review. Int J Appl Sci Technol 2:118–126

Frossard E, Bucher M, Mächler F et al (2000) Potential for increasing the content and bioavailability of Fe, Zn and Ca in plants for human nutrition. J Sci Food Agric 80:861–879. https://doi.org/10.1002/(SICI)1097-0010(20000515)80:7<861::AID-JSFA601>3.0.CO;2-P

Fuge R, Johnson CC (2015) Iodine and human health, the role of environmental geochemistry and diet, a review. Appl Geochem 63:282–302

Fuller TJ, Carter NW, Barcenas C, Knochel JP (1976) Reversible changes of the muscle cell in experimental phosphorus deficiency. J Clin Invest 57:1019–1024. https://doi.org/10.1172/JCI108343

Furrer AN, Chegeni M, Ferruzzi MG (2018) Impact of potato processing on nutrients, phytochemicals, and human health. Crit Rev Food Sci Nutr 58:146–168. https://doi.org/10.1080/10408398.2016.1139542

Hajŝlová J, Schulzová V, Slanina P et al (2005) Quality of organically and conventionally grown potatoes: four-year study of micronutrients, metals, secondary metabolites, enzymic browning and organoleptic properties. Food Addit Contam 22:514–534. https://doi.org/10.1080/02652030500137827

Hänsch R, Mendel RR (2009) Physiological functions of mineral micronutrients (Cu, Zn, Mn, Fe, Ni, Mo, B, Cl). Curr Opin Plant Biol 12:259–266
Haynes KG, Yencho GC, Clough ME et al (2012) Genetic variation for potato tuber micronutrient content and implications for biofortification of potatoes to reduce micronutrient malnutrition. Am J Potato Res 89:192–198
He FJ, MacGregor GA (2008) Beneficial effects of potassium on human health. Physiol Plant 133:725–735. https://doi.org/10.1111/j.1399-3054.2007.01033.x
Hopkins BG, Horneck DA, MacGuidwin AE (2014) Improving phosphorus use efficiency through potato rhizosphere modification and extension. Am J Potato Res 91:161–174. https://doi.org/10.1007/s12230-014-9370-3
Hoppo SD, Elliott DE, Reuter DJ (1999) Plant tests for diagnosing phosphorus deficiency in barley (Hordeum vulgare L.). Aust J Exp Agric 39:857–872. https://doi.org/10.1071/EA99029
Ikeda S, Yamashita Y (1994) Buckwheat as a dietary source of zinc, copper and manganese. Fagopyrum 14:29–34
Ilin Z, Durovka M, Markovic V (2002) Effect of irrigation and mineral nutrition on the quality of potato. In: Acta horticulturae. International Society for Horticultural Science, Leuven, pp 625–629
Ježek P, Hlušek J, Lošák T et al (2011) Effect of foliar application of selenium on the content of selected amino acids in potato tubers (Solanum tuberosum L.). Plant Soil Environ 57 (7):315–320. https://doi.org/10.17221/57/2011-PSE
Katan MB, De Roos NM (2004) Promises and problems of functional foods. Crit Rev Food Sci Nutr 44:369–377. https://doi.org/10.1080/10408690490509609
Kennedy G, Nantel G, Shetty P (2003) The scourge of "hidden hunger": global dimensions of micronutrient deficiencies. Food Nutr Agric 32:8–16
Khush GS (2002) The promise of biotechnology in addressing current nutritional problems in developing countries. Food Nutr Bull 23:354–357. https://doi.org/10.1177/156482650202300406
Khush GS, Lee S, Cho JI, Jeon JS (2012) Biofortification of crops for reducing malnutrition. Plant Biotechnol Rep 6:195–202
Koch M, Busse M, Naumann M et al (2019) Differential effects of varied potassium and magnesium nutrition on production and partitioning of photoassimilates in potato plants. Physiol Plant 166:921–935. https://doi.org/10.1111/ppl.12846
Kromann P, Valverde F, Alvarado S et al (2017) Can Andean potatoes be agronomically biofortified with iron and zinc fertilizers? Plant Soil 411:121–138. https://doi.org/10.1007/s11104-016-3065-0
La Frano MR, de Moura FF, Boy E et al (2014) Bioavailability of iron, zinc, and provitamin a carotenoids in biofortified staple crops. Nutr Rev 72:289–307. https://doi.org/10.1111/nure.12108
Llanos RM, Mercer JFB (2002) The molecular basis of copper homeostasis copper-related disorders. DNA Cell Biol 21:259–270
Lombardo S, Pandino G, Mauromicale G (2013) The influence of growing environment on the antioxidant and mineral content of "early" crop potato. J Food Compos Anal 32:28–35. https://doi.org/10.1016/j.jfca.2013.08.003
Lopez AD, Mathers CD, Ezzati M et al (2006) Global and regional burden of disease and risk factors, 2001: systematic analysis of population health data. Lancet 367:1747–1757
Madsen E, Gitlin JD (2007) Copper and Iron disorders of the brain. Annu Rev Neurosci 30:317–337. https://doi.org/10.1146/annurev.neuro.30.051606.094232
Marschner H (1995) Mineral nutrition of higher plants, 2nd edn. Academic press, London
Marschner H (2011) Marschner's mineral nutrition of higher plants. Academic press, Amsterdam
McGill CR, Kurilich AC, Davignon J (2013) The role of potatoes and potato components in cardiometabolic health: a review. Ann Med 45:467–473

Meenakshi JV, Johnson NL, Manyong VM et al (2010) How cost-effective is biofortification in combating micronutrient malnutrition? An ex ante assessment. World Dev 38:64–75. https://doi.org/10.1016/j.worlddev.2009.03.014
Mishra T, Raigond P, Thakur N et al (2020) Recent updates on healthy phytoconstituents in potato: a nutritional depository. Potato Res:1–21. https://doi.org/10.1007/s11540-019-09442-z
Murphy J, Riley JP (1962) A modified single solution method for the determination of phosphate in natural waters. Anal Chim Acta 27:31–36. https://doi.org/10.1016/S0003-2670(00)88444-5
Narwojsz A, Borowska EJ, Polak-Śliwińska M, Danowska-Oziewicz M (2020) Effect of different methods of thermal treatment on starch and bioactive compounds of potato. Plant Foods Hum Nutr 75:298–304. https://doi.org/10.1007/s11130-020-00808-0
Nassar AMK, Sabally K, Kubow S et al (2012) Some Canadian-grown potato cultivars contribute to a substantial content of essential dietary minerals. J Agric Food Chem 60:4688–4696
National Research Council (1989) Recommended dietary allowances. National Academies Press, Washington, DC
Navarre DA, Goyer A, Shakya R (2009) Nutritional value of potatoes. In: Advances in potato chemistry and technology, 1st edn. Elsevier, Amsterdam, pp 395–424
Nestel P, Bouis HE, Meenakshi JV, Pfeiffer W (2006) Biofortification of staple food crops. J Nutr 136:1064–1067. https://doi.org/10.1093/jn/136.4.1064
O'Neill KR (2005) Statistical derivation of nutrition label for raw potatoes from 2001--2002 United States Department of Agriculture (USDA) Agricultural Research Service (ARS) Nutrient Data Laboratory (NDL) Data for Raw Russet, White, and Red Potatoes by Weighting by Marke. Center for Food Safety and Applied Nutrition FDA, College Park, MD
Paget M, Amoros W, Salas E et al (2014) Genetic evaluation of micronutrient traits in diploid potato from a base population of Andean Landrace Cultivars. Crop Sci 54:1949–1959. https://doi.org/10.2135/cropsci2013.12.0809
Pandey R, Lal MK, Vengavasi K (2018) Differential response of hexaploid and tetraploid wheat to interactive effects of elevated [CO2] and low phosphorus. Plant Cell Rep 37:1231–1244. https://doi.org/10.1007/s00299-018-2307-4
Phillippy BQ, Lin M, Rasco B (2004) Analysis of phytate in raw and cooked potatoes. J Food Compos Anal 17:217–226. https://doi.org/10.1016/j.jfca.2003.08.001
Prasad AS, Miale A, Farid Z et al (1963) Biochemical studies on dwarfism, hypogonadism and anemia. Arch Intern Med 111:407–428. https://doi.org/10.1001/archinte.1963.03620280007003
Prasad R, Shivay YS, Kumar D (2014) Agronomic biofortification of cereal grains with iron and zinc. Adv Agron 125:55–91. https://doi.org/10.1016/B978-0-12-800137-0.00002-9
Puzina TI (2004) Effect of zinc sulfate and boric acid on the hormonal status of potato plants in relation to tuberization. Russ J Plant Physiol 51:209–215. https://doi.org/10.1023/B:RUPP.0000019216.92202.4a
Raghothama KG (1999) Phosphate acquisition. Plant Soil 274:37–49. https://doi.org/10.1007/s11104-004-2005-6
Ramos SJ, Rutzke MA, Hayes RJ et al (2011) Selenium accumulation in lettuce germplasm. Planta 233:649–660. https://doi.org/10.1007/s00425-010-1323-6
Rayman MP (2012) Selenium and human health. Lancet 379:1256–1268. https://doi.org/10.1016/S0140-6736(11)61452-9
Rosen CJ, Kelling KA, Stark JC, Porter GA (2014) Optimizing phosphorus fertilizer Management in Potato Production. Am J Potato Res 91:145–160. https://doi.org/10.1007/s12230-014-9371-2
Rout GR, Sahoo S (2015) Role of iron in plant growth and metabolism. Rev Agric Sci 3:1–24. https://doi.org/10.7831/ras.3.1
Rowe RCE (1993) Potato health management. APS Press, St. Paul, MN
Salisbury FB, Ross CW (1992) Plant physiology. Wadsworth Publishing Company, Belmont, CA
Saltzman A, Birol E, Bouis HE et al (2013) Biofortification: Progress toward a more nourishing future. Glob Food Sec 2:9–17

Sanchez-Castillo CP, Dewey PJS, Aguirre A et al (1998) The mineral content of Mexican fruits and vegetables. J Food Compos Anal 11:340–356. https://doi.org/10.1006/jfca.1998.0598
Schulze KJ (2012) Calcium. In: Encyclopedia of human nutrition. Academic Press, Cambridge, MA
Senbayram M, Gransee A, Wahle V, Thiel H (2015) Role of magnesium fertilisers in agriculture: plant–soil continuum. Crop Pasture Sci 66:1219. https://doi.org/10.1071/CP15104
Singh B, Pandey R (2003) Differences in root exudation among phosphorus-starved genotypes of maize and green gram and its relationship with phosphorus uptake. J Plant Nutr 26:2391–2401. https://doi.org/10.1081/pln-120025467
Sosa P, Guild G, Burgos G et al (2018) Potential and application of X-ray fluorescence spectrometry to estimate iron and zinc concentration in potato tubers. J Food Compos Anal 70:22–27. https://doi.org/10.1016/j.jfca.2018.03.004
Staiger C (2008) 2008: international year of the potato. Pharm J 281:743–744
Stern BR, Solioz M, Krewski D et al (2007) Copper and human health: biochemistry, genetics, and strategies for modeling dose-response relationships. J Toxicol Environ Heal B Crit Rev 10:157–222. https://doi.org/10.1080/10937400600755911
Taiz L, Zeiger E (2015) Plant physiology and development, 6th edn. Sinauer Assoc Inc, Sunderland, MA
Takeda E, Yamamoto H, Yamanaka-Okumura H, Taketani Y (2012) Dietary phosphorus in bone health and quality of life. Nutr Rev 70:311–321. https://doi.org/10.1111/j.1753-4887.2012.00473.x
True RH, Hogan JM, Augustin J et al (1978) Mineral composition of freshly harvested potatoes. Am Potato J 55:511–519. https://doi.org/10.1007/BF02852157
True RH, Hogan JM, Augustin J et al (1979) Changes in the nutrient composition of potatoes during home preparation: III. Minerals. Am Potato J 56:339–350. https://doi.org/10.1007/BF02853849
Van Der Straeten D, Fitzpatrick TB, De Steur H (2017) Editorial overview: biofortification of crops: achievements, future challenges, socio-economic, health and ethical aspects. Curr Opin Biotechnol 44:vii–x. https://doi.org/10.1016/j.copbio.2017.03.007
Vert G, Grotz N, Dédaldéchamp F et al (2002) IRT1, an Arabidopsis transporter essential for iron uptake from the soil and for plant growth. Plant Cell 14(6):1223–1233. https://doi.org/10.1105/tpc.001388
Vo TQ, Tran BTT, Nguyen NT et al (2020) Psychometric properties of the World Health Organization quality of life instrument, short form: validity in the Vietnamese healthcare context. Syst Rev Pharm 11:14–22
Walker CLF, Ezzati M, Black RE (2009) Global and regional child mortality and burden of disease attributable to zinc deficiency. Eur J Clin Nutr 63:591–597. https://doi.org/10.1038/ejcn.2008.9
Warman PR, Havard KA (1998) Yield, vitamin and mineral contents of organically and conventionally grown potatoes and sweet corn. Agric Ecosyst Environ 68:207–216. https://doi.org/10.1016/S0167-8809(97)00102-3
Welch RM (1995) Micronutrient nutrition of plants. Crit Rev Plant Sci 14:49–82. https://doi.org/10.1080/07352689509701922
Welch RM, Graham RD (2004) Breeding for micronutrients in staple food crops from a human nutrition perspective. J Exp Bot 55(396):353–364
White PJ, Broadley MR (2005) Biofortifying crops with essential mineral elements. Trends Plant Sci 10:586–593
White PJ, Broadley MR (2009) Biofortification of crops with seven mineral elements often lacking in human diets - Iron, zinc, copper, calcium, magnesium, selenium and iodine. New Phytol 182:49–84
White PJ, Brown PH (2010) Plant nutrition for sustainable development and global health. Ann Bot 105:1073–1080
Woolfe JA, Poats SV (1987) The potato in the human diet. Cambridge University Press, Cambridge
World Health Organization (2015) The global prevalence of anaemia in 2011. WHO, Geneva

Zasoski RJ, Burau RG (1977) A rapid nitric-perchloric acid digestion method for multi-element tissue analysis. Commun Soil Sci Plant Anal 8:425–436. https://doi.org/10.1080/00103627709366735

Zhou JR, Erdman JW (1995) Phytic acid in health and disease. Crit Rev Food Sci Nutr 35:495–508. https://doi.org/10.1080/10408399509527712

Zhu Z, Chen Y, Shi G, Zhang X (2017) Selenium delays tomato fruit ripening by inhibiting ethylene biosynthesis and enhancing the antioxidant defense system. Food Chem 219:179–184. https://doi.org/10.1016/j.foodchem.2016.09.138

Zimmermann MB, Jooste PL, Pandav CS (2008) Iodine-deficiency disorders. Lancet 372:1251–1262

Potato Vitamins

7

Maharishi Tomar, Reetu, and Sushil Sudhakar Changan

Abstract

Potato (*Solanum tuberosum* L.) is the world's largest non-grain food crop and ranked third overall after rice and wheat production. Potato is a highly nutritious, simple to digest, balanced food that contains proteins, carbohydrates, vitamins, minerals, and a high-quality dietary fiber. Vitamins belong to a diverse group of organic or related group of compounds. Ordinarily, they are not synthesized by humans but are primary requisite for maintaining proper functioning and overall health of the body. Vitamins content depends on a number of factors, with variety and environmental conditions being the most important ones. Up to 45% of the recommended daily allowance (RDA) for vitamin C, 8% niacin, 10% vitamin B_6 and 6% folate is provided by a single 150 g tuber. Vitamins are an essential nutritional element of potatoes and their improvement in making potato as a "full food" using advanced biotechnological, breeding approaches, method of estimation, the effect of growing conditions and their health benefits is mentioned in the proceeding chapter. Being established as a valuable nutritional source, development of better management practices, molecular and traditional breeding approach, and employment of potent biotechnological tools like genome editing through CRISPR and TALEN, the vitamin and phytonutrient content of potatoes can be substantially increased. Potatoes can possibly be part of the solution to increasing agricultural output to meet the expected increase in demand for global food security by 2050.

M. Tomar (✉) · Reetu
ICAR-Indian Grassland and Fodder Research Institute, Jhansi, Uttar Pradesh, India
e-mail: maharishi89@gmail.com

S. S. Changan
ICAR-Central Potato Research Institute, Shimla, Himachal Pradesh, India

P. Raigond et al. (eds.), *Potato*, https://doi.org/10.1007/978-981-15-7662-1_7

Keywords

Potato · Vitamin C · Vitamin B complex · Estimation methods

7.1 Introduction

Regardless of being ranked as the fourth most prevalent grown crop overall, one of the most cultivated tuber crops in the world, and the third most significant crop after rice and wheat, the overall nutritive value of a potato is still highly underestimated. Taking into consideration its short duration crop cycle it is appropriate to say that it generates more dry matter, edible protein, and palatable energy in a short period as compared to cereals such as wheat and rice. Potatoes are a source of extremely nutritious, easy to digest, healthy food that contain carbohydrates, minerals, proteins, vitamins, high-quality soluble, insoluble dietary fiber and healthy phytochemicals (Camire et al. 2009). Potato is also a vital source of substances in the modern diet that promotes overall health and wellness (Gibson and Kurilich 2013). Potato provides a prominent portion of daily requirement values (DRV) of various indispensable nutrients, e.g., potassium (26% of DRV; plays a critical role in maintaining acid-base equilibrium in the cell, also causes elimination of Na), vitamin C (28% of DRV; serves as an essential antioxidant molecule, integral for repair, maintenance, and growth of tissues and muscles in the body, enhances non-haem iron absorption), vitamin B_6 (27% of DRV; vital for most enzymatic reactions related to metabolism of amino acids, carbohydrates, and lipids in the body), dietary fiber (15% of DRV; necessary for regulating and managing body weight, have an essential role in cholesterol metabolism), magnesium (12% of DRV; serves as an essential cofactor for various metalloenzymes related to metabolism and cellular functions), and iron (10% of DRV; a necessary cofactor of several enzyme related to vital cellular activities and metabolic processes). Vitamins belong to a diverse group of organic or related group of compounds. Ordinarily, they are not synthesized by humans but are primary requisite for maintaining proper functioning and overall health of the body. Table 7.1 summarizes the major vitamins present in potato along with their most active and coenzymatic forms.

Among several other vitamins, potato contains vitamin C (ascorbic acid/L-ascorbate) which is water-soluble and is present in the highest concentration ranging from 84 to 145 mg/100 g dry weight (Camire et al. 2009). Upon examination of 98 vegetables, potato was found to be the most inexpensive source of vitamin C (Drewnowski and Rehm 2013). Deficiency of vitamin C causes scurvy (Neufeld 2017), its early symptoms include fatigue, sore legs and arms. If left untreated, it leads to the destruction of red blood cells, low immunity, liver spots, bleeding skin and gums, poor healing of wounds, and finally death (Shaath et al. 2016). Vitamin C is synthesized by tubers, some of which is later channelized into stems and leaves (Tedone et al. 2004). For animal system, vitamin C serves as a cofactor for several enzymes and has a prominent role in immune system (Carr and Maggini 2017), it modulates some epigenetic modifications (Young et al. 2015), bone development

Table 7.1 Chief vitamins in potato and their metabolically active forms

S. no	Nutrient (vitamins)	Active form
1	Vitamin C	Vitamin C/L-ascorbic acid/ascorbate/ascorbic acid
2	Vitamin B_6 (pyridoxin)	Pyridoxal 5′-phosphate (PLP)
3	Vitamin B_1 (thiamin)	Thiamine pyrophosphate (TPP)
4	Vitamin B_3 (niacin)	Nicotinamide adenine dinucleotide (NAD) and (NADP) nicotinamide adenine dinucleotide phosphate
5	Vitamin B_5 (pantothenic acid)	Pantothenic acid
6	Vitamin B_9 (folic acid)	Tetrahydrofolic acid and methyltetrahydrofolate
7	Vitamin E	α-, β-, γ-, δ-tocopherols

(Aghajanian et al. 2015), promotes skin health by stimulating collagen synthesis by acting as cofactor for hydroxylation reactions performed by prolyl-4-hydroxylase, lysyl-hydroxylase and preventing photodamage induced by UV (Pullar et al. 2017; Al-Niaimi and Chiang 2017) and enhances the non-haem iron bioavailability. According to the FAO, the range of recommended nutrient intake is from 25–45 mg/day (FAO/WHO 2001).

Plants are the chief source of dietary vitamin C, with concentration ranging from 5 mM in leaves (Wheeler et al. 1998) and 20–300 mM in chloroplasts (Smirnoff and Wheeler 2000). It has several substantial roles in the plant system like electron transport systems, photosynthesis, scavenging reactive oxygen species (ROS) produced during photorespiration and oxygen photoreduction, plant defense, control of stomatal dynamics, drought-oxidative stress, metal/metalloid stress tolerance, crucial cellular antioxidant, alleviation of abiotic stress, cell signalling, ozone sensitivity, regulation of cell death, disease resistance, and many more (Anjum et al. 2014; Venkatesh and Park 2014).

Dietary study on Japanese men as subjects demonstrated that their blood plasma levels of vitamin C increased considerably after the consumption of potato chips and mashed potatoes (Kondo et al. 2012). Love et al. (2004) analyzed vitamin C content in 75 potato genotypes and found a range of concentration from 11.5 to 29.8 mg/100 g fresh weight. Plants have various biosynthetic pathways for vitamin C production, among them, the characterization of enzymes for L-galactose pathway has been successfully done (Smirnoff 2011). Biosynthesis of L-ascorbate in plant systems occurs predominantly by Smirnoff–Wheeler pathway having L-galactose and D-mannose as primary precursors which are interconverted by the enzyme GDP-D-mannose-3,5-epimerase. Primarily D-glucose-6-phosphate is converted to GDP-L-galactose and GDP-D-mannose. GDP-L-galactose is then hydrolyzed by a two-step process to L-galactose-1-phosphate which is further oxidized to L-galactono1,4-lactone, and is finally oxidized by mitochondrial L-galactono-1,4-lactone dehydrogenase to form L-ascorbate (Wheeler et al. 1998). Although Smirnoff–Wheeler pathway appears to be the major pathway for L-ascorbate biosynthesis, same uronic acid intermediates like D-galacturonic acid can also synthesize L-ascorbate. In this pathway, D-galacturonate reductase (GalUR) catalyzes the reduction of D-galacturonic to L-galactonic acid, which is further converted to L-galactono-1,4-lactone. L-galactono-1,4-lactone dehydrogenase (GALDH) catalyzes the final step of oxidizing L-galactono-1,4-lactone to L-ascorbate (Agius et al. 2003).

Environmental factors have a significant impact on vitamin C content in potato tubers. Several studies have proved beyond doubt that the level of vitamin C in plants declines drastically during cold storage (Keijbets and Ebbenhorst-Seller 1990). Dale et al. (2003) found that there was a considerable decrease in vitamin C content in 33 potato genotypes when placed in cold storage for 15–17 weeks. Blauer et al. (2013) suggested that the transcript level of key genes controlling the rate-limiting steps of L-ascorbate biosynthetic pathway like GGP (GDP-L-galactose phosphorylase/guanylyltransferase) increased considerably as the tuber formation progressed 0.6–1.5 gm tuber stage. There was an increased expression of genes related to L-ascorbate biosynthesis during early bulking stage and tuberization. Further, there was a decline in L-ascorbate content during senescence, and a progressive continuous regression was noted throughout storage and maturation. A loss as high as 65% was recorded throughout 8.5 months storage. It was also indicated that the storage of tubers under low oxygen concentration reduced the loss of L-ascorbate. Tudela et al. (2002) demonstrated that the content of vitamin C in freshly cut potatoes remained almost unchanged when stored at 4 °C for 6 days in normal air, while a drastic reduction was observed when they were stored at −22 °C. After storing 12 genotypes in cold storage for 7 months, a 50% loss in vitamin C content was confirmed. Similar results were found by Külen et al. (2013), who also observed a drastic decline in vitamin C content in potato during cold storage. This decline had a gradual pattern, i.e., total vitamin C dropped by 24% after 2 months, 45% after 4 months, and 52% after 7 months.

One study revealed that wounding causes a substantial increase in vitamin C content of potatoes. Mondy and Leja (1986) observed that when the tubers were stored for 2 days followed by bruising and slicing, their vitamin C content increased by 400%, while a decrease of 347% was observed in bruised tubers. Vitamin C content of freshly cut potatoes increases when stored in air, while it decreases dramatically under frozen or modified atmospheric conditions (Tudela et al. 2002). Tosun and Yücecan (2008) studied the effects of washing, peeling, blanching, slicing, frying, drying, freezing, cooling, and storage on vitamin C content in potatoes. They observed a loss of 11.1% during peeling, 25.0% during washing, 30.5% during slicing, 40.9% during blanching, 46.4% during drying, 51.5% during frying and cooling, 56.8% during freezing, 59.6% during a period of 3 month storage, and 61.5% during 6 months of storage. Haase and Weber (2003) found that vitamin C content decreased drastically during the blanching and frying of potatoes while making French fries and chips. Critical external factors that cause a decrease in vitamin C content in tubers during cooking are oxidation in the presence of air, heating, and leaching. Methods of food preparation that include minimum cutting and shredding, retention of peel, using of minimal amount of water and reduced cooking time at low temperature results in an increase of vitamin C retention by several folds (Love and Pavek 2008).

Breeding varieties having superior vitamin C content is the first logical step to increase its availability in the human diet. Increase in ascorbate can be achieved by accessing the genetic variability already present among different potato cultivars (Okeyo and Kushad 1995). Various researchers suggested that vitamin C content in potato is heritable, thus traditional breeding approach can be employed as an essential tool to create new cultivars with high vitamin C levels.

Additionally, various molecular and biotechnological tools are employed for rapid and direct manipulation of vitamin levels in tubers. Overexpression of GDP L-galactose phosphorylase in potato tubers resulted in an increase in vitamin C by three-folds. When knockout mice that cannot synthesize vitamin C were fed potato chips, the mice had increased accumulation of vitamin C and a significant drop in the level of ROS concluding that fast-dry-processed potato chips are an excellent source of vitamin C (Kondo et al. 2014). In animals, L-ascorbic acid is synthesized by the oxidation of L-gulono-γ-lactone to L-ascorbic acid catalyzed by L-gulono-γ-lactone oxidase (Smirnoff 2001). Hemavathi et al. (2010) transformed potato plants by L-gulono-γ-lactone oxidase isolated from rat and overexpressed it by using CaMV35S constitutively active promoter resulting in enhancement of L-ascorbic acid accumulation by 40% and increase in tolerance to abiotic stress. Potato transgenic lines with 1.6–2 fold high L-ascorbic acid and increased abiotic stress were created by transforming and overexpressing tubers with strawberry GalUR (D-galacturonate reductase) gene controlled by CaMV 35S promoter (Hemavathi et al. 2009).

Overexpression of GalUR (D-galacturonic acid reductase), one of the chief gene in ascorbate biosynthetic pathway enhances the ascorbate content and increases the ability of transgenic plants to combat oxidative stress. GalUR overexpressing transgenic lines continued to tuberize even under the presence of 200 mM NaCl. These transgenics also had a higher ratio of reduced to oxidized glutathione (GSH: GSSG) (Upadhyaya et al. 2011). Dehydroascorbate reductase (DHAR) recycles the oxidized ascorbic acid, thus maintaining its optimum concentration in the cell. DHAR overexpressing transgenic potato lines exhibited high L-ascorbate levels in their tubers and leaves (Qin et al. 2011).

There are numerous methods for estimation of vitamin C. One of the classical approaches for its determination is a colorimetric technique by using Folin phenol reagent. It is quantified by color developed due to the interaction between the folic reagent and ascorbic acid at 760 nm (Jagota and Dani 1982). It can also be quantified spectrophotometrically through the reduction of 2,6-dichlorophenolindophenol dye by ascorbic acid (Egoville et al. 1988). High-performance liquid chromatography (HPLC) is used as a more precise method for its quantification in potatoes (Han et al. 2004) using C18 column. Spectrofluorimetric technique as determined by Liu et al. (2000) is also promising for accurate estimation of vitamin C. Structure of vitamin C is given in Fig. 7.1.

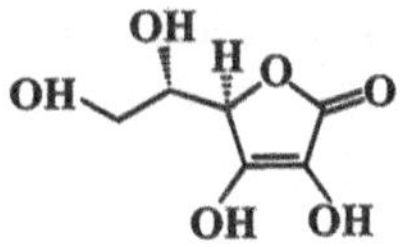

Fig. 7.1 Structure of vitamin C

7.2 Vitamin B Complex

Vitamin B family includes water-soluble vitamins with distinct chemical, physical, and functional properties. They play critical and intricate roles in cellular metabolism, functioning and act in their metabolically active coenzymatic forms for numerous anabolic and catabolic biochemical reactions. This group includes vitamin B_1 (thiamin), vitamin B_2 (riboflavin), vitamin B_3 (niacin or niacin amide), vitamin B_5 (pantothenic acid), vitamin B_6 (pyridoxine), vitamin B_7 (biotin), vitamin B_9 (folic acid), and vitamin B_{12} (cobalamins) (Combs Jr and McClung 2016).

Thiamin is an essential nutrient for humans (Bender 1999) as it can only be synthesized by plants (Goyer 2010), fungi (Bettendorff and Wins 2013), and bacteria (Gigliobianco et al. 2013). It functions as a cofactor for various enzymes in protein, energy, and lipid (Casteels et al. 2007) metabolic pathways. Its deficiency causes beriberi since it acts as a critical cofactor in pyruvate and alpha-ketoglutarate dehydrogenases, both necessary for the tricarboxylic acid cycle. Beriberi leads to failure of the autonomic nervous system and cardiovascular problems (Abdou and Hazell 2015). It appertains indirectly in specific crucial steps of the electron transport system, carbohydrate metabolism and Krebs cycle. Structurally it is composed of pyrimidine group and thiazole ring joined by a methylene bridge (Kraft and Angert 2017; Manzetti et al. 2014). Its active form is thiamine pyrophosphate (TPP) or thiamine diphosphate, which is synthesized from thiamin by the action of thiamine diphosphokinase (Martin 2001). TPP has an indispensable function as a cofactor for enzymes participating in α-ketoglutarate decarboxylation, also as a catalyst in pentose phosphate pathway and hexose monophosphate shunt (Tittmann 2009). It functions as a cofactor for pyruvate dehydrogenase complex that catalyzes the production of acetyl-CoA from pyruvate (Tittmann 2009) and for dehydrogenases involved in the degradation of branched chain amino acids like valine, leucine, and isoleucine (Binder et al. 2007). Besides playing as a central element in various energy metabolism reactions it also participates in oxidative decarboxylation (Li et al. 2012), protein formation, activation of neuroblastoma conduction channels, development and function of brain, interneuronal communication and activation of immune system (Gibson and Blass 2007). Generally, thiamin deficiency occurs due to high consumption of alcohol affecting its transport, metabolism, and storage (Rees and Gowing 2012).

According to the National Institutes of Health, the RDA (Recommended Dietary Allowance) of thiamin for adult men is 1.2 mg/day and for adult women is 1.1 mg/day. As per the findings of Goyer and Sweek (2011) several potato varieties contain approximately 20% of thiamin RDA in as few as 100 g of tubers. Being a heat

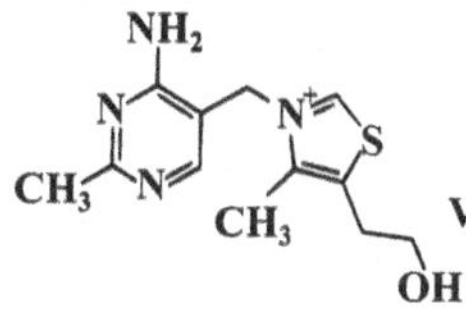

Fig. 7.2 Structure of vitamin B_1 (Thiamin)

sensitive vitamin (Kandutsch and Baumann 1953) thiamin levels are considerably lower in processed potatoes. Thiamin present in 100 g of oil fried French fries is 0.08 mg, for chips is 0.17 mg, for baked potatoes is 0.067 mg, and for boiled and cooked potatoes is 0.106 mg indicating that the procedure for processing, cooking time, and temperature determines the amount of thiamin preserved in the product. Hundred percent thiamin retention was observed in the case of chilled, baked, and reheated potatoes and 90% in rehydrated, cooked, chilled, and reheated potato granules (Augustin et al. 1980).

In recent years, substantial advancement has been made in illustrating the pathway of thiamin biosynthesis, promoting the development of concrete schemes for biofortification, with specific emphasis to genetic engineering. The presently available data sets substantiate the capacity of potatoes for thiamin fortification through bioengineering or breeding approach. High-performance liquid chromatography at 254 nm using C18 column (Cho et al. 2000), ultraviolet-visible (UV-Vis) spectrophotometry (López-de-Alba et al. 2006) is used for precise and rapid estimation of thiamin. Fluorimetry also serves as a promising technique for analysts (Sarkiyayi and Ikioda 2010). Recently more accurate methods based on spectrofluorometry are used to determine thiamin (Hashemi et al. 2019). Structure of vitamin B_1 is given in Fig. 7.2.

It is a water-soluble vitamin occurring in two different forms, i.e., nicotinamide and nicotinic acid. They belong to pyridinecarboxylic acid group of compounds. An adequate amount of niacin in the body can either be ensured through its biosynthesis from the aromatic amino acid tryptophan, or it can directly be absorbed from the food. Several reports signify the potency of niacin to prevent serious ailments like diabetes, hypercholesterolemia, hypotension, issues with vasodilation, protein synthesis, skin aging, and atherosclerosis (Rolfe 2014). It also helps in burn-wound recovery, prevention of skin cancer, and cataracts (Benavente et al. 2009). Niacin is the precursor for coenzymes (NAD) nicotinamide adenine dinucleotide and (NADP) nicotinamide adenine dinucleotide phosphate. NAD serves as a two-electron carrier during glucose oxidation, ultimately synthesizing ATP (adenosine triphosphate) the molecular unit for energy transfer, during oxidative phosphorylation in the cellular mitochondria (Depeint et al. 2006a, b). The level of NAD regulates glucose oxidative pathway and ATP generation, i.e., a decrease in NAD inhibits glucose oxidation, while a high cellular glucose concentration reduces NADPH/NADP and NAD/NADH ratios (Van den Enden et al. 1995).

According to Berger et al. (2004) both NAD and NADP are involved in a battery of numerous cellular signalling pathways, protein modifications, and calcium mobilization. Biological effects of coenzymes NADP-NADPH and NAD-NADH are

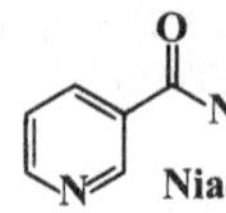

Fig. 7.3 Structure of vitamin B_3 (Niacin)

determined by their redox state (Wahlberg et al. 2000). NADPH functions as a cofactor in sterol and fatty acid biosynthesis, photosynthesis, and carbohydrate metabolism (Wahlberg et al. 2000). Niacin deficiency causes pellagra which is symptomized by diarrhea (watery stool), dermatitis (skin inflammation), and dementia (loss of memory and judgment) (Fu et al. 2014). Niacin deficiency is ubiquitous in counties where the majority of the population depends on maize as their staple food. Retention of niacin requires complex seed processing method termed as "nixtamalization" (Gwirtz and Garcia-Casal 2014).

Wills et al. (1984) was unable to observe any significant variation in niacin content among numerous tested genotypes of potato. RDA values for niacin are 14 mg/day for men and 16 mg/day for women. It is estimated that 100 g boiled potatoes contain 1.44 mg thiamin, which can fulfil 10% of RDA. Niacin content is significantly low, 0.28 mg/100 g in French fries, 3.38 mg/100 g in potato chips, and 1.41 mg/100 g in baked potatoes. Page and Hanning (1963) studied the effect of cooking on niacin availability and found that the loss of vitamin B_6 was 4.8% for boiled and 8.8% for baked potatoes. Similarly, loss of nicotinic acid was 1.5% for boiled and 4.2% for baked potatoes concluding that 100 g baked or boiled potatoes can supply 1/10 of adult's daily requirement of vitamin B6 and nicotinic acid. Retention of niacin in oven-baked chips and boiled, peeled potatoes was substantially lower in comparison to raw samples (Augustin et al. 1978). Nisha et al. (2009) studied the degradation of niacin over a range of temperature (50–120 °C). The study concluded that first-order kinetics was followed in the degradation process with an increase in the rate constant in proportion to temperature. Hundred percent niacin retention was observed in the case of chilled, baked, and reheated potatoes and 90% in rehydrated, cooked, chilled, and reheated potato granules (Augustin et al. 1980). HPLC is considered as a precise and rapid method for qualitative and quantitative estimation of vitamin B_3 using C18 column (Seal and Chaudhuri 2017). Several other techniques like microemulsion electrokinetic capillary chromatography (Aurora-Prado et al. 2010) and reversed-phase chromatography (Singh et al. 2013) are also successfully employed to estimate niacin content. Structure of vitamin B_3 is given in Fig. 7.3.

Pantothenic acid is the precursor for a 4'phosphopantetheine moiety present in acyl carrier protein (ACP) and coenzyme A (CoA). ACP and CoA serve as cofactors for enzymes that are integral for crucial energy-yielding metabolic pathways for all living systems especially fatty acid, amino acid, and carbohydrate metabolism (Depeint et al. 2006a, b). Deficiency of this vitamin has not been reported so far in humans even though animal tissues do not produce it, owing to its synthesis by gastrointestinal microbiota and its presence in most food products (Wojtczak and Slyshenkov 2003). However, under medical conditions like anemia, dermatitis, encephalopathy, and convulsions, it can turn into a limiting factor (Bender 1999).

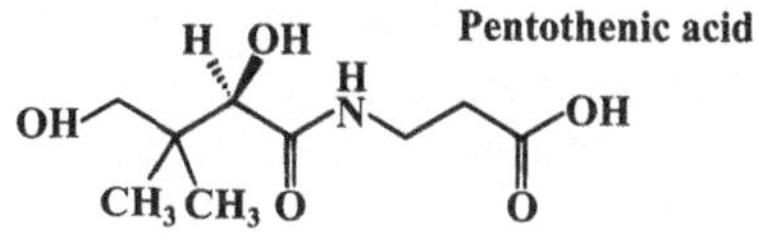

Fig. 7.4 Structure of vitamin B5 (Pantothenic Acid)

Beneficial effects of pantothenic acid include protection against radiation injury (Dombradi et al. 1964), skin wound healing (Grenier et al. 1982), maintenance of lung epithelium (Wigand et al. 1991), and cornea of the eye (Egger et al. 1999). It is also known to cure injuries in internal organs like liver and heart caused by oxidative stress and atherosclerosis (Kumerova et al. 1992). Due to the lack of its malnutrition only AI (adequate intake) values have been established instead of RDA which is close to 5 mg/day for both adult men and women (Tahiliani and Beinlich 1991).

Considerable work has not been done on the content of vitamin B_5 in potatoes. Nonetheless, Riaz et al. (2009) stated that pantothenic acid is moist heat stable but labile under dry heating conditions. Vitamin B_5 content in French fries was found to be 0.5 mg/100 g, 0.4 mg/100 g for potato chips, 0.38 mg/100 g for baked, and 0.52 mg/100 g for boiled potato. Acrylamide is a toxic, mutagenic, and carcinogen organic compound produced through Maillard reaction during food processing. Pantothenic acid was found to inhibit the acrylamides generated in fried potato strips (Zeng et al. 2009). Vitamin B_5 can be detected rapidly by gas chromatography–mass spectrometry (GC-MS) (Juhász et al. 2014). Techniques like HPLC—diode array detector—mass spectrometry and reverse-phase liquid chromatography–fluorimetry are also used for its quantification along with other water-soluble vitamins (Santos et al. 2012). Structure of vitamin B_5 is given in Fig. 7.4.

Vitamin B_6 family constitutes of three derivatives of 3-hydroxy-2-methylpyridine (pyridoxal, pyridoxamine, and pyridoxol) along with their 5′-phosphates (Velisek and Cejpek 2007). They occur as pyridoxamine, pyridoxal, and pyridoxal-5-′-phosphate in their bioactive forms. It functions as a coenzyme for close to 140 metabolic reactions (Mooney and Hellmann 2010). These metabolic reactions involve amino acid anabolism and catabolism (Stolz and Vielreicher 2003) through transaminations (erythrocyte aspartate aminotransferase), transsulfuration (cystathionine γ-lyase, cystathionine β-synthase) (Zhang et al. 2009) and also participates in selenoamino acid metabolism (selenocysteine γ-lyase, selenocysteine β-lyase). They are involved in single-carbon metabolism, synthesis of niacin from tryptophan, gluconeogenesis, utilization of glycogen in animal tissues through glycogen phosphorylase. They administer various neurological functions by synthesizing several key neurotransmitters (γ-aminobutyric acid, dopamine, serotonin, norepinephrine) and sphingolipids, modulate the immune response through histamine biosynthesis by histidine decarboxylase, participate in hemoglobin synthesis through δ-aminolevulinic acid synthase, and perform several cardiovascular functions. They also function as potent oxidizing agents and crucial factors that govern resistance of plants against pathogens (Zhang et al. 2015). Deficiency of vitamin B_6 is hard to detect, but early symptoms include dermatitis and stomatitis which closely resemble riboflavin and niacin insufficiency. Prolonged deficiency can cause neurological problems leading to peripheral neuropathy and dementia (Spinneker et al. 2007).

According to Mooney et al. (2013), there is a notable variation in vitamin B_6 content among several potato cultivars. He also studied the difference in vitamin content among mature and immature tubers and established that its concentration range in mature tubers was 18 μg B_6/g DW (dry weight) to 27 μg B_6/g DW and for immature tubers was 16 μg/g DW to 22 μg/g DW. These results indicate that there is not much variation in B_6 content among different developmental stages, implying that its good quantity is present in both developmental stages. RDA of B_6 is 2.0 mg/day for men and 1.6 mg/day for women (Manore 2000). Potato is an adequate source of this vitamin and its concentration is not susceptible to heat and processing methods. Its concentration is 0.36 mg/100 g in French fries, 0.78 mg/100 g in potato chips, and 0.31 mg/100 g in baked and boiled potatoes.

Vitamin B_6 serves as a cofactor for α-glucan phosphorylase of plastids, which serves as a critical enzyme engaged in initial fragmentation of starch into soluble carbohydrates in potato tubers (Kossmann and Lloyd 2000). When potatoes were stored under low temperatures (4–8 °C), they tend to aggregate carbohydrates like fructose, glucose, and sucrose, and this process is termed as cold sweetening (Chen et al. 2012). These carbohydrates augment the tuber's resistance against tissue damage due to cold shock but limit their commercial value for the processing industry, attributable to the formation of acrylamide, and unacceptable brown discoloration (Tareke et al. 2002). The leading cause of this discoloration is Maillard reaction, which takes place between carbonyl group and amino compounds (Becalski et al. 2003). The concentration of vitamin B_6 governs the adequacy of potato for consumption in two ways. First, a higher concentration of vitamin B_6 accelerates the process of cold induces sweetening in potato through starch catabolism by modulating the activity of α-glucan phosphorylase (Manore 2000). Second, out of all vitamins, the structural features of B_6 vitamin i.e. pyridoxamine gives it the ability to scavenge dicarbonyls generated by sugar degradation and Maillard reaction during processing (Arribas-Lorenzo and Morales 2009). Accordingly, modification in B_6 content can change the acceptability of potato for various consumption and processing uses. High vitamin B_6 is not only beneficial for the consumer but also for the plant itself as it increases the overall growth, yield, and stress tolerance (Zhang et al. 2015; Moccand et al. 2014).

Transgenic potato plants were developed through their transformation and overexpression with crucial enzyme PDXII (glutaminase activity) from *Arabidopsis thaliana* in vitamin B_6 biosynthetic pathway. One hundred and seven to one hundred and fifty increase in vitamin B_6 accumulation was observed in transgenic tubers compared to untransformed plants (Bagri et al. 2018). UV-spectrophotometry is an efficient method to simultaneously estimate Vitamin B_1 and B_6 using the H-point standard addition method (Dena and Ammar 2019). HPLC is also used for its estimation using C18 column (Datta et al. 2019). Structure of vitamin B_6 is given in Fig. 7.5.

Vitamin B_9 is an essential vitamin, synthesized by plants from p-aminobenzoate, amino acid glutamate, and 6-hydroxymethyldihydropterin (da Silva et al. 2014). In

Fig. 7.5 Structure of vitamin B_6 (Pyridoxine)

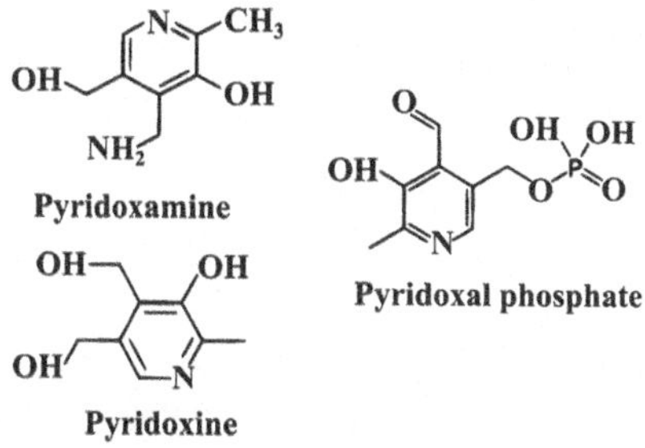

cells, folate is reduced to 7,8-dihydrofolate (DHF) then to 5,6,7,8-tetrahydrofolate (THF), which functions as an active form of this vitamin (Tibbetts and Appling 2010). THF serves as an elemental cofactor in every living organism where it functions as one-carbon (C1) unit carrier in enzymatic reactions that synthesizes amino acids (serine, methionine, glycine), thymidylate, purines, formylmethionyl–transfer RNA (for translation initiation) and pantothenate (vitamin B_5) (Hanson and Gregory III 2011). It has an indispensable role in DNA and sulfur-iron cluster metabolism (Waller et al. 2010). THF functions as a cofactor for serine hydroxymethyltransferase and glycine decarboxylase complex (mitochondrial multienzyme complex) that are central for photorespiratory serine-glycine interconversion (Rajinikanth et al. 2007).

Folate deficiency is a global dietary issue as the vitamin affects processes like oxidation and reduction (Delchier et al. 2014). Lack of folate causes chromosome-breaking due to disproportionate incorporation of uracil, DNA hypomethylation, and deletions in mitochondrial DNA leading to genomic instability (Fenech 2012). It also leads to serious afflictions like neural tube defects (NTDs) like spina bifida during pregnancy (Greenberg et al. 2011), certain types of cancers, and diseases like megaloblastic anemia and heart disorder, programmed cell death or apoptosis (Koury et al. 2000).

Goyer and Navarre (2007) demonstrated notable variation in this vitamin content in potatoes ranging from 521 $\pm$ 96 to 1373 $\pm$ 230 ng/g dry tuber weight, while 7 months tuber storage caused an increase in its content. Variation of vitamin B_9 ranging from 221 $\pm$ 19 to 2336 $\pm$ 285 ng/g dry tuber weight was also observed by Robinson et al. (2015). Among 54 cultivated potato cultivars folate ranged from 388 to 2098 ng/g DW (Goyer and Sweek 2011). RDA of folate for both men and women are 400 μg, and its food supplements are recommended during pregnancy to reduce the risks of congenital disabilities. Around 2.5% of the recommended RDA is present in 100 g boiled potatoes, while 6.5% RDA is present in 100 g baked potatoes. Processed potato products are significantly deficient in folate when compared to unprocessed one (Goyer and Sweek 2011). Since potato is an excellent quality food consumed by many nations abundantly, it can be projected as an exemplary day-to-day source of nourishment. Goyer and Navarre (2009) determined the folate concentration in potato cultivars and found that higher folate levels were present in young tubers and its accumulation gradually declined by 2.6–3.4 folds till

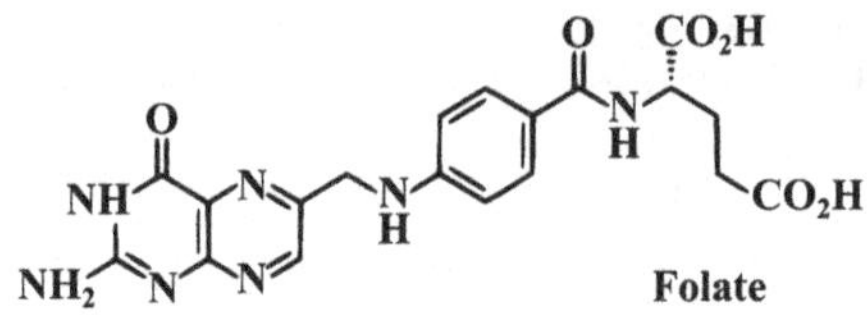

Fig. 7.6 Structure of vitamin B_9 (Folic Acid)

the harvesting time. The assay of various genes associated with folate metabolism like γ-glutamyl hydrolase (GGH1, GGH2) also revealed that variation in its concentration may be regulated post-transcriptionally. Microbiological assay using *Lactobacillus rhamnosus* employing tri-enzyme extraction technique is used to quantify folate content in samples (Goyer and Navarre 2007). Several modern techniques like liquid chromatography–mass spectrometry (LC-MS) (Chen et al. 2017) are also used for its rapid and accurate determination. Structure of vitamin B_9 is given in Fig. 7.6.

They are a group of lipophilic antioxidant compounds, comprising of tocotrienols and tocopherols chiefly synthesized by photosynthetic organisms. Vitamin E encompasses eight diverse forms (α-, β-, γ-, δ-tocotrienols and α-, β-, γ-, δ-tocopherols). Vitamin E activity of these isoforms in-vitro varies substantially, with α-tocopherol having the maximum molar activity and the highest prevalence in nature (Fitzpatrick et al. 2012). Structurally both tocotrienols and tocopherols have a typical chroman-6-ol ring, and the position of methyl groups present on this ring governs the identity and nature of α, γ, δ or β tocotrienols or tocopherols isomers (DellaPenna and Mène-Saffrané 2011). All these isomeric forms have an aromatic ring which can reduce ROS by donating hydrogen atoms (Kanellis and Manganaris 2014). Tocochromanols can chemically or physically quench highly reactive singlet oxygen species ($^1O_2$) which can damage nearly all biological molecules, including polyunsaturated fatty acids (PUFA). The process of physical quenching of $^1O_2$ is executed through charge transfer and thermal dissipation, returning singlet oxygen to $^3O_2$, its ground state without causing any structural and functional damage to tocochromanol. On the other hand, during chemical quenching, the ROS triggers the opening in the chromanol ring system of tocochromanol converting it chemically to its respective tocopherolquinone, which plays a significant role in electron transport reactions in both animals and plants (Munné-Bosch et al. 2005). In plants photosystem II is the primary source of $^1O_2$ generation, therefore, these quenching mechanisms for checking the damage caused by ROS and $^1O_2$ play a more imperative part in plant system than animal metabolic reactions (Triantaphylidès and Havaux 2009). In plants, the biosynthesis and accumulation of tocopherols increase under oxidative stress to counter oxidative damage by limiting the generation of ROS (Lushchak and Semchuk 2012). They share interlinked functions with phytohormones and positively affect seed germination and plant growth (Horvath et al. 2006).

Vitamin E deficiency is uncommon; it generally occurs during severe malnutrition state or defects in α-tocopherol transfer protein or abnormality in dietary fat absorption (Muller 2010). Its deficiency culminates in neurological malfunctions like ataxia, muscle weakness (myopathy), and retina damage (retinopathy) (Chowers et al. 2001). Suboptimal level of vitamin E in plasma is closely related to cardiovascular disease, some types of cancers, and declined immunity (Aslam et al. 2004). Potato tubers contain about 0.7 μg/g fresh weight of vitamin E, comprising of 90% α and 10% γ and β tocopherols (DellaPenna and Mène-Saffrané 2011). Minimum dietary levels of vitamin E are debatable since it does not serve as a cofactor for any known metabolic reaction. Taking into account the above-mentioned issues, dietary reference intake of 15 mg/day was determined (Maras et al. 2004).

As per the studies conducted by Hofius et al. (2004), a potato gene StSXD1 encodes an enzyme having tocopherol cyclase (TC) activity. TC catalyzes ring cyclization reaction which produces α, β, γ, and δ tocopherols. Studies revealed that StSXD1 is localized in the chloroplast and its silencing through RNAi causes an acute deficit of tocopherol concentration in transgenic potato lines. These transgenic lines also had a high accumulation of 2,3-dimethyl-5-phytyl-1,4-hydroquinone, a substrate for TC and obstruction in the process of photoassimilate export. This indirectly has a negative influence on the plant's photosynthetic capacity and results in callose deposition. Two chief enzymes p-hydroxyphenylpyruvate dioxygenase (HPPD) (Garcia et al. 1999), which catalyzes the production of homogentisic acid from p-hydroxyphenylpyruvate and homogentisate phytyl transferase (HPT) which catalyzes the conversion of homogentisic acid to 2-methyl-6-phytyl-1,4-benzoquinone (Cheng et al. 2003) mainly govern tocopherol biosynthesis. With this information of potato, transgenic lines were created by transformation with At-HPPD and At-HPT genes from *Arabidopsis thaliana*. At-HPPD overexpressing lines displayed a 266% increase in α-tocopherol accumulation, while the transgenic lines that overexpressed At-HPT demonstrated a 106% increase. α-tocopherol levels in the tubers were still substantially less than leaves and seeds (Crowell et al. 2008). Vitamin E is most preferably estimated through HPLC (Crowell et al. 2008). Structure of vitamin E is given in Fig. 7.7.

7.3 Conclusion

According to Anderson (2010) potatoes have the potential to provide energy quantified as 216 MJ/ha/day which is greater than rice and corn. They are chemically complex and contain an adequate amount of several essential vitamins. In developing poor countries potato serves as an essential energy and nutrient source providing high amounts of complex carbohydrates, minerals, vitamins, proteins, and dietary fibers. The content and composition of these nutritional elements vary, depending on genotype and environment. Effects of nutritional factors on health results through their bioavailability which in turn depend on the interaction between several food matrix components. Being established as a valuable nutritional source, development of better management practices, molecular and traditional breeding approach, and employment of potent biotechnological tools like genome editing through CRISPR and TALEN, the vitamin and phytonutrient content of potatoes can be substantially increased. Various features of nutrient and vitamin bioavailability should be

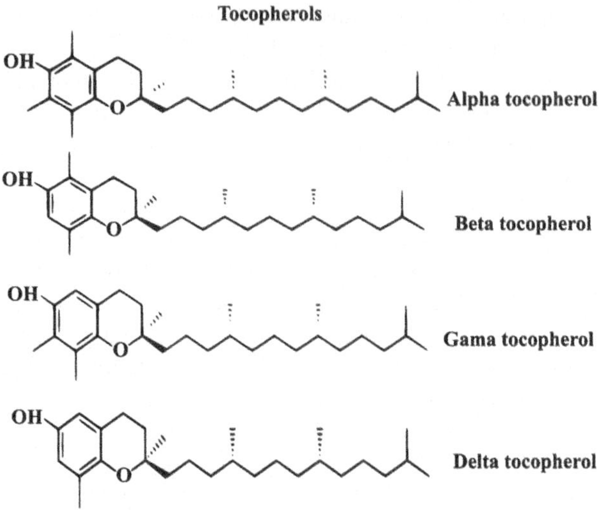

Fig. 7.7 Structure of vitamin E (Tocochromanols)

assessed using greater metabolomics approach. Potatoes can be a solution to increase agricultural output for meeting the projected food demand in 2050.

References

Abdou E, Hazell AS (2015) Thiamine deficiency: an update of pathophysiologic mechanisms and future therapeutic considerations. Neurochem Res 40(2):353–361

Aghajanian P, Hall S, Wongworawat MD et al (2015) The roles and mechanisms of actions of vitamin C in bone: new developments. J Bone Miner Res 30(11):1945–1955

Agius F, González-Lamothe R, Caballero JL et al (2003) Engineering increased vitamin C levels in plants by overexpression of a D-galacturonic acid reductase. Nat Biotechnol 21(2):177–181

Al-Niaimi F, Chiang NY (2017) Topical vitamin C and the skin: mechanisms of action and clinical applications. J Clin Aesthet Dermatol 10(7):14–17

Anderson P (10 Sept 2010) The contribution of potatoes to global food security. International Potato Center, Lima

Anjum NA, Gill SS, Gill R et al (2014) Metal/metalloid stress tolerance in plants: role of ascorbate, its redox couple, and associated enzymes. Protoplasma 251(6):1265–1283

Arribas-Lorenzo G, Morales FJ (2009) Effect of pyridoxamine on acrylamide formation in a glucose/asparagine model system. J Agric Food Chem 57(3):901–909

Aslam A, Misbah SA, Talbot K et al (2004) Vitamin E deficiency induced neurological disease in common variable immunodeficiency: two cases and a review of the literature of vitamin E deficiency. J Clin Immunol 112(1):24–29

Augustin J, Johnson SR, Teitzel C et al (1978) Changes in the nutrient composition of potatoes during home preparation: II. Vitamins. Am Potato J55(12):653–662

Augustin J, Marousek GI, Tholen LA et al (1980) Vitamin retention in cooked, chilled, and reheated potatoes. J Food Sci 45(4):814–816

Aurora-Prado MS, Silva CA, Tavares MF et al (2010) Rapid determination of water-soluble and fat-soluble vitamins in commercial formulations by MEEKC. Chromatographia 72 (7–8):687–694

Bagri DS, Upadhyaya DC, Kumar A et al (2018) Overexpression of PDX-II gene in potato (Solanum tuberosum L.) leads to the enhanced accumulation of vitamin B6 in tuber tissues and tolerance to abiotic stresses. Plant Sci 272:267–275

Becalski A, Lau BP, Lewis D et al (2003) Acrylamide in foods: occurrence, sources, and modeling. J Agric Food Chem 51(3):802–808

Benavente CA, Jacobson MK, Jacobson EL (2009) NAD in skin: therapeutic approaches for niacin. Curr Pharm Des 15(1):29–38

Bender DA (1999) Optimum nutrition: thiamin, biotin and pantothenate. Proc Nutr Soc 58 (2):427–433

Berger F, Ramírez-Hernández MH, Ziegler M (2004) The new life of a centenarian: signalling functions of NAD (P). Trends Biochem Sci 29(3):111–118

Bettendorff L, Wins P (2013) Biochemistry of thiamine and thiamine phosphate compounds. In: The encyclopedia of biological chemistry. Academic Press, Cambridge, pp 202–209

Binder S, Knill T, Schuster J (2007) Branched-chain amino acid metabolism in higher plants. Physiol Plant 129(1):68–78

Blauer JM, Kumar GM, Knowles LO et al (2013) Changes in ascorbate and associated gene expression during development and storage of potato tubers (Solanum tuberosum L.). Postharvest Biol Technol 178:76–91

Camire ME, Kubow S, Donnelly DJ (2009) Potatoes and human health. Crit Rev Food Sci Nutr 49 (10):823–840

Carr A, Maggini S (2017) Vitamin C and immune function. Nutrients 9(11):1211

Casteels M, Sniekers M, Fraccascia P et al (2007) The role of 2-hydroxyacyl-CoA lyase, a thiamin pyrophosphate-dependent enzyme, in the peroxisomal metabolism of 3-methyl-branched fatty acids and 2-hydroxy straight-chain fatty acids. Biochem Soc Trans 35(5):876–880

Chen X, Song B, Liu J et al (2012) Modulation of gene expression in cold-induced sweetening resistant potato species Solanum berthaultii exposed to low temperature. Mol Genet Genom Med 287(5):411–421

Chen L, Ducker GS, Lu W et al (2017) An LC-MS chemical derivatization method for the measurement of five different one-carbon states of cellular tetrahydrofolate. Anal Bioanal Chem 409(25):5955–5964

Cheng Z, Sattler S, Maeda H et al (2003) Highly divergent methyltransferases catalyze a conserved reaction in tocopherol and plastoquinone synthesis in cyanobacteria and photosynthetic eukaryotes. Plant Cell 15(10):2343–2356

Cho CM, Ko JH, Cheong WJ (2000) Simultaneous determination of water-soluble vitamins excreted in human urine after eating an overdose of vitamin pills by a HPLC method coupled with a solid phase extraction. Talanta 51(4):799–806

Chowers I, Banin E, Merin S et al (2001) Long-term assessment of combined vitamin A and E treatment for the prevention of retinal degeneration in abetalipoproteinaemia and hypobetalipoproteinaemia patients. Eye 15(4):525–530

Combs GF Jr, McClung JP (2016) The vitamins: fundamental aspects in nutrition and health. Academic press, Cambridge

Crowell EF, McGrath JM, Douches DS (2008) Accumulation of vitamin E in potato (Solanum tuberosum) tubers. Transgenic Res 17(2):205–217

da Silva RP, Kelly KB, Al Rajabi A (2014) Novel insights on interactions between folate and lipid metabolism. Biofactors 40(3):277–283

Dale MF, Griffiths DW, Todd DT (2003) Effects of genotype, environment, and postharvest storage on the total ascorbate content of potato (*Solanum tuberosum*) tubers. J Agric Food Chem 51 (1):244–248

Datta S, Sinha BK, Bhattacharjee S (2019) Nutritional composition, mineral content, antioxidant activity and quantitative estimation of water soluble vitamins and phenolics by RP-HPLC in some lesser used wild edible plants. Heliyon 5(3):e01431

Delchier N, Ringling C, Maingonnat JF et al (2014) Mechanisms of folate losses during processing: diffusion vs. heat degradation. Food Chem 157:439–447

DellaPenna D, Mène-Saffrané L (2011) Vitamin E. Adv Bot Res 59:179–227

Dena AS, Ammar AA (2019) H-point standard addition for simultaneous reagent-free spectrophotometric determination of B1 and B6 vitamins. Spectrochim Acta A Mol Biomol Spectrosc 206:491–497

Depeint F, Bruce WR, Shangari N et al (2006a) Mitochondrial function and toxicity: role of the B vitamin family on mitochondrial energy metabolism. Chem Biol 163(1–2):94–112

Depeint F, Bruce WR, Shangari N et al (2006b) Mitochondrial function and toxicity: role of B vitamins on the one-carbon transfer pathways. Chem Biol 163(1–2):113–132

Dombradi GA, Szepesvári E, Szórády I (1964) Possible mode of action of pantothenic acid in radioprotection. Acta Physiol Lat Am 14:327–329

Drewnowski A, Rehm CD (2013) Vegetable cost metrics show that potatoes and beans provide most nutrients per penny. PLoS One 8(5):e63277

Egger SF, Huber-Spitzy V, Alzner E et al (1999) Corneal wound healing after superficial foreign body injury: vitamin A and dexpanthenol versus a calf blood extract. Ophthalmologica 213 (4):246–249

Egoville MJ, Sullivan JF, Kozempel MF et al (1988) Ascorbic acid determination in processed potatoes. Am Potato J65(2):91–97

FAO/WHO (2001) Human vitamin and mineral requirements. Report of a joint FAO/WHO expert consultation, Bangkok, Thailand. Food and Nutrition Division. FAO, Rome, pp 235–247

Fenech M (2012) Folate (vitamin B_9) and vitamin B_{12} and their function in the maintenance of nuclear and mitochondrial genome integrity. Mutat Res/Fundam Mol Mech Mutagen 733 (1–2):21–33

Fitzpatrick TB, Basset GJ, Borel P et al (2012) Vitamin deficiencies in humans: can plant science help? Plant Cell 24(2):395–414

Fu L, Doreswamy V, Prakash R (2014) The biochemical pathways of central nervous system neural degeneration in niacin deficiency. Neural Regen Res 9(16):1509–1513

Garcia I, Rodgers M, Pepin R et al (1999) Characterization and subcellular compartmentation of recombinant 4-hydroxyphenylpyruvate dioxygenase from Arabidopsis in transgenic tobacco. Plant Physiol 119(4):1507–1516

Gibson GE, Blass JP (2007) Thiamine-dependent processes and treatment strategies in neurodegeneration. Antioxid Redox Signal 9(10):1605–1620

Gibson S, Kurilich AC (2013) The nutritional value of potatoes and potato products in the UK diet. Nutr Bull 38(4):389–399

Gigliobianco T, Gangolf M, Lakaye B et al (2013) An alternative role of FoF1-ATP synthase in Escherichia coli: synthesis of thiamine triphosphate. Sci Rep 3:1071

Goyer A (2010) Thiamine in plants: aspects of its metabolism and functions. Phytochemistry 71 (14–15):1615–1624

Goyer A, Navarre DA (2007) Determination of folate concentrations in diverse potato germplasm using a trienzyme extraction and a microbiological assay. J Agric Food Chem 55(9):3523–3528

Goyer A, Navarre DA (2009) Folate is higher in developmentally younger potato tubers. J Sci Food Agric 89(4):579–583

Goyer A, Sweek K (2011) Genetic diversity of thiamin and folate in primitive cultivated and wild potato (Solanum) species. J Agric Food Chem 59(24):13072–13080

Greenberg JA, Bell SJ, Guan Y et al (2011) Folic acid supplementation and pregnancy: more than just neural tube defect prevention. Rev Obstet Gynecol l4(2):52–59

Grenier JF, Aprahamian M, Genot C et al (1982) Pantothenic acid (vitamin B_5) efficiency on wound healing. Acta Vitaminol Enzymol 4(1–2):81–85

Gwirtz JA, Garcia-Casal MN (2014) Processing maize flour and corn meal food products. Ann N Y Acad Sci 1312(1):66–75

Haase NU, Weber L (2003) Ascorbic acid losses during processing of French fries and potato chips. J Food Eng 56(2–3):207–209

Han JS, Kozukue N, Young KS et al (2004) Distribution of ascorbic acid in potato tubers and in home-processed and commercial potato foods. J Agric Food Chem 52(21):6516–6521

Hanson AD, Gregory JF III (2011) Folate biosynthesis, turnover, and transport in plants. Annu Rev Plant Biol 62:105–125

Hashemi SH, Yahyavi H, Kaykhaii M et al (2019) Spectrofluorometrical determination of vitamin B_1 in different matrices using Box-Behnken designed pipette tip solid phase extraction by a carbon nanotube sorbent. Chem Select 4(11):3052–3057

Hemavathi, Upadhyaya CP, Young KE et al (2009) Over-expression of strawberry D-galacturonic acid reductase in potato leads to accumulation of vitamin C with enhanced abiotic stress tolerance. Plant Sci 177(6):659–667

Hemavathi, Upadhyaya CP, Akula N et al (2010) Enhanced ascorbic acid accumulation in transgenic potato confers tolerance to various abiotic stresses. Biotechnol Lett 32(2):321–330

Hofius D, Hajirezaei MR, Geiger M et al (2004) RNAi-mediated tocopherol deficiency impairs photoassimilate export in transgenic potato plants. Plant Physiol 135(3):1256–1268

Horvath G, Wessjohann L, Bigirimana J et al (2006) Accumulation of tocopherols and tocotrienols during seed development of grape (*Vitis vinifera* L. cv. Albert Lavallée). Plant Physiol Biochem 44(11–12):724–731

Jagota SK, Dani HM (1982) A new colorimetric technique for the estimation of vitamin C using Folin phenol reagent. Anal Biochem 127(1):178–182

Juhász Z, Dancs G, Marincs F et al (2014) Vitamin C, B_5, and B_6 contents of segregating potato populations detected by GC-MS: a method facilitating breeding potatoes with improved vitamin content. Plant Breed 133(4):515–520

Kandutsch AA, Baumann CA (1953) Factors affecting the stability of thiamine in a typical laboratory diet: one figure. J Nutr 49(2):209–219

Kanellis A, Manganaris G (2014) Antioxidants and bioactive compounds in fruits. In: Nath P, Bouzayen M, Mattoo AK, Pech JC (eds) Fruit ripening: physiology, signalling and genomics, vol 16. CABI, Wallingford, pp 99–126

Keijbets MJ, Ebbenhorst-Seller G (1990) Loss of vitamin C (L-ascorbic acid) during long-term cold storage of Dutch table potatoes. Potato Res 33(1):125–130

Kondo Y, Higashi C, Iwama M et al (2012) Bioavailability of vitamin C from mashed potatoes and potato chips after oral administration in healthy Japanese men. Br J Nutr 107(6):885–892

Kondo Y, Sakuma R, Ichisawa M et al (2014) Potato chip intake increases ascorbic acid levels and decreases reactive oxygen species in SMP30/GNL knockout mouse tissues. J Agric Food Chem 62(38):9286–9295

Kossmann J, Lloyd J (2000) Understanding and influencing starch biochemistry. Crit Rev Plant Sci 19(3):171–226

Koury MJ, Price JO, Hicks GG (2000) Apoptosis in megaloblastic anemia occurs during DNA synthesis by a p53-independent, nucleoside-reversible mechanism. Blood 96(9):3249–3255

Kraft CE, Angert ER (2017) Competition for vitamin B_1 (thiamin) structures numerous ecological interactions. Q Rev Biol 92(2):151–168

Külen O, Stushnoff C, Holm DG (2013) Effect of cold storage on total phenolics content, antioxidant activity and vitamin C level of selected potato clones. J Sci Food Agric 93(10):2437–2444

Kumerova AO, Utno LJ, Lipsberga ZE et al (1992) Myocardial protection by pantothenic acid derivatives in heart model with experimental ischemia and reperfusion. Bull Exp Biol Med 113 (4):489–492

Li T, Huo L, Pulley C et al (2012) Decarboxylation mechanisms in biological system. Bioorg Chem 43:2–14

Liu Z, Wang Q, Mao L et al (2000) Highly sensitive spectrofluorimetric determination of ascorbic acid based on its enhancement effect on a mimetic enzyme-catalyzed reaction. Anal Chim Acta 413(1–2):167–173

López-de-Alba PL, López-Martínez L, Cerdá V et al (2006) Simultaneous determination and classification of riboflavin, thiamine, nicotinamide and pyridoxine in pharmaceutical formulations, by UV-visible spectrophotometry and multivariate analysis. J Brazil Chem Soc 17(4):715–722

Love SL, Pavek JJ (2008) Positioning the potato as a primary food source of vitamin C. Am J Potato Res 85(4):277–285

Love SL, Salaiz T, Shafii B et al (2004) Stability of expression and concentration of ascorbic acid in North American potato germplasm. HortScience 39(1):156–160

Lushchak VI, Semchuk NM (2012) Tocopherol biosynthesis: chemistry, regulation and effects of environmental factors. Acta Physiol Plant 34(5):1607–1628

Manore MM (2000) Effect of physical activity on thiamine, riboflavin, and vitamin B-6 requirements. Am J Clin Nutr 72(2):598–606

Manzetti S, Zhang J, van der Spoel D (2014) Thiamin function, metabolism, uptake, and transport. Biochemistry 53(5):821–835

Maras JE, Bermudez OI, Qiao N et al (2004) Intake of α-tocopherol is limited among US adults. J Am Diet Assoc 104(4):567–575

Martin PR (2001) Molecular mechanisms of thiamine utilization. Curr Mol Med 2:197–207

Moccand C, Boycheva S, Surriabre P et al (2014) The pseudoenzyme PDX1. 2 boosts vitamin B6 biosynthesis under heat and oxidative stress in Arabidopsis. J Biol Chem 289(12):8203–8216

Mondy NI, Leja M (1986) Effect of mechanical injury on the ascorbic acid content of potatoes. J Food Sci 51(2):355–357

Mooney S, Hellmann H (2010) Vitamin B_6: killing two birds with one stone? Phytochemistry 71 (5–6):495–501

Mooney S, Chen L, Kühn C et al (2013) Genotype-specific changes in vitamin B_6 content and the PDX family in potato. J Food Sci 2013:2013. https://doi.org/10.1155/2013/389723

Muller DP (2010) Vitamin E and neurological function. Mol Nutr Food Res 54(5):710–718

Munné-Bosch S, Shikanai T, Asada K (2005) Enhanced ferredoxin-dependent cyclic electron flow around photosystem I and α-tocopherol quinone accumulation in water-stressed ndhB-inactivated tobacco mutants. Planta 222(3):502–511

Neufeld M (2017) Scurvy: the disease of discovery by Jonathan Lamb. Bull Hist Med 91 (4):811–812

Nisha P, Singhal RS, Pandit AB (2009) A study on degradation kinetics of niacin in potato (*Solanum tuberosum* L.). J Food Compos Anal 22(6):620–624

Okeyo JA, Kushad MM (1995) Composition of four potato cultivars in relation to cold storage and reconditioning. HortTechnology 5(3):250–253

Page E, Hanning FM (1963) Vitamin B6 and niacin in potatoes. Retention after storage and cooking. J Am Diet Assoc 42:42–45

Pullar J, Carr A, Vissers M (2017) The roles of vitamin C in skin health. Nutrients 9(8):866

Qin A, Shi Q, Yu X (2011) Ascorbic acid contents in transgenic potato plants overexpressing two dehydroascorbate reductase genes. Mol Biol Rep 38(3):1557–1566

Rajinikanth M, Harding SA, Tsai CJ (2007) The glycine decarboxylase complex multienzyme family in Populus. J Exp Bot 58(7):1761–1770

Rees E, Gowing LR (2012) Supplementary thiamine is still important in alcohol dependence. Alcohol Alcohol 48(1):88–92

Riaz MN, Asif M, Ali R (2009) Stability of vitamins during extrusion. Crit Rev Food Sci Nutr 49 (4):361–368

Robinson B, Sathuvalli V, Bamberg J (2015) Exploring folate diversity in wild and primitive potatoes for modern crop improvement. Genes 6(4):1300–1314

Rolfe HM (2014) A review of nicotinamide: treatment of skin diseases and potential side effects. J Cosmet Dermatol 13(4):324–328

Santos J, Mendiola JA, Oliveira MB (2012) Sequential determination of fat-and water-soluble vitamins in green leafy vegetables during storage. J Chromatogr A 1261:179–188

Sarkiyayi S, Ikioda H (2010) Estimation of thiamine and ascorbic acid contents in fresh and dried Hibiscus sabdariffa (Roselle) and Lactuca sativa (Tettuce). Adv J Food Sci Technol 2(1):47–49

Seal T, Chaudhuri K (2017) High performance liquid chromatography method for the estimation of water soluble vitamin in five wild edible fruits consumed by the tribal people of north-eastern region in India. Int J Curr Microbiol App Sci 6(10):2900–2913

Shaath T, Fischer R, Goeser M (2016) Scurvy in the present times: vitamin C allergy leading to strict fast food diet. Dermatol Online J 22(1):10

Singh NK, Sharma A, Sharma M (2013) Estimation of naturally produced water-soluble vitamins in different asavas and arishtas using liquid chromatography. J Food Drug Anal 1(3):81–84

Smirnoff N, Wheeler GL (2000) Ascorbic acid in plants: biosynthesis and function. Crit Rev Plant Sci 19:267–290

Smirnoff N (2001) L-ascorbic acid biosynthesis. Vitam Horm 61:241–266

Smirnoff N (2011) Vitamin C: the metabolism and functions of ascorbic acid in plants. In: Rébeillé F, Douce R (eds) Advances in botanical research, vol 59. Academic Press, Cambridge, pp p107–p177

Spinneker A, Sola R, Lemmen V et al (2007) Vitamin B6 status, deficiency and its consequences-an overview. Nutr Hosp 22(1):7–24

Stolz J, Vielreicher M (2003) Tpn1p, the plasma membrane vitamin B_6 transporter of Saccharomyces cerevisiae. J Biol Chem 278(21):18990–18996

Tahiliani AG, Beinlich CJ (1991) Pantothenic-acid in health and disease. Vitam Horm 46:165–228

Tareke E, Rydberg P, Karlsson P (2002) Analysis of acrylamide, a carcinogen formed in heated foodstuffs. J Agric Food Chem 50(17):4998–5006

Tedone L, Hancock RD, Alberino S (2004) Long-distance transport of L-ascorbic acid in potato. BMC Plant Biol 4(1):16

Tibbetts AS, Appling DR (2010) Compartmentalization of mammalian folate-mediated one-carbon metabolism. Annu Rev Nutr 30:57–81

Tittmann K (2009) Reaction mechanisms of thiamin diphosphate enzymes: redox reactions. FEBS J276(9):2454–2468

Tosun BN, Yücecan S (2008) Influence of commercial freezing and storage on vitamin C content of some vegetables. Int J Food Sci 43(2):316–321

Triantaphylidès C, Havaux M (2009) Singlet oxygen in plants: production, detoxification and signaling. Trends Plant Sci 14(4):219–228

Tudela JA, Espın JC, Gil MI (2002) Vitamin C retention in fresh-cut potatoes. Postharvest Biol Technol 26(1):75–84

Upadhyaya CP, Venkatesh J, Gururani MA et al (2011) Transgenic potato overproducing L-ascorbic acid resisted an increase in methylglyoxal under salinity stress via maintaining higher reduced glutathione level and glyoxalase enzyme activity. Biotechnol Lett 33(11):2297

Van den Enden MK, Nyengaard JR, Ostrow E et al (1995) Elevated glucose levels increase retinal glycolysis and sorbitol pathway metabolism. Implications for diabetic retinopathy. Investig Ophthalmol Vis Sci 36(8):1675–1685

Velisek J, Cejpek K (2007) Biosynthesis of food constituents: vitamins. 1. Fat-soluble vitamins-a review. Czech J Food Sci 25:1–16

Venkatesh J, Park SW (2014) Role of L-ascorbate in alleviating abiotic stresses in crop plants. Bot Stud 55(1):38

Wahlberg G, Adamson U, Svensson J (2000) Pyridine nucleotides in glucose metabolism and diabetes: a review. Diabetes Metab Res Rev 16(1):33–42

Waller JC, Alvarez S, Naponelli V et al (2010) A role for tetrahydrofolates in the metabolism of iron-sulfur clusters in all domains of life. Proc Natl Acad Sci 107(23):10412–10417

Wheeler GL, Jones MA, Smirnoff N (1998) The biosynthetic pathway of vitamin C in higher plants. Nature 393:365–369

Wigand ME, Göde U, Linger F et al (1991) Normal wound healing of the paranasal sinuses: clinical and experimental investigations. Eur Arch Oto-Rhino-Laryngol 248(7):390–394

Wills RB, Lim JS, Greenfield H (1984) Variation in nutrient composition of Australian retail potatoes over a 12 month period. J Sci Food Agric 35(9):1012–1017

Wojtczak L, Slyshenkov VS (2003) Protection by pantothenic acid against apoptosis and cell damage by oxygen free radicals-the role of glutathione. Biofactors 17(1–4):61–73

Young JI, Züchner S, Wang G (2015) Regulation of the epigenome by vitamin C. Annu Rev Nutr 35:545–564

Zeng X, Cheng KW, Jiang Y et al (2009) Inhibition of acrylamide formation by vitamins in model reactions and fried potato strips. Food Chem 116(1):34–39

Zhang Z, Kebreab E, Jing M et al (2009) Impairments in pyridoxine-dependent sulphur amino acid metabolism are highly sensitive to the degree of vitamin B_6 deficiency and repletion in the pig. Animal 3(6):826–837

Zhang Y, Jin X, Ouyang Z et al (2015) Vitamin B_6 contributes to disease resistance against *Pseudomonas syringae* pv. tomato DC3000 and Botrytis cinerea in Arabidopsis thaliana. J. Plant Physiol 175:21–25

Phenolics

8

Bandana, Vineet Sharma, Nitasha Thakur, Pinky Raigond, and Brajesh Singh

Abstract

Potato is a wholesome and versatile food that is not only rich in carbohydrtaes but also have variety of nutritional compounds. Due to its potential to produce more dry matter, protein and minerals per unit area, potato can be a better staple crop compared to cereals. Because of the beneficial effects on health, compounds present in colored (yellow, red, and purple flesh) potatoes such as phenols and carotenoids are very important and therefore they are highly desirable in the human diet. Phytonutrients are antioxidants that reduce or inhibit oxidative processes in human body and food products. Free radicals or reactive oxygen species are responsible for the degenerative reactions and are associated with many chronic diseases during metabolism. Antioxidants such as polyphenols, anthocyanins and ascorbic acid, etc., help the body to neutralize free radicals. Due to more per capita consumption of potatoes compared to other vegetables, breeders are focusing on development of new potato varieties rich in antioxidants, viz. phenols, carotenoids, and ascorbic acid so as to increase the availability of antioxidants to large populations worldwide. In this chapter an attempt was made to retrace the information on potato phenols that differ with the cultivar, growing location, cultural practices adopted during cultivation, maturity at harvest, subsequent storage, processing parameters, different cooking methods, and other related factors.

Keywords

Antioxidant · Nutritional security · Phenols · Processing · Phytonutrient · Quality

Bandana (✉) · V. Sharma
ICAR-Central Potato Research Institute Campus, Meerut, Uttar Pradesh, India
e-mail: bandana.biochem@gmail.com

N. Thakur · P. Raigond · B. Singh
Crop Physiology, Biochemistry & Post-Harvest Technology Division, ICAR-Central Potato Research Institute, Shimla, Himachal Pradesh, India

P. Raigond et al. (eds.), *Potato*, https://doi.org/10.1007/978-981-15-7662-1_8

8.1 Introduction

Potatoes contain carbohydrates, dietary fiber, high-quality proteins, vitamins, minerals, and polyphenols. Carotenoids are also present in low concentration in white-/cream-fleshed potatoes and in high concentration in yellow potatoes. Potatoes are popular component of diet not only because of their low cost but also because of their nutritional quality. The major antioxidants present in potato are ascorbic acid and polyphenols. Antioxidative potential of potato was understood lately and now it is becoming popular as low-cost antioxidant-rich commodity. Antioxidants present in food play an important role in amelioration of lipid peroxidation in both animal and plant tissues. Antioxidants not only retain the quality of food but also reduce the risk of disease development (Matsuzoe et al. 1999). Potatoes contribute significantly to the antioxidant activity of the diet, since their consumption is quite high than many other vegetables. The most important polyphenolic antioxidants in potatoes are chlorogenic acid, cryptochlorogenic acid, caffeic acid, *L*-tyrosine, scopolin, and ferulic acid (Andre et al. 2007). In case of potato, chlorogenic acids (CGA) consist up to 90% of the total polyphenol content. Antioxidative property of potato is mainly due to phenolics that contribute almost 58–82% of the total antioxidant property. Potatoes contain high concentration of antioxidants, among which phenolics contribute to 58–82% of the total antioxidant activity (Reddivari et al. 2007). Compared to other sources of total phenolic content such as fruits and vegetables like carrots, onions, or tomatoes, potatoes contribute significantly due to their high consumption rates. Potential nutritional and therapeutic properties of potato polyphenols are gaining attention of nutritionists. Colored-flesh potatoes are a good source of dietary polyphenols such as carotenoids, phenolic compounds, flavonoids, and anthocyanins that help to reduce the risk of several diseases such as cancer, age related neuronal degeneration, and cardiovascular diseases (Brown 2005). Purple-flesh potatoes contain 69–350 mg anthocyanins/kg of fresh weight (Fossen et al. 2003). The concentration of these compounds depends on the intensity of tuber color (Brown et al. 2003). Antioxidant activity showed a positive correlation with the content of total polyphenols and anthocyanins. Potato germplasm shows huge variation in terms of the phenolic content. Red and purple potatoes contain acylated anthocyanins. Antioxidant concentration in pigmented potatoes is 2–3 times higher compared to white-fleshed ones. Red potato tubers display glycosides of pelargonidin and peonidin, whereas purple potatoes are sources of glycosides of malvidin and petunidin. Colored potatoes (red- and purple-fleshed) can replace the synthetic colorants and can be used as fortificants in the food products. Potato peel as well as flesh contains significant concentration of phenolic compounds, though the concentration is much higher in peels (Ezekiel et al. 2013). One of the most abundant flavonoids in potato is catechin with 0–204 mg/100 g dry weight range. Potato also contains flavonols like quercetin and kaempferol rutinose. Potato skin is relatively high in activity and thus has potential for development of "healthier" snack products (Lister and Wilson 2001). Potatoes are always cooked/processed before consumption. Concentration and quality of potato phytonutrients is influenced by cooking methods. Phytonutrients undergo various changes such as destruction, release, and structural transformation during cooking which results into loss of phytonutrients, particularly water soluble and heat labile compounds. Loss varies with the cooking treatment which can either increase or decrease the phytonutrient concentration in potatoes (Lin and Chang 2005).

8.2 Total Phenolics in Potato

After apples and oranges, potatoes are the third most important source of phenols (Chun et al. 2005) and are a high-quality source of polyphenol compounds (Manach et al. 2004). Polyphenols comprises over 8000 recognized substances, which can be categorized into groups, such as phenolic acids, flavonoids, stilbenes, lignins, and coumarins (Ross and Kasum 2002). These compounds have a wide range of health promoting properties (Bravo 1998). Besides this, phenolic compounds influence the organoleptic properties of potatoes. Caffeic acid derivatives like chlorogenic acid are the main phenolic constituents in potatoes (Mattila and Hellstrom 2007). Yellow-fleshed potatoes are known to contain high concentration of carotenoids and major carotenoids present in these potatoes are lutein, violaxanthin, zeaxanthin, and antheraxanthin. Red-, blue-, or purple-fleshed potatoes mainly contain anthocyanins in high concentration. Red- and purple-fleshed potatoes are reported to contain high concentrations of phenolic compounds, viz. chlorogenic, caffeic, ferulic, p-coumaric, and protocatechuic acids. Traces of flavonoids such as rutin, myricetin, quercetin, kaempferol, and naringenin are also present (Fig. 8.1). Anthocyanins present in potatoes provide protection (biotic and abiotic stresses) against potato blight (Poiatti et al. 2009). Phenolic compounds, namely coumarins, pterocarpans,

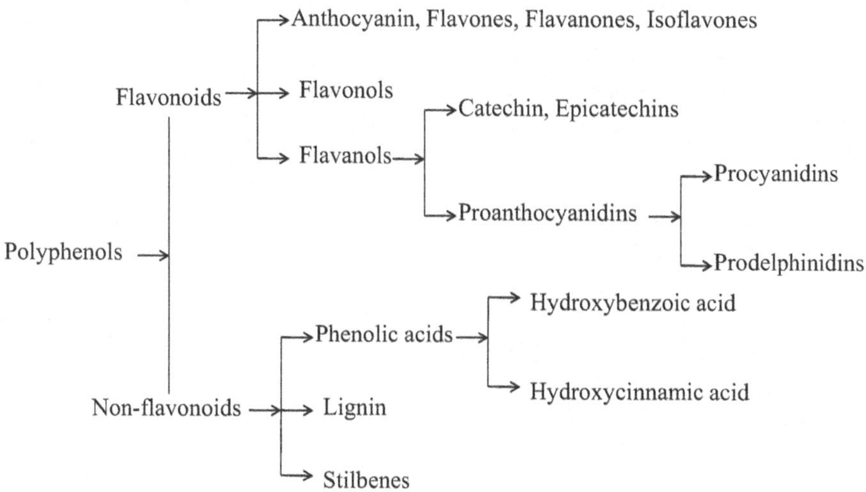

Fig. 8.1 Polyphenols classification (adopted from Thakur et al. 2020)

flavonols, and isoflavones are an important group of secondary metabolites that have antimicrobial activity and hence are involved in resistance to pathogens.

Caffeoyl-esters are present in high concentration in tubers. Chlorogenic acid confers several health benefits and is reported to have anti-viral and anti-bacterial properties. Chlorogenic acid is reported to reduce the risk of type II diabetes and is reported as a glycemic index lowering agent (Legrand and Scheen 2007; Bassoli et al. 2008).

Ah-Hen et al. (2012) reported that phenolic content in peeled potatoes ranged from 191 to 1864 mg/100 g when compared to unpeeled samples which contained higher amount of phenolic content varying from 345 to 2852 mg/100 g. Among 20 native Andean potato cultivars with different flesh colors, total phenolics ranged from 162.19 to 510.20 mg GAE/100 g DW in purple-fleshed potatoes, while 152.40 to 261.49 mg GAE/100 g DW in red-fleshed and 113.37–114.63 in yellow-fleshed cultivars (Giusti et al. 2014). Total phenolics in potato somaclones ranged from 61 to 122 GAE/150 g fresh weight (FW) (Nassar et al. 2012). Phenolic compounds in colored potato cultivar Purple Majesty were 209 mg GAE/100 g FW (Lemos et al. 2015). In Indian potato varieties, Dalamu et al. (2015) reported phenolics in whole tuber in range from 22.53 to 85.85 mg GAE/ 100 g FW and in flesh from 20.26 to 63.05 mg GAE/100 g FW.

Polyphenols are more concentrated in potato peel than the cortex and pith. In potatoes, distribution of phenolics is mostly between the cortex and peel (Friedman 1997). Potato peel and adjoining tissues contain almost 50% of phenolic concentration and it decreases in concentration toward the pith or center of potato tubers. Phenols were present in higher concentration significantly in peels of indigenous varieties Kufri Surya (194.0 mg/100 g FW) and Kufri Pukhraj (163.5 mg/100 g fresh weight) (Bandana et al. 2016). Total anthocyanin and total phenol concentrations in purple- and red-fleshed potatoes peel were reported to be 0.9–1.6-fold higher than in flesh (Reyes et al. 2005). Nara et al. (2006) have observed higher antioxidative potential in potato peel as compared to flesh through DPPH radical scavenging activity. Chlorogenic acid is the major potato peel phenolic acid, i.e. almost 50.31%. Other phenolic compounds present in potato peel are gallic acid (41.67%), protocatechuic (7.81%), and caffeic (0.21%). Extract of potato peel exhibits very strong antioxidant activity which is at par with synthetic antioxidants. Potato antioxidants have free radical scavenging effects which help to reduce the cholesterol accumulation in the blood serum and enhance the resistance of vascular walls, hence decrease the risk of coronary heart diseases.

8.3 Major Phenolics: Chlorogenic Acid

Chlorogenic acids are formed from *trans*-cinnamic acids and (−)-quinic acid. Three isomeric forms of chlorogenic acid, i.e. 3-, 4-, and 5-caffeoylquinic acids are present in potatoes. Chlorogenic acid has multiple health promoting properties including antioxidant, anti-inflammatory, anticarcinogenic, antimicrobial, analgesic, neuroprotective, and cardioprotective effects. Researchers reported different

Table 8.1 Chlorogenic acid range in different potato cultivars

Potato varieties	Chlorogenic acid range (μg/g DW)	References
S. tuberosum cv. Sikli and Timo	455–600	Mattila and Hellstrom (2007)
S. tuberosum	1500	Lukaszewicz et al. (2004)
S. tuberosum	421–2185	Zhu et al. (2010)
Potato(unspecified)	420–3183	Xu et al. (2009)
Potato (unspecified)	6.5	Chiou et al. (2007)
Potato (unspecified)	33–6370	Im et al. (2008)
S. tuberosum cv. Norkotah and Ranger	1000–2200	Shakya and Navarre (2006)
S. tuberosum	343–1249	Hajslova et al. (2005)
S. tuberosum	39–65	Blessington et al. (2010)

proportion of chlorogenic acid in total phenols of potato. Riciputi et al. (2018) reported 49.3–61.0%, whereas Friedman (1997) reported up to 90% of the total potato phenolic content to be chlorogenic acid. The concentration and stability of phenolics is influenced by several factors such as agro-technical processes, climatic conditions, and post-harvest manipulations (Zarzecka et al. 2019). Chlorogenic acid in different potato cultivars is summarized in Table 8.1. In addition, genotype, storage conditions after harvest, processing/cooking methods affect the concentration of phenolics significantly (Stushnoff et al. 2008; Blessington et al. 2010; Lachman et al. 2012). Though mainly chlorogenic acid is present in raw, microwaved, or fried potatoes, p-coumaric, ellagic, caffeic and ferulic acids were also identified along with traces of sinapic acid, protocatechuic acid, and vanillic acids (Ezekiel et al. 2013; Mader et al. 2009; Rytel et al. 2014). Colored potatoes retained three to four times more chlorogenic acid (800–4730 mg/kg DW) compared to white- and yellow-fleshed potatoes with concentration ranging from 200 to 760 mg/kg DW (Navarre et al. 2010). The effect of cooking on chlorogenic acid content in potato varieties of some references has been summarized in Table 8.2. Dao and Friedman (1992) evaluated chlorogenic acid in seven potato varieties and results suggest that the UV method may be advantageous over HPLC and chlorogenic acid ranged from 10 to 19 mg/100 g of fresh weight. They further reported that oven baked potatoes contained 0%, boiled potatoes 35%, and microwaved potatoes 55% of the original concentration of chlorogenic acid. Evers and Deußer (2012) reported that chlorogenic, ferulic acid and rutin levels were not significantly affected by boiling, whereas increase in concentration of neochlorogenic acid, cryptochlorogenic acid, vanillin, and catechin was observed. Blessington et al. (2010) observed increase in the average concentration of 5-O-CQA, caffeic acid, vanillic acid, p-coumaric acids, and epicatechin; and decrease in quercetin after cooking. Barba et al. (2008) reported that the losses depend on the cooking method and conventional boiling caused the maximum losses that varied between 85.6% and 94.8%, whereas losses in microwave baking was low, i.e. 49.6

Table 8.2 Effect of different cooking methods on chlorogenic acid in potato

Sr. no.	Potato/potato varieties	Processing method	Chlorogenic acid range	References
1	Potato	Boiled potatoes	3.5 (mg/100 g FW)	Dao and Friedman (1992)
		Microwaved potatoes	5.5	
2	Bintje	Raw	0.61 (mg/g DW)	Navarre et al. (2010)
		Microwaved	2.19	
		Steamed	3.23	
		Boiled	2.86	
		Baked	3.67	
	Piccolo	Raw	1.70	
		Microwaved	1.74	
		Steamed	2.49	
		Boiled	2.30	
		Baked	2.78	
	Purple Majesty	Raw	4.16	
		Microwaved	4.35	
		Steamed	4.81	
		Boiled	5.29	
		Baked	6.20	
3	Average of red potato cultivars (HB Red, Rote Emma)	Fresh non-peeled	1711	Lachman et al. (2013)
		Fresh peeled	1048 (mg/kg DM)	
		Boiled peeled	1095	
		Microwaved non-peeled cut	702	
		Baked non-peeled cut	1032	
	Average of red potato cultivars (HB Red, Rote Emma)	Fresh non-peeled	2149	
		Fresh peeled	1496	
		Boiled peeled	1148	
		Microwaved non-peeled cut	537	
		Baked non-peeled cut	603	
	Yellow potato cultivar	Fresh non-peeled	314	
		Fresh peeled	102	
		Boiled peeled	67	
		Microwaved non-peeled cut	49	
		Baked non-peeled cut	74	

(continued)

Table 8.2 (continued)

Sr. no.	Potato/potato varieties	Processing method	Chlorogenic acid range	References
4	Purple-flesh potatoes	Raw	463 (mg/100 g DW)	Tiam et al. (2016)
		Boiled	370	
		Baked	285	
		Steamed	495	
		Microwaved	365	
		Fried	306	
		Air-fried	274	
		Stir-fried	229	
5	Light yellow potato	Raw	840 (mg/kg DW)	Silveira et al. (2017)
		Microwaved	483	
		Fried	151	
	Red potato	Raw	2214	
		Microwaved	1275	
		Fried	445	
	Red-white potato	Raw	2152	
		Microwaved	901	
		Fried	526	
	Purple potato	Raw	2541	
		Microwaved	1237	
		Fried	343	
	Purple-white potato	Raw	2277	
		Microwaved	765	
		Fried	362	

and 77.3%. After frying and microwaving, chlorogenic acid content in purple-fleshed potatoes decreased by 33.8% and 21%, respectively (Tiam et al. 2016).

8.4 Most Preferred Methods for Estimation of Total and Individual Phenolics

Polyphenols are reported to have antioxidative, anti-aging, anti-bacterial, and antimutagenic properties. Though role of phenols in human diet and health has been well explored and established, there are several limitations in their extraction efficiency. To overcome difficulties and limitations related to screening, extraction, separation, and purification of these compounds, new modern methods were developed. It is well understood that the extraction step is one of the most crucial steps for estimation of any compound. Several extraction methods are described in literature such as ultrasound-assisted extraction method (UAE), microwave-assisted extraction (MAE), pressurized fluid extraction (PFE), supercritical CO_2 extraction ($SC\text{-}CO_2$). Besides this, enzyme-assisted extraction (EAE) is considered superior for polyphenols extraction.

Since maceration was widely used in the past, however, presently other more feasible and reliable methods are being used. In one of the method, sample is mixed in appropriate solvent and incubated at room temperature with continuous agitation. After incubation, the supernatant is collected through filtration. Though this is a simple method, but is more time and solvent consuming.

In routine laboratory use, colorimetric reactions are measured in the UV/VIS spectrophotometric method, which are easy to perform, rapid, and low cost (Pelozo et al. 2008). Polyphenols present in plant extracts react with specific redox reagents like Folin–Ciocalteu reagent to form a blue colored complex (constituted by a phosphotungstic-phosphomolybdenum complex) that are measured by visible-light spectrophotometer (Schofield et al. 2001).

For extraction of phenols from potato peels ultrasonic assisted extraction method can be used with different solvents (ethanol, methanol, hexane, acetone, and water). After filtration and centrifugation the content is stored in freezer. In another method of solvent extraction, peels are ground and extracted in methanol using magnetic stirrer and incubated at room temperature overnight. After filtration content is evaporated in rotary evaporator below 40 °C.

In percolation method, sample is placed in cylindrical close container and solvent is passed drop-wise through the sample. Liquid/extract is collected at the bottom filtration device. This method has same issues as with maceration, which included more time and solvent consumption, solubility of polyphenols, particle size of sample, and contact time between solvent and sample. Polyphenols are commonly extracted with water or organic solvents such as methanol, ethanol, acetone, n-hexane, chloroform, propanol, and ethyl acetate. The miscibility of organic solvents significantly affects the polyphenol extraction yield. The extraction of polyphenols and other phytochemicals must be standardized with different solvent systems, because the diverse structures of polyphenols interfere in the extraction process. Because of variations in polyphenol structures, selection of solvent to develop a standard method for extraction of all types of phenols is difficult. Several factors such as (1) selectivity (2) polarity (3) boiling temperature (4) reactivity (5) low viscosity (6) stability to heat, oxygen, and light (7) safety (8) if possible, suitability for reuse, and (9) compatible with legislation for food applications, need to be considered while selection of a specific solvent.

Phenol extraction is carried out using soxhlet procedure (Sticher 2008; Kaufmann and Christen 2002) by placing the powdered samples in cellulose bags, which are then placed in extraction chamber. Collecting flask is located beneath the reflux condenser and after addition of the solvent the solvent condenses once it reaches certain level of temperature. Due to reflux, the liquid extract is collected to the flask. The Soxhlet extraction requires less time and solvent than maceration or percolation methods. Soxhlet extraction is reported to be widely used because of its convenience.

8.4.1 Advanced Methods of Extraction

Due to simple procedures and low economic cost, classical extraction methods are being practiced in many laboratories. The major disadvantages of maceration, percolation, and Soxhlet extraction are low efficiency, consumption of high volumes of organic solvents, and evaporation to concentrate the extracts which leads to more time consumption. A number of methods such as microwave-assisted extraction (MAE), ultrasound-assisted extraction (UAE), supercritical CO_2 extraction (SC-CO_2), pressurized fluid extraction (PFE), enzyme-assisted extraction (EAE), or even combined approaches have been developed during the last few years to overcome the problems faced in the classical extraction methods. These novel extraction methods have several advantages over classical methods such as less requirement of time, solvents, low toxicity due to low solvent volumes, high extraction yields, and improved reproducibility (Samarin et al. 2012).

8.5 Effect of Growing Conditions, Storage as Well as Cooking on Phenolics

The content of total antioxidants and phenolic compounds such as chlorogenic acid is significantly influenced by both intrinsic factors, such as variety, flesh color (Andre et al. 2007) and extrinsic factors, such as cultivation conditions, location, year of cultivation and fertilization (Hamouz et al. 2010). Nutrient content is dependent on a number of factors. Potato quality parameters are also affected by cultivation methods (conventional and organic cultivation) and fertilization (Hajslova et al. 2005). High chlorogenic acid was reported in organic potatoes. Potato tubers grown at higher potassium dose were reported to have lower polyphenol content. Because of this, potatoes grown in more potassium fertilization may have less enzymatic browning. Organic potatoes contained higher concentrations of chlorogenic acid (208 $\pm$ 114 mg/kg) as compared to tubers harvested from conventional farms (144–80 mg/kg). The organic potatoes contained higher concentration of ascorbic acid (AA), chlorogenic acid, higher dry matter and starch content than the conventional ones. Comparison of organically grown potato tubers with conventionally grown potato tubers shows no significant difference in their soluble and hydrolyzable polyphenolic content. Potato varieties grown at same altitude in three locations of North Indian plains with different temperatures showed no significant differences in the total phenolic content. Total anthocyanin and phenolic content of potatoes increased 2.5 and 1.4 times, respectively, when grown at cooler temperatures, long days with higher solar radiation. Temperature during crop growth affects the phytochemical content.

Potatoes are generally stored in cold storage to provide supply to the consumers throughout the year. Storage at low temperatures alters potatoes' phytochemical composition to some extent. After storage of potatoes for 0, 30, 60, and 90 days at room temperature, 15 and 4 °C, respectively, potatoes were analyzed for total individual phenols revealing chlorogenic acid as the most abundant phenolic acid and cinnamic acid in low concentrations. Para-coumaric acid decreased at 4 °C, whereas all the other phenolic acids increased with storage. Before storage, chlorogenic acid was most abundant followed by gallic acid, sinapic acid, ellagic acid, caffeic acid, vanillic acid, para-coumaric acid, ferulic acid, syringic acid, protocatechuic acid, salicylic acid, and cinnamic acid. Whereas after storage, though chlorogenic acid remained the most abundant but rest of the trend changed to sinapic > gallic > ellagic > caffeic > vanillic > ferulic > para-coumaric > syringic (Galani et al. 2017).

Phytonutrient content of potatoes is influenced by environmental conditions and developmental stage. Immature potatoes contain high concentrations of phytonutrients such as folate and chlorogenic acid than mature potatoes. Immature potatoes contain more total carotenoids content in tubers that decreases with tuber maturity. "Baby potatoes" or immature potatoes contain considerable concentrations of phytonutrients compared to mature potatoes. Anthocyanins and total phenolic concentrations decrease with tuber growth and maturity; however, total yield per ha of these compounds increases with time. Potatoes harvested at maturity have high anthocyanin and total phenolic content and low glycoalkaloids, hence have high nutritional quality. Anthocyanin content of tubers from plants grown with two different doses of nitrogen fertilizers, i.e. 100 and 200 kg/ha showed no significant difference. Phenolic acids profiling of potatoes stored at different temperatures showed that gallic acid ranged from 4.42 to 17.53 μg/g FW initially and its content increased with storage at all the 3 temperatures (room temperature, 15 °C, and 4 °C) and content was up to 45.7059 μg/g FW, but the highest increase was observed at room temperature. After 30 or 60 DOS, a decrease in gallic acid content was observed in some varieties at 15 °C and 4 °C. Varieties Kufri Badshah and Kufri Himsona contained higher gallic acid contents, whereas processing variety Kufri Chipsona-3 contained low content. The initial content of Pro (protocatechuic acid) was recorded between 0.0261 and 0.5677 μg/g FW, whereas at 4 °C, the highest concentration (1.05 μg/g FW) was reported in DSP 287. Pro content was high in varieties DSP 287 and Kufri Badshah, whereas content was low in Kufri Sutlej and Kufri Sadabahar. Before storage, chlorogenic acid ranged from 5.98 to 28.88 μg/g FW. Chlorogenic acid increased with storage at all the 3 temperatures, with maximum increase reported at room temperature, with a peak value 62.51 μg/g FW reported in Kufri Badshah. After 30 and 60 days at (15 and 4 °C) in a few varieties, decline was observed. Caffeic acid content initially ranged between 0.32 and 1.54 μg/g FW. At 4 °C, Caffeic acid increased significantly from 0 to 30 DOS, whereas decrease was reported at 90 DOS. Variety DSP 287 exhibited the highest caffeic acid content (6.38 μg/g FW) at 60 DOS at 4 °C (Galani et al. 2017).

Most of vegetables are processed before consumption which may have favorable or unfavorable impact on the food flavor and texture, palatability, and affect the quality of bioactive compounds. Changes occurring during processing may have beneficial or harmful effects on the human health. Though thermal processing is known to inactivate microorganisms, decrease the anti-nutritional factors/ compounds and enhance food digestibility, modify bioavailability of phenolics, but it has negative impact on most of bioactive compounds. Most processing methods such as fresh-cut, blanching, drying, pasteurization, use of electric fields and membranes affect the phytonutrient bioavailability. Phenolics are present in both bound and free forms in plants. During cooking the bound phenolics are released due to rupture of cell wall, which in turn may increase the phenol bioavailability. After cooking the retention of phytonutrients varies considerably between vegetables due to difference in matrix. Boiling may decrease the total phenolic content, while steaming and stir frying may promote the release of these compounds. Phenolics being highly soluble in water may leach during boiling, whereas their extraction increases after boiling due to softening of tissues. Contradictory reports are available on the impact of processing on phenolic content concentration, where some reports indicate the increase, whereas other showed decrease in phenolics after processing. Faller and Fialho (2009) reported that concentration of hydrolyzable phenolic compound remains constant with boiling and microwaving as compared to the soluble phenolics. Tudela et al. (2002) reported that the cooking methods reduce the phenolic content in potato but minimum loss was noticed in boiling. Lachman et al. (2013) observed that concentration of chlorogenic acid decreased with cooking treatment (microwaving, boiling, baking) but total anthocyanin concentration was not affected. Tuber flesh color exhibits positive correlation with antioxidant retention after cooking. Frying and microwaving reduce total polyphenols in five potato cultivars of different flesh colors. The total polyphenols in dark-fleshed potatoes were significantly higher (2880.5–3241.6 mg GAE/kg DW) than those of light-fleshed cultivars (Silveira et al. 2017). Frying of red- and purple-fleshed potatoes is reported to reduce anthocyanin content by 38–70% (Kita et al. 2013). According to Navarre et al. (2010) cooking methods, viz. baking, steam, stir-fry, microwave, and boiling increased the concentration of total phenolics, flavonols, and chlorogenic acid considerably. Baking enhanced antioxidant activity by 1.2 folds in cultivar "Bintje", chlorogenic acid and rutin enhanced to 1.36- and 2-fold in cultivar "Purple Majesty". In case of cultivar "Purple Majesty" cooking decreased polyphenols, whereas anthocyanin increased by all cooking methods except for baking (Lemos et al. 2015). Effect of cooking was evaluated on total antioxidant potential and major antioxidants such as anthocyanins, phenolics, flavonols, and lutein in 14 potato varieties with varying flesh color. The results showed that antioxidants decreased in all the varieties as indicated by lower DPPH radical scavenging activity except for the colored (red-, purple-fleshed) potatoes that retained considerable concentrations of antioxidants after cooking (Perla et al. 2012).

8.6 Health Benefits Associated with Phenolics

Phenols and anthocyanins have shown maximum impact on diabetes management among all the antioxidants. Potatoes extract rich in chlorogenic acid has been reported to prevent diabetes type II and cardiovascular diseases (Paynter et al. 2006). Chlorogenic acid is considered as novel insulin sensitizer that can be substituted with metformin drug used for treatment of diabetes type II as it possesses similar effect as that of drug. Chlorogenic acid found to inhibit rate limiting enzyme hepatic glucose-6-phosphatase in gluconeogenesis process (Arion et al. 1997). Obese, hyperlipidemic, and insulin resistant rats when administered with chlorogenic acid showed slow absorption of glucose in the gut and increased insulin sensitivity without affecting insulin release (Rodriguez de Sotillo and Hadley 2002). Purple potato extract was most effective for inhibition of α-amylase and α-glucosidase inhibitory activities. Raigond et al. (2017) reported that among 46 Indian potato varieties, α-amylase inhibitory activity was obtained only in one variety, i.e. Kufri Frysona with 20.5% inhibitory activity. Kumar et al. (2011) further observed that α-glucosidase inhibitory activity was also shown by compounds such as alkaloids, phenolics, curcuminoids, terpenoids, and anthocyanin. Reports on inhibitory effect of 5-O-caffeoylquinic acid on starch hydrolytic enzymes were also supported by in vivo animal studies. Glycemic peaks decreased by 22 and 17% after 10 and 15 min of oral administration of 5-O-caffeoylquinic acid at 3.5 mg/kg body weight (Bassoli et al. 2008). Antioxidant activity in potato tubers has been reported extensively (Singh and Rajini 2008). Pigmented potatoes such as purple- and red-fleshed ones are better source of antioxidants compared to cream-/white-fleshed potatoes. Consumption of colored potatoes (yellow- and purple-fleshed) for 6 weeks reduced inflammation and DNA damage in men (Kaspar et al. 2010). Increase in consumption of anthocyanin rich potatoes decreases the concentration of C-reactive protein, that are known as biomarker for disease progression, in plasma. Consumption of colored potatoes is reported to reduce the multiple markers of inflammation and oxidative stress in humans (Kaspar et al. 2011). Consumption of peeled and without peeled potatoes was found to significantly affect lipid metabolism and antioxidant capacity in rats (Robert et al. 2008). Administration of purple potato flakes to rats showed that they increase the antioxidant potential in the serum as well as liver of cholesterol-fed rats. Whereas red potato flakes promoted the hepatic superoxide dismutase mRNA in rats (Ezekiel et al. 2013). Consumption of cooked unpeeled potatoes is reported to improve the lipid metabolism and antioxidant concentration in cholesterol-fed rats (Robert et al. 2006). Diabetic rats when administered with 10% potato peels in diet were able to manage some aspects of oxidative stress (Singh et al. 2005). Dried peels of potatoes bound bile acid and carcinogen benzo(a) pyrene as indicated by in-vitro studies when potato peel were processed by extrusion cooking method (Camire et al. 1993).

8.7 Genetic Modifications to Improve Phenol Content

Potato may contain more genetic diversity than any other crop. Currently, breeding approaches are being practiced with the aim to develop antioxidant rich potatoes containing high concentrations of phenolic content and carotenoids content (Brown 2005). Modern genomics and biotechnological strategies, including candidate gene approaches, QTL detection and genetic transformation, are important tools to identify genomic regions and genes exhibiting key role in synthesis or accumulation of phenolic acids. However, increasing the phenolic acid content genetically may affect other agronomic traits important for acceptability of a variety. Caffeoylquinic acids

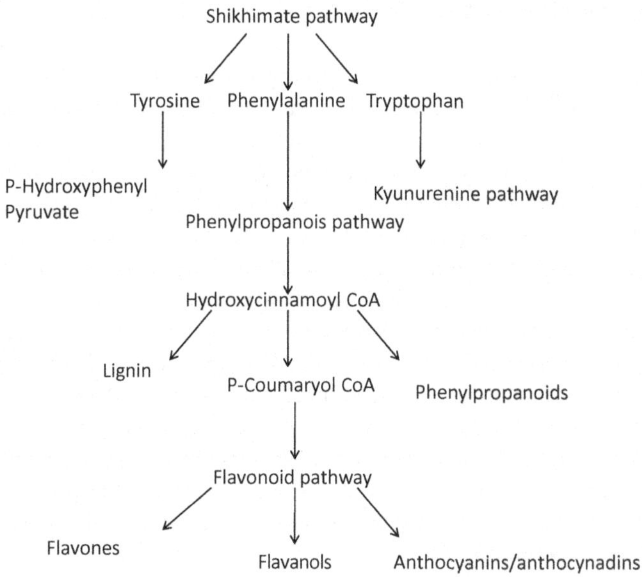

Fig. 8.2 Metabolic pathway for phenolics

(CQAs) are the major form of phenolic acid occupying almost 80% of the phenolic compounds in potatoes. Expression of an exogenous gene flavonol-specific transcriptional activator (AtMYB12: derived from *Arabidopsis thaliana*) reported to increase CQAs and total flavonoid content more than 3-folds (Li et al. 2014). Marker-free approach was utilized to facilitate the downstream regulatory approvals. Identification of varieties/germplasm rich in phenolic acid has the potential as breeding materials. Biosynthesis of phenolic acid in plants takes place through Shikimic acid pathway using mainly L-phenylalanine as precursor. L-phenylalanine is catalyzed by enzyme phenylalanine ammonia-lyase (PAL) for synthesis of cinnamic acid. Further cinnamic acid is transformed into different phenolic acids through the catalytic action of enzymes such as hydroxylases and methyltransferase (Fig. 8.2). Genes of interest can be used as tools for selecting new high-quality genotypes with stably fixed trait. The transgenic plants with overexpressed key biosynthesis pathway enzymes expressed genes encoding chalcone synthase (CHS), chalcone isomerase (CHI), and dihydroflavonol reductase (DFR) exhibited significantly increased phenolic acids and anthocyanins. The plants rich in flavonoids exhibit improved antioxidant capacity; however, the relationship between antioxidant capacity and flavonoids content is quite complex, suggesting the role of other compounds in the antioxidant potential of the plants (Lukaszewicz et al. 2004).

8.8 Conclusion/Future Directions

Potatoes have great potential as source of total phenolics and may also be considered as good source of bioactive compounds. The identification of potato cultivars with high phenolic content might open new market niches for cultivation of new species rich in phenols. Nowadays, due to inadequate intake of bioactive compounds people are becoming more prone to the degenerative diseases. In the last few decades the demand for new effective strategies is, therefore, aimed for prevention of these diseases. One approach to improve the health of large masses is to increase the phytonutrient content in most consumed crops and products. Potato as sources of nutraceuticals could be recommended where the incidence of significant oxidative stress-induced diseases occurred frequently. Phenols are metabolized and have a wide range of bioavailability but not yet thoroughly defined. More research needs to focus on development of nutrient rich/biofortified potatoes. Such potatoes can significantly improve the health status of consumers due to high per capita consumption of potatoes. Further research is required to determine the potential of waste streams for extraction of nutraceuticals and to substantiate the potential health benefits of potato phenolics.

References

Ah-Hen K, Fuenzalida C, Hess S et al (2012) Antioxidant capacity and total phenolic compounds of twelve selected potato landrace clones grown in Southern Chile. Chil J Agric Res 72:3. https://doi.org/10.4067/S0718-58392012000100001

Andre CM, Oufir M, Guignard C et al (2007) Antioxidant profiling of native Andean potato tubers (*Solanum tuberosum* L.) reveals cultivars with high levels of beta-carotene, alphatocopherol, chlorogenic acid, and petanin. J Agric Food Chem 55:10839–10849

Arion WJ, Canfield WK, Ramos FC et al (1997) Chlorogenic acid and hydroxynitrobenzaldehyde: new inhibitors of hepatic glucose 6-phosphatase. Arch Biochem Biophys 339:315–322

Bandana, Sharma V, Kaushik SK et al (2016) Variation in biochemical parameters in different parts of potato tubers for processing purposes. J Food Sci Technol 53:2040–2046

Barba A, Calabretti A, D'Amore M et al (2008) Phenolic constituents levels in cv. Agria potato under microwave processing. LWT Food Sci Technol 41(10):1919–1926. https://doi.org/10.1016/j.lwt.2008.02.004

Bassoli BK, Casolla P, Murad B et al (2008) Chlorogenic acid reduces the plasma glucose peak in the oral glucose tolerance test: effects on hepatic glucose release and glycemia. Cell Biochem Funct 26:320–328

Blessington T, Nzaramba MN, Scheuring DC et al (2010) Cooking methods and storage treatments of potato: effects on carotenoids, antioxidant activity and phenolics. Am J Potato Res 87:479–491

Bravo L (1998) Polyphenols: chemistry, dietary sources, metabolism and nutritional significance. Nutr Rev 56(11):317–333

Brown CR (2005) Antioxidants in potato. Am J Potato Res 82:163–172

Brown C, Wrolstad R, Durst R et al (2003) Breeding studies in potatoes containing high concentrations of anthocyanins. Am J Potato Res 80:241–249

Camire ME, Zhao J, Violette DA (1993) In vitro binding of bile acids by extruded potato peels. J Agric Food Chem 41:2391–2394

Chiou A, Salta FN, Kalogeropoulos N et al (2007) Retention and distribution of polyphenols after pan-frying of French fries in oils enriched with olive leaf extract. J Food Sci 72(8):S574–S584

Chun OK, Kim DO, Smith N et al (2005) Daily consumption of phenolics and total antioxidant capacity from fruit and vegetables in the American diet. J Sci Food Agric 85:1715–1724

Dalamu, Singh B, Barwal S, Raigond P et al (2015) Assessment of phytochemical diversity in Indian potato cultivars. Indian J Hortic 72:447–450

Dao L, Friedman M (1992) Chlorogenic acid content of fresh and processed potatoes determined by ultraviolet spectroscopy. J Agric Food Chem 40:2152–2156

Evers D, Deußer H (2012) Potato antioxidant compounds: Impact of cultivation methods and relevance for diet and health, nutrition, well-being and health. In: Jaouad B (ed) Nutrition, well-being, and health. InTech Open, Shanghai, pp 95–118

Ezekiel R, Singh N, Sharma S et al (2013) Beneficial phytochemicals in potato-a review. Food Res Int 50:487–496

Faller ALK, Fialho E (2009) The antioxidant capacity and polyphenol content of organic and conventional retail vegetables after domestic cooking. Food Res Int 42:210–215. https://doi.org/10.1016/j.foodres.2008.10.009

Fossen T, Ovstedal DO, Slimestad R et al (2003) Anthocyanins from a Norwegian potato cultivar. Food Chem 81:433–437

Friedman M (1997) Chemistry, biochemistry and dietary role of potato polyphenols. J Agric Food Chem 45:1523–1540

Galani JHY, Mankad PM, Avadh KS et al (2017) Effect of storage temperature on vitamin C, total phenolics, UPLC phenolic acid profile and antioxidant capacity of eleven potato (*Solanum tuberosum*) varieties. Hortic Plant J 3:73–89

Giusti MM, Polit MF, Ayvaz H et al (2014) Characterization and quantitation of anthocyanins and other phenolics in native Andean potatoes. J Agric Food Chem 62:4408–4416. https://doi.org/10.1021/jf500655n

Hajslova J, Schulzova V, Slanina P et al (2005) Quality of organically and conventionally grown potatoes: four-year study of micronutrients, metals, secondary metabolites, enzymic browning and organoleptic properties. Food Addit Contam 22:514–534

Hamouz K, Lachman J, Hejtmankova A et al (2010) Effect of natural and growing conditions on the content of phenolics in potatoes with different flesh colour. Plant Soil Environ 56:368–374

Im HW, Suh BS, Lee SU et al (2008) Analysis of phenolic compounds by high-performance liquid chromatography and liquid chromatography/mass spectrometry in potato plant flowers, leaves, stems, and tubers and in home-processed potatoes. J Agric Food Chem 56(9):3341–3349

Kaspar K, Park J, Brown C et al (2010) Pigmented potato consumption alters oxidative stress and inflammatory damage in men. J Nutr 141:108–111

Kaspar KL, Park JS, Brown CR (2011) Pigmented potato consumption alters oxidative stress and inflammatory damage in men. J Nutr 141:108–111

Kaufmann B, Christen P (2002) Recent extraction techniques for natural products: microwave assisted extraction and pressurised solvent extraction. Phytochem Anal 13:105–113

Kita A, Bąkowska-Barczak A, Hamouz K et al (2013) The effect of frying on anthocyanin stability and antioxidant activity of crisps from red-and purple-fleshed potatoes (*Solanum tuberosum* L.). J Food Compos Anal 32:169–175. https://doi.org/10.1016/j.jfca.2013.09.006

Kumar S, Narwal S, Kumar V (2011) α-Glucosidase inhibitors from plants: a natural approach to treat diabetes. Pharmacogn Rev 5(9):19–29

Lachman J, Hamouz K, Orsak M et al (2012) Impact of selected factors – cultivar, storage, cooking and baking on the content of anthocyanins in coloured-flesh potatoes. Food Chem 133:1107–1116

Lachman J, Hamouz K, Musilová J (2013) Effect of peeling and three cooking methods on the content of selected phytochemicals in potato tubers with various colour of flesh. Food Chem 138:1189–1197

Legrand D, Scheen AJ (2007) Does coffee protect against type 2 diabetes? Rev Med Liege 62:554–559

Lemos MA, Aliyu MM, Hungerford G (2015) Influence of cooking on the levels of bioactive compounds in Purple Majesty potato observed via chemical and spectroscopic means. Food Chem 173:462–467. https://doi.org/10.1016/j.foodchem.2014.10.064

Li A, Li S, Zhang Y et al (2014) Resources and biological activities of natural polyphenols. Nutrients 6:6020–6047

Lin CH, Chang CY (2005) Textural change and antioxidant properties of broccoli under different cooking treatments. Food Chem 90(1-2):9–15. https://doi.org/10.1016/j.foodchem.2004.02.053

Lister E, Wilson P (2001) Measurement of total phenolics and ABTS assay for antioxidant activity (personal communication). Crop Research Institute, Lincoln

Lukaszewicz M, Matysiak-Kata I, Skala J et al (2004) Antioxidant capacity manipulation in transgenic potato tuber by changes in phenolic compounds content. J Agric Food Chem 52:1526–1533

Mader J, Rawel H et al (2009) Composition of phenolic compounds and glycoalkaloids a-solanine and a-chaconine during commercial potato processing. J Agric Food Chem 57:6292–6297

Manach C, Scalbert A, Morand C et al (2004) Polyphenols: food sources and bioavailability. Am J Clin Nutr 79:727–747

Matsuzoe N, Yamaguchi M, Kawanobu S et al (1999) Effect of dark treatment of the eggplant on fruit skin color and its anthocyanin components. J Jpn Soc Hortic Sci 68:138–145

Mattila P, Hellstrom J (2007) Phenolic acids in potatoes, vegetables and some of their products. J Food Compos Anal 20:152–160

Nara K, Miyoshi T, Honma T et al (2006) Antioxidative activity of bound-form phenolics in potato peel. Biosci Biotechnol Biochem 70:1489–1491. https://doi.org/10.1271/bbb.50552

Nassar AM, Sabally K, Kubow S et al (2012) Some Canadian-grown potato cultivars contribute to a substantial content of essential dietary minerals. J Agric Food Chem 60:4688–4696. https://doi.org/10.1021/jf204940t

Navarre DA, Shakya R et al (2010) The effect of different cooking methods on phenolics and vitamin C in developmentally young potato tubers. Am J Potato Res 87:350–359

Paynter NP, Yeh HC, Voutilainen S et al (2006) Coffee and sweetened beverage consumption and the risk of type 2 diabetes mellitus: the atherosclerosis risk in communities study. Am J Epidemiol 164:1075–1084

Pelozo MIG, Cardoso MLC et al (2008) Spectrophotometric determination of tannins and caffeine in preparations from Paullinia cupana var. sorbilis. Braz Arch Biol Technol 51:447–451

Perla V, Holm DG, Jayanty SS (2012) Effects of cooking methods on polyphenols, pigments and antioxidant activity in potato tubers. LWT Food Sci Technol 45:161–171. https://doi.org/10.1016/j.lwt.2011.08.005

Poiatti V, Dalmas AD, Fernando R et al (2009) Defense mechanisms of *Solanum tuberosum* L. in response to attack by plant-pathogenic bacteria. Biol Res 42:205–215

Raigond P, Singh B, Dutt S et al (2017) Potential of Indian potatoes for the management of hyperglycemia. Indian J Hortic 74(1):103–108

Reddivari L, Hale AL, Miller JC Jr (2007) Determination of phenolic content, composition and their contribution to antioxidant activity in specialty potato selections. Am J Potato Res 84:275–282

Reyes LF, Miller JC, Cisneros-Zevallos L (2005) Antioxidant capacity, anthocyanins and total phenolics in purple-and red-fleshed potato (*Solanum tuberosum* L.) genotypes. Am J Potato Res 82:271. https://doi.org/10.1007/BF02871956

Riciputi Y, Diaz-de-Cerio E, Akyol H et al (2018) Establishment of ultrasound-assisted extraction of phenolic compounds from industrial potato by-products using response surface methodology. Food Chem 269:258–263

Robert L, Narcy A, Rock E (2006) Entire potato consumption improves lipid metabolism and antioxidant status in cholesterol-fed rat. Eur J Nutr 45:267–274

Robert L, Narcy A, Rayssiguier Y et al (2008) Lipid metabolism and antioxidant status in sucrose vs. potato-fed rats. J Am Coll Nutr 27:109–116

Rodriguez de Sotillo D, Hadley M (2002) Chlorogenic acid modifies plasma and liver concentrations of: cholesterol, triacylglycerol, and minerals in (fa/fa) Zucker rats. J Nutr Biochem 13:717–726

Ross JA, Kasum CM (2002) Dietary flavonoids: bioavailability, metabolic effects and safety. Annu Rev Nutr 22:19–34

Rytel E, Tajner Czopek A et al (2014) Content of polyphenols in coloured and yellow fleshed potatoes during dices processing. Food Chem 161:224–229

Samarin AM, Poorazarang H, Hematyar N et al (2012) Phenolics in potato peels: extraction and utilization as natural antioxidants. World Appl Sci J 18:191–195

Schofield P, Mbugua DM et al (2001) Analysis of condensed tannins: a review. Anim Feed Sci Technol 91:21–40

Shakya R, Navarre DA (2006) Rapid screening of ascorbic acid, glycoalkaloids, and phenolics in potato using high-performance liquid chromatography. J Agric Food Chem 54:5253–5260

Silveira A, Falagán N, Aguayo E et al (2017) Compositional changes on colored and light-yellow-fleshed potatoes subjected to two cooking processes. CyTA J Food 15(2):241–248. https://doi.org/10.1080/19476337.2016.1243155

Singh N, Rajini P (2008) Antioxidant-mediated protective effect of potato peel extract in erythrocytes against oxidative damage. Chem Biol Interact 173:97–104

Singh N, Kamath V, Rajini PS (2005) Attenuation of hyperglycemia and associated biochemical parameters in STZ-induced diabetic rats by dietary supplementation of potato peel powder. Clin Chim Acta 353:165–175

Sticher O (2008) Natural product isolation. Nat Prod Rep 25:517–554

Stushnoff C, Holm D, Thompson MD et al (2008) Antioxidant properties of cultivars and selections from the Colorado potato breeding program. Am J Potato Res 85:267–276

Thakur N, Raigond P, Singh Y et al (2020) Recent updates on bioaccessibility of phytonutrients. Trends Food Sci Technol 97:366–380

Tiam J, Che J, Lv F et al (2016) Domestic cooking methods affect the phytochemical composition and antioxidant activity of purple-fleshed potatoes. Food Chem 197:1264–1270. https://doi.org/10.1016/j.foodchem.2015.11.049

Tudela JA, Cantos E, Espín JC et al (2002) Induction of antioxidant flavonol biosynthesis in fresh-cut potatoes: effect of domestic cooking. J Agric Food Chem 50:5925–5931. https://doi.org/10.1021/jf020330y

Xu X, Li W, Lu Z et al (2009) Phenolic content, composition, antioxidant activity, and their changes during domestic cooking of potatoes. J Agric Food Chem 57(21):10231–10238

Zarzecka K, Gugała M, Sikorska A et al (2019) The effect of herbicides and biostimulants on polyphenol content of potato (*Solanum tuberosum* L.) tubers and leaves. J Saudi Soc Agric Sci 18:102–106

Zhu F, Cai YZ, Ke J (2010) Compositions of phenolic compounds, amino acids and reducing sugars in commercial potato varieties and their effects on acrylamide formation. J Sci Food Agric 90 (13):2254–2262

Potato Carotenoids

9

Sushil Sudhakar Changan, Mark A. Taylor, Pinky Raigond, Som Dutt, Dharmendra Kumar, Milan Kumar Lal, Manoj Kumar, Maharishi Tomar, and Brajesh Singh

Abstract

Potato (*Solanum tuberosum* L.) is the world's third most important crop in terms of human consumption following wheat and rice. It has been used as a primary nutritional source in many diets and for preparation of a variety of processed products in all countries of the world whether developing or developed. Potato tubers are considered as a rich source of bioactive compounds, which are highly desirable in healthy human diet. Although potato is a nutrient rich food, there is ample scope for improving its nutritional quality and making it more nutritious food. Among various phytonutrients, carotenoids are major lipophilic constituents contributing to total antioxidant activity of potato. Potato with improved carotenoid content and composition are essential to fulfil the carotenoid requirement of malnourished populations of many countries around the world, which is important to alleviate vitamin A deficiency and other health-related disorders. The objective of the chapter is to summarize various aspects related to potato tuber carotenoid, viz. health promoting properties of carotenoids and specially their contents and composition in different potato varieties affected by tuber flesh colour (purple-, yellow-, white- and red-fleshed) and the effect of various factors on total and individual carotenoid levels, such as genotype,

S. S. Changan (✉) · P. Raigond · S. Dutt · D. Kumar · M. K. Lal · B. Singh
Crop Physiology, Biochemistry & Post-Harvest Technology Division, ICAR-Central Potato Research Institute, Shimla, Himachal Pradesh, India
e-mail: changan.sushil7@gmail.com

M. A. Taylor
Cellular and Molecular Sciences, The James Hutton Institute, Dundee, UK

M. Kumar
ICAR-Central Institute for Research on Cotton Technology, Mumbai, Maharashtra, India

M. Tomar
ICAR-Indian Grassland and Fodder Research Institute, Jhansi, Uttar Pradesh, India

P. Raigond et al. (eds.), *Potato*, https://doi.org/10.1007/978-981-15-7662-1_9

breeding methodology, tuber development, effect of year, locality, storage, thermal processing—cooking, frying, etc.

Keywords

Solanum tuberosum · Carotenoids · Phytonutrients · Processing · Storage

9.1 Introduction

Worldwide, potato is the third most consumed staple food crop after rice and wheat. Generally, potato tubers contain several phytochemicals, including various phenolic compounds like chlorogenic acid, anthocyanins in coloured potatoes and a wide range of carotenoids (Akyol et al. 2016; Fogelman et al. 2019). Hence, potato is considered as an important source of bioactive health promoting compounds, which are most essential in human diets, although the concentrations of different phytochemicals are affected by cooking and other processes (Ezekiel et al. 2013; Lachman et al. 2013). In particular, the wide range of nutrient content among native potato germplasm can be used in conventional breeding programs to increase tuber nutritional value (Tatarowska et al. 2019; Bradshaw 2019). The colour of potato tubers depends on carotenoid concentration and composition (Hejtmánková et al. 2013) and total carotenoid content is positively correlated with tuber yellow intensity. Major factors affecting carotenoid levels are cultivar, climate, growing conditions, storage and cooking processes.

Carotenoids are the most important class of natural organic isoprenoid pigments that are synthesized by photosynthetic organisms (plants and algae), some non-photosynthetic fungi and bacteria (Valcarcel et al. 2015). All have a tetraterpenoid structure (40 carbon atoms), a long chain of conjugated double bonds in the centre of the molecule and have symmetry around the central double bond. The pathway of carotenoid synthesis is shown in Fig. 9.1. Carotenoids have two major roles in algae and plants: help in photosynthesis by absorbing light energy and protect chlorophyll from photo-damage. As most animals are unable to produce carotenoids, they depend on other foods (Varela et al. 2015a, b). More than 750 carotenoid structures have been reported, which can be categorized into two major groups; (1) Carotenes, which are purely hydrocarbons and contain no oxygen, including β-carotene and lycopene and (2) Xanthophylls, having oxygen as a functional group, including lutein, astaxanthin, violaxanthin and zeaxanthin (Rao and Rao 2007). Carotenoids are most important antioxidants having various health-promoting functions like provitamin A activity, immune system enhancement and protection against cardiovascular diseases and atherosclerosis (Bonierbale et al. 2009; Eggersdorfer and Wyss 2018). Due to the health benefits of carotenoids consumption, considerable attention is currently being given to screening and development of food crops with enhanced concentrations of total and individual carotenoids (Andersson et al. 2017; Amah et al. 2019).

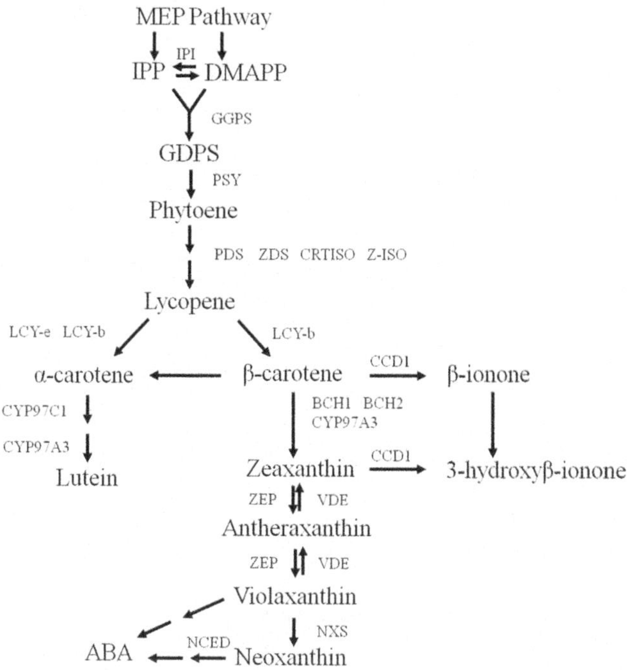

Fig. 9.1 Carotenoids biosynthesis pathway in potato. *IPI* isopentenyl pyrophosphate isomerase, *GGPS* geranylgeranyl pyrophosphate synthase, *PSY* phytoene synthase, *PDS* phytoene desaturase, *ZDS* f-carotene desaturase, *CRTISO* carotenoid isomerase, *LCY-ε* lycopene ε-cyclase, *LCY β* lycopene β-cyclase, *CHY-ε* ε-ring hydroxylase, *CHY-β* β-carotene hydroxylase, *VDE* violaxanthin de-epoxidase, *ZEP* zeaxanthin epoxidase, *NXS* neoxanthin synthase, *CCD* carotenoid cleavage dioxygenase, *NCED* 9-cis-epoxycarotenoids dioxygenase (source: Dutt et al. 2019)

9.2 Content of Total and Individual Carotenoids in Potato Tubers

Some yellow- and orange-fleshed potatoes are a rich source of carotenoids, which are lipophilic compounds synthesized from isoprenoid precursors in the plastids (Dellapenna and Pogson 2006). The major carotenoids found in potato are lutein, violaxanthin, zeaxanthin and neoxanthin, whereas β-carotene is reported in trace amounts. The structure of major potato carotenoids is shown in Fig. 9.2. The tuber flesh colour, yellow and orange, is due to the presence of lutein and zeaxanthin, respectively. Yellow or orange tuber-fleshed potato cultivars contained higher amount of carotenoids compared with white-fleshed potato cultivars. Brown (2005) reported total carotenoids content in the range of 50–350 μg/100 g fresh weight (FW) and 800–2000 μg/100 g FW in white- and yellow-fleshed potato cultivars, respectively. The total carotenoid content was higher in tuber skin (maximum 2800 μg/100 g dry weight (DW)) as compared to flesh (900 μg/100 g DW).

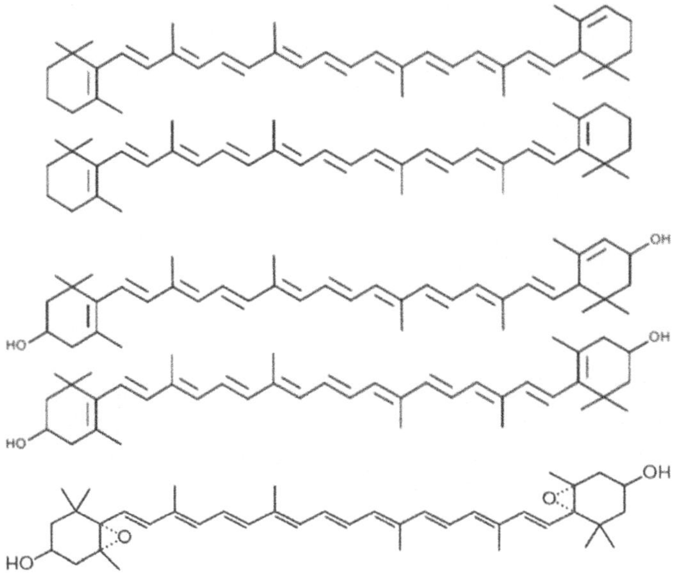

Fig. 9.2 Structure of major carotenoids present in potato

Yellow-skinned or fleshed tetraploid cultivars possessed higher carotenoid contents compared to paler or white tissues, whereas no such relationship found for other colours (Valcarcel et al. 2015). Some diploid potato lines contain up to 22 times higher carotenoid concentrations as compared to potatoes with white-flesh (Haynes et al. 2011). Tuber carotenoid content is affected by genotype and growing environment (Hamouz et al. 2016). According to Breithaupt and Bamedi (2002), the total carotenoid content in potato tubers could be considered as a parameter to screen for white- or yellow-fleshed cultivars. The total carotenoid contents for different cultivars ranged from 5 to 1550 μg/100 g DW, while most of the cultivars had less than 1000 μg/100 g DW, with an average value of about 435 μg/100 g DW. Carotenoid content of different cultivars reported by various research groups is shown in Table 9.1. Morris et al. (2004) observed more than 20-fold range in total carotenoid concentrations in potato germplasm. Earlier studies have reported that violaxanthin and lutein are the predominant carotenoids, whereas zeaxanthin is present at very low levels in raw potato tubers (Iwanzik et al. 1983; Lu et al. 2001). Iwanzik et al. (1983) reported a range of 27–74 μg/100 g FW for total carotenoids in white-fleshed potato cultivars, whereas Brown et al. (1993) reported a higher zeaxanthin content of 2000 μg/100 g FW in diploid potatoes derived from *Solanum phureja* and *S. stenotomum*. Burgos et al. (2009) reported significant amounts of antheraxanthin and zeaxanthin in deep yellow-fleshed potatoes, whereas the carotenoid profile of yellow-fleshed potatoes contains violaxanthin, antheraxanthin, zeaxanthin and lutein, and that of cream-fleshed potatoes contain lutein, violaxanthin and b-carotene. The carotenoid pattern of four white-fleshed and four yellow-fleshed German potato cultivars was investigated by Breithaupt and

Table 9.1 Carotenoids content reported in potato tubers (mg/kg DW* or FW**)

Carotenoid	Content	Potato cultivars	References
Total carotenoids			
	0.58–1.75	Yellow cultivars	Breithaupt and Bamedi (2002)
	26.2*/5.69*	Yellow/red/purple	Brown (2005)
	5.60–35.0*	Transgen. Desirée	Ducreux et al. (2005)
	3.0–36.0*	Andean landraces	Andre et al. (2007)
	5.67	Inca-no-hitomi orange	Kobayashi et al. (2008)
	26	Papa Amarilla cvs	Brown et al. (2005)
	1.03–21.4	*S. phureja* accession	Bonierbale et al. (2009)
	28.0*	Skin of tubers	Campbell et al. (2010)
	9.0*	Flesh of tubers	
	14.8 ± 2.22*	Red Laura	Burmeister et al. (2011)
	0.50–15.5*	Different cultivars	Fernandez-Orozco et al. (2013)
	1.51 ± 0.31*	Boiled Shetl. Black	Tierno et al. (2015)
	1.10–12.2*	Different cultivars	Hamouz et al. (2016)
	16–29*	Orange-fleshed group	Sulli et al. (2017)
	5.57–20.20**	Different cultivars	Tatarowska et al. (2019)
	1.5**	Andean sunrise	Fogelman et al. (2019)
Individual carotenoids			
All-trans-Lutein	1.12–17.7	Andean landraces	Andre et al. (2007)
	2.92–6.66**	Different cultivars	Tatarowska et al. (2019)
	0.55–1.89	*S. phureja* accession	Bonierbale et al. (2009)
	3.27–9.50*	Raw tubers	Clevidence (2005)
	3.89–9.50*	Boiled tubers	
All-trans-Violaxanthin	Trace–2.78	*S. phureja* accession	Bonierbale et al. (2009)
All-trans-Zeaxanthin	18	Andean landraces	Andre et al. (2007)
	12.9	*S. phureja*	Burgos et al. (2009)
	>10.0	*S. phureja*	Bonierbale et al. (2009)
	Trace–12.9	*S. phureja*	
	Trace–40*	Accession raw/boiled tubers	Clevidence (2005)
All-trans-β-Carotene	2	Andean landraces	Andre et al. (2007)
	>0.1	*S. phureja* accession	Bonierbale et al. (2009)

(continued)

Table 9.1 (continued)

Carotenoid	Content	Potato cultivars	References
Lutein-5,6-epoxide	Identified	Commercial, bred, old, and native cultivars	Fernandez-Orozco et al. (2013)
9-cis-Lutein	Identified		
13-cis-Lutein	Identified		
Mutatoxanthin	Identified		
Neochrome	Identified		
All-trans-β-Cryptoxanthin	Identified		

Adapted and updated from Lachman et al. (2016) and Dutt et al. (2019)

Bamedi (2002). The carotenoid content was dominated by zeaxanthin, antheraxanthin, violaxanthin and lutein, which were found in different ratios, while neoxanthin, β-carotene and β-cryptoxanthin were found in trace amounts. Antheraxanthin was the only carotenoid epoxide found in native extracts. The total concentration of the four major carotenoids was 175 μg/100 g FW, while the total carotenoid esters were 41–131 μg/100 g FW. Andean potato germplasm pool is a rich and varied source of carotenoids. A range of 300–3600 μg/100 g DW for total carotenoids among 74 Andean landraces was reported by Andre et al. (2007). In another study, 23 Andean potato cultivars were screened and identified genotypes having a high concentration of zeaxanthin (18 μg/g DW), lutein (1.12–17.69 μg/g DW) and β-carotene (2 μg/g DW) (Andre et al. 2007). Burgos et al. (2009) analysed 23 accessions of *S. Phureja* for carotenoid content in tubers and reported two accessions containing a very high zeaxanthin concentration (1290 μg/100 g FW). Total and individual carotenoid concentrations were analysed by Bonierbale et al. (2009) in 152 accessions of *S. phureja* germplasm. They sorted two cultivars containing very high zeaxanthin content (>1000 μg/100 g FW) and other 43 accessions having relatively high β-carotene concentrations (>10 μg/100 g FW). The total carotenoid, violaxanthin, zeaxanthin, antheraxanthin, lutein and β-carotene concentrations ranged from 103 to 2135 μg/100 g FW, 0 to 278 μg/100 g FW, 0 to 1290 μg/100 g FW, 3 to 354 μg/100 g FW, 55 to 189 μg/100 g FW and 0 to 18 μg/100 g FW, respectively.

9.2.1 Methods for Estimation of Total and Individual Carotenoids

Various methods have been used for the identification and quantification of total and individual carotenoids in food and plant samples including thin layer chromatography, UV/Visible spectroscopy, high pressure liquid chromatography (HPLC) with a photodiode array detector (HPLC-PDA), HPLC in combination with mass spectrometry including MALDI-TOF, liquid chromatography—tandem nuclear magnetic resonance (LC-NMR), liquid chromatography–mass spectrometry (LC-MS) and Raman resonance spectroscopy. But, the most commonly employed method for identification and quantification of carotenoids is HPLC with UV–Visible absorption detection system (Kopec et al. 2012). High performance liquid chromatography (HPLC) is the most powerful analytical tool that advanced carotenoid research especially for complex samples, followed by solvent extraction/spectrophotometry, chromatography with supercritical fluid, etc. (Luterotti and Kljak 2010). The photodiode array (PDA) is the most commonly used detector for HPLC carotenoid analysis, whereas other detectors, viz. fluorescence, electrochemical detectors (ECD), nuclear magnetic resonance (NMR) and mass spectrometers (MS) can be used for analysis (Kopec et al. 2012). Although both normal and reverse phase HPLC can be used for carotenoid separation, due to poor separation of non-polar carotenoids normal phase HPLC is not generally applicable for carotenoid separation. In contrast, reverse phase HPLC significantly increases the interaction between analyte and non-polar stationary phase resulting in enhanced resolution of carotenoids (Sander et al. 2000). Mostly C18 columns are more preferred, with isocratic or gradient mode, for carotenoid analysis (Khachik et al. 1997). C18 columns are often used for separation of different carotenoids, although C30 columns provide better separation for geometrical and positional isomers, e.g. all-trans-lutein and all-trans-zeaxanthin (Hart and Scott 1995; Updike and Schwartz 2003). Hence, C30 column are mostly used to distinguish between beta-carotene and beta-carotene cis-isomers (Emenhiser et al. 1995), and lycopene and cis-lycopene isomers (Frohlich et al. 2007). The recently developed UHPLC technique has been used most commonly for carotenoid analysis in various matrices (Li et al. 2012). UHPLC refers to ultra-high performance liquid chromatography, which improves speed, resolution and sensitivity of analysis. Better resolution, separation and sensitivity are obtained by UHPLC compared to HPLC as well as higher sample throughput and reduced consumption of solvent (Taleuzzaman et al. 2015). C18 columns have been used for carotenoid separation despite their poor performance to resolve carotenoid isomers, because of unavailability of C30 stationary phase columns for UHPLC. In earlier studies, C30 HPLC columns were compared with C18 UHPLC columns. Though C30 HPLC column gives better resolution of carotenoids, especially for geometrical isomers compared to C18 UHPLC columns, this advantage was limited by the longer run time needed for C30 HPLC column to elute the analyte, i.e. 100 min versus 23 min for C18 UHPLC column (Bijttebier et al. 2014). Maurer et al. (2014) reported the separation of 11 major carotenoids from vegetable crops within 13.50 min using UHPLC-UV C18 column. However, this method could be applied to detect only few geometrical isomers.

Liquid chromatography–mass spectrometry and LC-MS-MS are widely used techniques for the characterization and identification of carotenoids due to their high sensitivity, specificity and selectivity. The LC-MS is important technique for analysis of carotenoids obtained from various natural sources, as these compounds are often contaminated with biological matrices and generally present in trace amounts. The major advantage of the LC-MS tool is that it enables not only quantification, but also the elucidation of structure, on the basis of the molecular mass and fragmentation. Different techniques are used to ionize the carotenoids such as atmospheric pressure chemical ionization (APCI), electrospray ionization (ESI) and atmospheric pressure photoionization (APPI). It is possible to obtain characteristic transitions of all of the carotenoids tested for their unequivocal identification using each ionization technique. However, APCI is a more powerful technique to ionize the carotenoids compared to ESI or APPI (Rivera et al. 2011). LC-MS/APCI was used for identification and quantification of carotenoids in potato varieties. In these samples, the carotenoid composition was dominated by violaxanthin, antheraxanthin, lutein, and zeaxanthin, whereas beta-cryptoxanthin, beta-carotene and neoxanthin generally were present in trace amounts (Breithaupt and Bamedi 2002; Weller and Breithaupt 2003).

Comprehensive two-dimensional LC (LC × LC) and supercritical fluid chromatography (SFC) have also been applied to enhance the chromatographic separations of complex carotenoid mixtures (Dugo et al. 2008). Though both techniques have better capability to perform separations of complex carotenoid, they generally require expensive and more specialized instruments, and longer analysis times, i.e. twice that of HPLC. Special precautions are needed to avoid carotenoid degradation during longer analysis time (Gupta et al. 2015).

Nuclear magnetic resonance (NMR) is a powerful technique for the structural identification of carotenoids (Tode et al. 2009; LaFountain et al. 2013), but in the past decade it has been interfaced with HPLC to accelerate carotenoid profiling. The LC-NMR technique has improved sensitivity due to higher magnetic fields of superconductive magnets and advanced techniques such as the solvent suppression method (Smallcombe et al. 1995). NMR is mostly used to obtain information about conformational geometry and is thus a powerful tool for structural analysis. Application of LC-NMR for carotenoids analysis could reduce the time required to examine unknown compounds and determine the component ratio in the given sample (Tode et al. 2009). Generally, C30 column has been preferred in LC-NMR for carotenoid analyses due to their greater shape selectivity compared to C18 columns (Sander et al. 1994). This feature enhances the sharpness of chromatographic peaks and subsequently, a higher concentration of analyte in the NMR flow cell, which increases sensitivity (Dachtler et al. 2001). Tode et al. (2009) used HPLC-NMR using a C18 column to identify the major carotenoids in tomato juice, palm oil, and satsuma mandarin orange juice. Although LC-MS is a powerful method for detection of trace amounts of carotenoids in a variety of matrices, it cannot distinguish between stereoisomers as they yield the same mass spectra and fragmentation patterns. Hence, HPLC-NMR is a preferred method for both separation and unambiguous identification of carotenoid stereoisomers.

9.3 Effect of Growing Conditions, Storage as Well as Cooking on Carotenoids

Cultivation of potato under different environmental conditions may alter the nutritional profile of tubers. It allows environment to be used as a tool to study fluctuations in tuber nutrient content. Earlier studies analysed gene expression and metabolite pools in potato tubers cultivated in diverse locations to understand how tuber metabolism is altered and whether environment affects the nutritional profile of a potato tuber. Smaller variation was observed in total carotenoid content, but major differences found in individual carotenoid content, i.e. violaxanthin, lutein and zeaxanthin were the predominant carotenoids in potato tubers from Alaska, Texas and Florida, respectively (Payyavula et al. 2012). Othman (2009) observed highly significant differences in carotenoid pigments among the cultivars and the locations, suggesting that carotenoid biosynthesis is a complex and an environment specific process. Light influenced the concentration of zeaxanthin, whereas drought stress and nutrient availability influenced the biosynthesis of neoxanthin and violaxanthin. Apart from environment factors, cultivar selection appeared to be the most influential factor for carotenoid content and profile (Othman 2009). The locality and average temperature during the growing season were key factors for total carotenoid contents in potato tubers (Hamouz et al. 2016). The year of cultivation had a significant effect on total carotenoids content observed by Kotikova et al. (2007). Another report also revealed that highly significant differences were observed both between years and cultivars for both the *S. tuberosum* cultivars and the *S. phureja* lines (Griffiths et al. 2007). In both species, 2003 grown tubers showed the lowest values with total carotenoid content in 2003 being on average 56 and 32% lower as compared to 2002 grown *S. tuberosum* cultivars and *S. phureja* lines, respectively. On comparing 2 years, 2002 and 2003, for carotenoid content, it appears that the higher solar irradiation (June–August 2002, 1629.1 MJ/m^2; June–August 2003, 1864.9 MJ/m^2) and drought conditions (June–August 2002, 294.4 mm; June–August 2003, 80.4 mm) may be the reason for the 33% less total tuber carotenoid content (Griffiths et al. 2007). These results indicate that the nutritional quality of potato tubers may vary significantly with season. Further detailed investigation utilizing plant material grown under controlled environmental conditions is necessary to prove this hypothesis.

To preserve quality and to fulfil consumer's demand throughout the year, potatoes are required to be stored at 4 °C for the fresh market and at 4–10 °C for processing. Potato phytonutrients are often influenced by storage conditions. Bhushan and Thomas (1990) reported that the concentrations of carotenoids increased in potato tubers stored at 4 and 25–30 °C compared to those stored at 15 and 20 °C suggesting that specific temperature conditions are required for biosynthesis, accumulation and stability of carotenoids in stored potatoes. During tuber development and storage *S. phureja* accession showed a high carotenoid-accumulation compared to two *S. tuberosum* cultivars (Pentland Javelin and Desirée) (Morris et al. 2004). After 9 months storage at 4 °C the concentrations of zeaxanthin and antheraxanthin decreased, while lutein concentration increased; however, there was a small reduction in total carotenoid content. Blessington et al. (2010) reported that non-stored potato tubers were lower in carotenoid content (expressed as lutein) as compared to stored potato tubers in four cultivars. In contrary to these reports, Griffiths et al. (2007) studied effect of post-harvest storage on the carotenoid content of 38 *S. phureja* potato lines and stated that post-harvest storage significantly decreases the carotenoid content of the potato tubers, and further reducing the storage temperature leads to lowering the carotenoid content. At both storage

(4 °C and 10 °C) conditions violaxanthin, antheraxanthin and zeaxanthin levels, average of all lines, were significantly decreased during storage, while the relative concentration of lutein and neoxanthin was increased during storage (Griffiths et al. 2007). Morris et al. (2004) also reported a decrease in total carotenoid content in a *S. phureja* accession and *S. tuberosum* cultivar, Desiree, after storage at 4 °C for 9 months. Post-harvest storage and disease conditions remarkably influenced the total carotenoid levels, neoxanthin, lutein, violaxanthin, zeaxanthin and β-carotene, in potatoes. The intensity of these effects depends on the cultivar, time of storage, storage conditions. Results showed that long term storage of potato tubers resulted in neoxanthin, violaxanthin and zeaxanthin accumulation, whereas decrease of lutein, β-carotene and total carotenoid content (Othman 2009). This revealed that lutein was the most stable carotenoid and least likely to be degraded, whereas the levels of the other carotenoids derived from beta-carotene were significantly reduced during storage.

The most of the dietary components are cooked/processed, prior to consumption, using different methods according to the recipes and the culinary techniques of various regions. During household potato cooking, the application of heating includes a various processes, i.e. boiling, steaming, baking, frying and roasting by using traditional and microwave oven techniques (Palermo et al. 2014). The physical, chemical and biological changes that occur during cooking process may lead to appearance, nutritional, sensory and textural modifications of food and that can be beneficial or harmful to human health. Generally, cooking kills the microorganisms and inactivates anti-nutritional factors, and also enhances the food digestibility and the nutrients bioavailability. Apart from this, cooking leads to the formation of various structural and desirable compounds, which confer crispiness, texture, flavour and antioxidants to the cooked food. Household cooking can result in adverse effects like certain nutrient losses and undesirable compound formation (e.g. acrylamide) (VanBoekel et al. 2010). Worldwide, potatoes are generally consumed only after cooking. The cooking methods are important factors because they not only affect the physical structure and chemical composition of the potato but also alter the characteristics of bioactive compounds present in normal diet (Rothwell et al. 2015). Carotenoids are more sensitive to heat, oxygen and light due to the presence of highly unsaturated structures. Thus, carotenoids content is depending on the cooking conditions, such as temperature and time (Blessington et al. 2010). Generally, cooking reduces the carotenoid concentration in potatoes; while, in some cases, carotenoids bonded with proteins (e.g. protein–xanthophyll complex) in potatoes are dissociated by heat during cooking, resulting in increase of carotenoid concentration in cooked potatoes (Burmeister et al. 2011). Kotíková et al. (2016) stated that after boiling the total carotenoid content reduced by 92% as compared to baking (by 88%) in purple-, yellow- or red-fleshed potatoes. Lutein reported as the most stable carotenoid against heat treatment (reduced by 24–43%) compared to β-carotene (reduced by 78–83%); whereas other carotenoids were degraded completely. Mostly frying in oil or boiling in water lead to degrade the carotenoid content due to the thermal effect or the lipophilic properties of carotenoids (Tierno et al. 2015; Kotíková et al. 2016; Tian et al. 2016). The effect of cooking on stability of potato carotenoids has been studied by several researcher groups. Tian et al. (2016) reported that cooking methods significantly affect the carotenoid content in purple-fleshed potatoes. The total carotenoid content was significantly reduced after baking, boiling and steaming and treatments by 51.99%, 20.15% and 34.89%, respectively, whereas the frying, air-frying, stir-frying and microwaving treatments reduced the content by 75.66%, 72.00%, 76.16% and 66.30%, respectively (Tian et al. 2016). Stahl and Sies (2005) stated that carotenoids were more prone degradation when treated with oil and heating because of their thermal sensitivity and liposolubility. Burmeister et al. (2011) compared the carotenoid content of raw and boiled tubers of four potato varieties in which they found lutein and violaxanthin as the major carotenoids and zeaxanthin in very low concentration, and also concluded that heat treatment transformed all-trans carotenoids to 9-cis and 13-cis isomeric forms or degraded them. Blessington et al. (2010) reported that boiled potato samples have lower carotenoid content as compared to the other samples cooked by baking, microwaving and frying. Another study reported the effects of cooking on seven Andean potato cultivars with different intensities of yellow-coloured flesh concluding that boiling significantly reduced violaxanthin and antheraxanthin content (Burgos et al. 2012). The total carotene content in uncooked potato was 0.18 mg/100 g FW, while in cooked (steamed) sample it was 0.12 mg/100 g FW. Boiling of potatoes results in 0–17% and 0–54% retention of violaxanthin and antheraxanthin, respectively. Although present in small quantity, cooking significantly affects the carotenoid content of potato (Bembem and Sadana 2013). Clevidence (2005) reported the decrease of zeaxanthin and lutein content after cooking in some cultivars due to degradation during cooking. However, in some cultivars the lutein and zeaxanthin content of a yellow-fleshed potato was higher following steaming and microwave cooking, indicating that cooking altered the plant cell structure and liberated these compounds.

9.4 Importance of Carotenoids in Human Nutrition and Their Health Benefits

Higher animals are unable to biosynthesize carotenoids, thus carotenoids are provided only through the diet. Carotenoid derivatives are essential components for many biological functions. In plants, carotenoids are primarily required for photosynthesis, photoprotection machinery and the biosynthesis of phytohormones (Sun et al. 2018). Hundreds of carotenoids have been reported in nature, but only 40 are important in a human diet, and around 20 carotenoids have been identified in human body. The six major carotenoids commonly found in serum are β-cryptoxanthin, lutein, lycopene, zeaxanthin, α-carotene and β-carotene. In animals, carotenoids promote health through their anti-oxidative and anti-inflammatory activity and ability to enhance immune system response. Thus, carotenoids help to prevent chronic diseases, such as cancer and cardiovascular disease, preserve visual function and delay ageing (Eggersdorfer and Wyss 2018). The carotenoids are converted into vitamin A (retinol and retinal) by the action of

Table 9.2 Health benefits of carotenoids

Health benefits	Carotenoids	References
Vision	α-Carotene, β-carotene, β-cryptoxanthin	Van Vliet et al. (1996)
UV skin protection	β-Carotene, lycopene	Seddon et al. (1994), Stahl et al. (2000), Heinrich et al. (2003), Köpcke and Krutmann (2008), and Stahl and Sies (2012)
Fertility	β-Carotene, lutein	Mínguez-Alarcón et al. (2012)
Immune function enhancement	β-Carotene	Lindley (1998), Jonasson et al. (2003), and Palmer et al. (2012)
Eye health improvement	Lutein, zeaxanthin	Mares (2016)
Reducing risk of cancer	Lycopene	Wan et al. (2014)
Brain development	β-Carotene, lutein	Grodstein et al. (2007), Johnson et al. (2008), and Walk et al. (2017)
Heart health	Lycopene	Cheng et al. (2019)
Cell communication (morphogenesis and cell differentiation)	All carotenoids	Palozza and Krinsky (1992)

carotene dioxygenase. The carotenoids having at least one unmodified β-ionone ring can be cleaved to form provitamin A structure. Thus, β-carotene, α-carotene, γ-carotene and β-cryptoxanthin are classified as provitamin A active carotenoids. β-carotene is the most studied carotenoid among other 50 carotenoids that have provitamin A activity, and it shows the highest provitamin A activity. Apart from provitamin A activity, carotenoids also have potent antioxidant activity through singlet oxygen quenching and free radical deactivation mechanism which play important roles in the prevention of cancer, cardiovascular diseases and macular degeneration (Müller et al. 2016). Low levels of β-carotene in plasma increase the risk of cataract development (Knekt et al. 1992). Also, deficiency of carotenoids or lycopene in diet leads to age related macular degeneration (AMD) (Seddon et al. 1994; Mares-Perlman et al. 1995). Zeaxanthin and lutein are the major components essential for the macular pigment biosynthesis. AMD is the major cause of blindness in the elderly. High intake of zeaxanthin and lutein through the diet can help to reduce the risk of AMD development (Seddon et al. 1994; Krinsky et al. 2003). Earlier studies reported that proper intake of carotenoids could reduce the risk of cataract and AMD (Van den Berg et al. 2000). Lycopene has a role in reducing the amount of DNA damage in white blood cells and prostate tissues, thus it has a protective effect against prostate cancer (Gann and Khachik 2003). The most important biological functions of various carotenoids are listed in Table 9.2. Apart from this, carotenoids also play important roles in various cellular and organelle functions. Beta-carotene also has role in inhibition of inflammatory gene expression in lipopolysaccharide stimulated macrophages with a possible positive anti-obesity role (Williams et al. 2013). Various other activities in which carotenoids play important role are cell differentiation, anti-inflammation, embryonic development, gap junction communication, bone metabolism, anti-angiogenesis, immune modulation, antioxidant activity, anti-proliferation, haematopoiesis, skin health and apoptosis.

9.5 The Genetic Control of Potato Tuber Carotenoid Content

Although significant genotype x environment interactions have been reported for potato tuber carotenoid content, heritabilities of carotenoid content are generally high (Haynes 2010). In fact it has long been mooted that tuber yellow flesh is primarily controlled by a single dominant allele at the Y (Yellow) locus (Bonierbale et al. 1988; Thorup et al. 2000). Good evidence from several studies has shown that the Y locus maps on chromosome 3 and is likely to encode an isoform of β-carotene hydroxylase (CHY) (Brown et al. 2006; Kloosterman et al. 2010). Other enzyme activities were implicated in tuber carotenogenesis by transgenic studies which demonstrated elevated tuber carotenoid content on down-regulation of the gene encoding zeaxanthin epoxidase (ZEP) (Römer et al. 2002). An inverse relationship between *Zep* expression level and tuber carotenoid content re-enforced this suggestion (Morris et al. 2004) and recently it was shown that within a range of potato diploid genotypes with orange tuber flesh, all were homozygous for one specific *Zep* allele (Wolters et al. 2010). Whilst alleles for *Chy* and *Zep* clearly have a major impact on tuber carotenoid levels, a comprehensive survey of loci affecting tuber carotenoid content using a QTL approach in combination with a genetical genomics approach identified a further major locus and multiple minor loci (Campbell et al. 2014).

9.6 Genetic Modifications to Improve Carotenoids Content in Potato

There are multiple examples of genetic manipulation that have led to increased carotenoid concentrations in various plant species and tissues. Various genes encoding the enzymes of the carotenoid biosynthetic pathway have been cloned and characterized such as isopentenyl diphosphate/dimethylallyl diphosphate synthase isomerase, geranylgeranyl diphosphate synthase, phytoene desaturase, phytoene synthase, ζ-carotene desaturase, lycopene α- and β-cyclase, neoxanthin synthase and β-carotene hydroxylase both from plant and microbial sources (Cunningham and Gantt 1998; Al-Babili et al. 2000; Hirschberg 2001; Herbers 2003; Fraser and Bramley 2004). Total carotenoid content has been increased by over-expression of phytoene synthase in carrot (Hauptmann et al. 1997), canola, tomato (Fraser et al. 2002), Arabidopsis (Lindgren et al. 2003) and potato (Ducreux et al. 2005). The increase in total carotenoid content was sixfold in potato, 1.6-fold in tomato and 50-fold in canola. The manipulation of potato tubers through transgenic approach has been used successfully to enhance carotenoid concentrations and also increase significantly the accumulation of the spectrum of carotenoids (Römer et al. 2002; Ducreux et al. 2005). Römer et al. (2002) genetically modified two potato cultivars. The zeaxanthin to violaxanthin conversion was inhibited by transforming with sense and antisense constructs which encoding zeaxanthin epoxidase. Both techniques (antisense and co-suppression) led to increase in accumulation of zeaxanthin in potato tubers. In transgenic lines, zeaxanthin concentration was increased 4–130-fold reaching values up to 40–78 μg/g DW. This genetic manipulation also results in dramatically decrease of violaxanthin concentration, whereas in some cases antheraxanthin (monoepoxy intermediate) accumulated. Most of the transformants with higher zeaxanthin levels also showed enhanced total carotenoid concentrations (up to 5.7-fold) and some of them exhibited diminished lutein accumulation. The increased accumulation of total carotenoids indicated that the genetic modification regulates the carotenoid biosynthetic pathway in potato tubers. Ducreux et al. (2005) revealed that expression of *Erwinia uredovora* crtB gene encoding phytoene synthase in potato resulted in increased levels of total carotenoids. The tuber of *S. tuberosum* L. cultivar Desiree normally contains 5.6 μg carotenoid/g DW, while tubers of *S. phureja* cultivar "*Mayan Gold*" contain carotenoid content of typically 20 μg/g DW. In transgenic crtB Desiree lines, carotenoid levels in developing tubers reached up to 35 μg/g DW and the composition of carotenoids changed radically compared with controls. The levels of beta-carotene in the transgenic tubers reached up to 11 μg/g DW, which was significantly higher than control tubers and lutein accumulated to a level up to 19-fold higher than empty-vector transformed controls. The transformation of *S. phureja* (cv. Mayan Gold) with crtB gene also resulted in an increase in total carotenoid content up to 78 μg/g DW in the transgenic lines. The major carotenoids were violaxanthin, lutein, antheraxanthin and beta-carotene in these transgenic tubers. Interestingly, there was no increase in expression levels of the major carotenoid biosynthetic genes in the transgenic tubers, despite the large increase in carotenoid accumulation. Diretto et al. (2007b) expressed three *Erwinia* genes encoding phytoene synthase (CrtB), phytoene desaturase/carotene isomerase (CrtI) and lycopene beta-cyclase (CrtY), with a tuber specific promoter, that convert geranylgeranyl diphosphate (GGPP) into β-carotene. This led to the development of tubers with a deep yellow colour without any adverse effects on leaf morphology. Total carotenoids and β-carotene content was increased about 20-fold (114 μg/g DW) and 3600-fold (47 μg/g DW), respectively. This golden potato contains the highest carotenoid and β-carotene of any major staple food crop. Consumption of 250 g of this potato tuber flesh is adequate to provide 50% of the RDA of Vitamin A. Diretto et al. (2007b) stated that the consumption of this tuber (47 μg/g DW) is better than Golden Rice 2 with 31 μg/g DW of β-carotene. In 2006, Diretto et al. (2007b) have silenced the first step in the epsilon-beta branch of carotenoid biosynthesis pathway by introducing an antisense fragment of lycopene epsilon cyclase (LCY-e) gene under the control of the patatin promoter, via *Agrobacterium*-mediated transformation, results in increase of total carotenoids up to 2.5-fold and β-carotene content up to 14-fold, whereas silencing of the non-heme β-carotene hydroxylases CHY1 and CHY2 leads to increase in total carotenoids up to 4.5-fold and β-carotene up to 38-fold (Diretto et al. 2007a). The data suggested that lycopene epsilon cyclase (LCY-e) and β-carotene hydroxylases (CHY1 and CHY2) are key regulatory steps in potato tuber carotenogenesis. However, there was no change in neoxanthin and violaxanthin contents, whereas zeaxanthin levels decreased. RNAi technique was used to silence the β-carotene hydroxylase (bch) gene that showed a significant increase in accumulation of β-carotene and lutein in the tubers (Van Eck et al. 2007). Morris et al. (2006) reported that overexpression of a bacterial 1-deoxy-D-xylulose 5-phosphate synthase gene in potato tubers led to twofold increases in total carotenoid and six to sevenfold increase in phytoene content. The introduction of the *Or* gene under the control of a potato granule-bound starch synthase (GBSS) promoter into a potato yielded the orange-yellow flesh transgenic tubers (Van Eck et al. 2007). The total carotenoid accumulation was sixfold higher than the non-transformed control tubers. The extra orange bodies structures in chromoplasts imparted orange-yellow flesh to transgenic tubers, which provide a metabolic sink to facilitate accumulation of carotenoids. From all these studies, it was evident that metabolic engineering of potato led to increase of β-carotene, phytoene, lutein and zeaxanthin accumulation (Ducreux et al. 2005; Diretto et al. 2007a, b). Further transgenic studies aimed to engineer potato to produce the high value ketocarotenoid astaxanthin. Astaxanthin is an important high valued carotenoid produced by several fungi, a few green algae and some bacteria but only a few plants from the genus *Adonis* (Cunningham and Gantt 2005). Astaxanthin has been industrially used in poultry farming and aquaculture as a feed supplement. Consumption of astaxanthin (ketocarotenoids) is also prominently associated with a several health benefits, tested in over 65 clinical studies (Yuan et al. 2011). Commercially astaxanthin is produced from algal production systems or by chemical synthesis. In various studies, a metabolic engineering approach has been used to produce astaxanthin in transgenic plants. Early attempts limited success has been achieved to produce transgenic potato tubers biofortified with astaxanthin (Morris et al. 2006; Gerjets and Sandmann 2006). However, the selection of desired parental genotypes for transgenic development and stacking genes of carotenoid biosynthetic pathway with the cauliflower *Or* gene resulted in sixfold higher accumulation of astaxanthin in tuber then has been achieved previously (Campbell et al. 2015). Bub et al. (2008) studied zeaxanthin bioavailability from zeaxanthin-rich genetically modified potatoes in humans and stated that zeaxanthin-rich potatoes consumption significantly increased chylomicron zeaxanthin content suggesting it as an important dietary source of zeaxanthin. Although earlier studies proved that overexpression of genes of the carotenoid biosynthetic pathway resulted in increased accumulation of carotenoids in tubers, the modification of sink capacity also proven as a novel approach to increase carotenoids content in storage tissues of food crops. Application of both manipulating tools can be more effective strategy to enhance carotenoids content quantitatively and qualitatively in order to satisfy the requirement for human nutrition and health. Hence, to better understand tubers carotenogenesis regulation more detailed study of the diversity of carotenoid pigments in potatoes and environmental factors influencing their accumulation is required.

9.7 Future Prospects

Carotenoids are antioxidants with a huge pharmaceutical potential. The deficiency of carotenoid in human diet is a major problem throughout the world. Potato being an important constituent of our diets is expected to play vital role in alleviating the carotenoid malnutrition. Hence, improvements in the levels of beneficial carotenoid of potato would be the panacea for world's population to overcome carotenoid deficiency. The available genetic diversity of wild-potato species must be utilized to develop nutritionally superior traits. A full potential of biotechnological tools along with conventional plant breeding programs must be put in use to develop potatoes with superior nutritional levels. Further investigations to identifying key molecular regulators of carotenoid biosynthesis pathway may help to have deeper insights for manipulation of potato nutritional profile.

References

Akyol H, Riciputi Y, Capanoglu E et al (2016) Phenolic compounds in the potato and its byproducts: an overview. Int J Mol Sci 17(6):835

Al-Babili S, Hugueney P, Schledz M et al (2000) Identification of a novel gene coding for neoxanthin synthase from Solanum tuberosum. FEBS Lett 485(2-3):168–172

Amah D, Alamu E, Adesokan M et al (2019) Variability of carotenoids in a Musa germplasm collection and implications for provitamin A biofortification. Food Chem: X 30(2):100024

Andersson MS, Saltzman A, Virk PS, Pfeiffer WH (2017) Progress update: crop development of biofortified staple food crops under Harvest Plus. Afr J Food Agric Nutr Dev 17 (2):11905–11935

Andre C, Oufir M, Guignard C et al (2007) Antioxidant profiling of native Andean potato tubers (*Solanum tuberosum* L.) reveals cultivars with high levels of β-carotene, α-tocopherol, chlorogenic acid, and petanin. J Agric Food Chem 55:10839–10849

Bembem K, Sadana B (2013) Effect of cooking methods on the nutritional composition and antioxidant activity of potato tubers. Int J Food Nutr Sci 2(4):26–30

Bhushan B, Thomas P (1990) Effects of gamma-irradiation and storage temperature on lipoxygenase activity and carotenoid disappearance in potato tubers (*Solanum tuberosum* L.). J Agric Food Chem 38(7):1586–1590

Bijttebier S, D'Hondt E, Noten B et al (2014) Automated analytical standard production with supercritical fluid chromatography for the quantification of bioactive C17-polyacetylenes: a case study on food processing waste. Food Chem 165:371–378

Blessington T, Nzaramba MN, Scheuring DC et al (2010) Cooking methods and storage treatments of potato: effects on carotenoids, antioxidant activity, and phenolics. Am J Potato Res 87 (6):479–491

Bonierbale MW, Plaisted RL, Tanksley SD (1988) RFLP maps based on a common set of clones reveal modes of chromosomal evolution in potato and tomato. Genetics 120:1095–1103

Bonierbale M, Grüneberg W, Amoros W et al (2009) Total and individual carotenoid profiles in *Solanum phureja* cultivated potatoes: II. Development and application of near-infrared reflectance spectroscopy (NIRS) calibrations for germplasm characterization. J Food Compos Anal 22 (6):509–516

Bradshaw JE (2019) Improving the nutritional value of potatoes by conventional breeding and genetic modification. In: Quality breeding in field crops. Springer, Cham, pp 41–84

Breithaupt DE, Bamedi A (2002) Carotenoids and carotenoid esters in potatoes (*Solanum tuberosum* L.): new insights into an ancient vegetable. J Agric Food Chem 50:7175–7181

Brown CR (2005) Antioxidants in potato. Am J Potato Res 82:163–172

Brown CR, Edwards CG, Yang CP et al (1993) Orange flesh trait in potato: inheritance and carotenoid content. J Am Soc Hortic Sci 118(1):145–150

Brown CR, Culley D, Yang C et al (2005) Variation of anthocyanins and carotenoid contents and associated antioxidant values in potato breeding lines. J Am Soc Hortic Sci 130:174–180

Brown CR, Kim TS, Ganga Z et al (2006) Segregation of total carotenoid in high level potato germplasm and its relationship to beta-carotene hydroxylase polymorphism. Am J Potato Res 83:365–372

Bub A, Möseneder J, Wenzel G et al (2008) Zeaxanthin is bioavailable from genetically modified zeaxanthin-rich potatoes. Eur J Nutr 47:99–103

Burgos G, Salas E, Amoros W et al (2009) Total and individual carotenoid profiles in *Solanum phureja* of cultivated potatoes: I concentrations and relationships as determined by spectrophotometry and HPLC. J Food Compos Anal 22:503–508

Burgos G, Amoros W, Salas E et al (2012) Carotenoid concentrations of native Andean potatoes as affected by cooking. Food Chem 133(4):1131–1137

Burmeister A, Bondiek S, Apel L et al (2011) Comparison of carotenoid and anthocyanin profiles of raw and boiled *Solanum tuberosum* and *Solanum phureja* tubers. J Food Compos Anal 24 (6):865–872

Campbell R, Ducreux LJM, Morris WL et al (2010) The metabolic and developmental roles of carotenoid cleavage dioxygenase 4 from potato (*Solanum tuberosum* L). Plant Physiol 154:656–664

Campbell R, Pont SD, Morris JA et al (2014) Genome-wide QTL and bulked transcriptomic analysis reveals new candidate genes for the control of tuber carotenoid content in potato (*Solanum tuberosum* L.). Theor Appl Genet 127(9):1917–1933

Campbell R, Morris WL, Mortimer CL et al (2015) Optimising ketocarotenoid production in potato tubers: effect of genetic background, transgene combinations and environment. Plant Sci 234:27–37

Cheng HM, Koutsidis G, Lodge JK et al (2019) Lycopene and tomato and risk of cardiovascular diseases: a systematic review and meta-analysis of epidemiological evidence. Crit Rev Food Sci Nutr 59(1):141–158

Clevidence B, Haynes K, Rao D et al (2005) Effect of cooking method on xanthophyll content of yellow-fleshed potato. US Jpn Nat Resour Protein Panel 34:280–284

Cunningham FX, Gantt E (2005) A study in scarlet: enzymes of ketocarotenoid biosynthesis in the flowers of *Adonis aestivalis*. Plant J 41:478–492

Cunningham FX Jr, Gantt E (1998) Genes and enzymes of carotenoid biosynthesis in plants. Annu Rev Plant Biol 49(1):557–583

Dachtler M, Glaser T, Kohler K, Albert K (2001) Combined HPLC– MS and HPLC– NMR on-line coupling for the separation and determination of lutein and zeaxanthin stereoisomers in spinach and in retina. Anal Chem 73(3):667–674

DellaPenna D, Pogson BJ (2006) Vitamin synthesis in plants: tocopherols and carotenoids. Annu Rev Plant Biol 57:711–738

Diretto G, Al-Babili S, Tavazza R et al (2007a) Metabolic engineering of potato carotenoid content through tuber-specific overexpression of a bacterial mini-pathway. PLoS One 2(4):e350

Diretto G, Welsch R, Tavazza R et al (2007b) Silencing of beta-carotene hydroxylase increases total carotenoid and beta-carotene levels in potato tubers. BMC Plant Biol 7(1):11

Ducreux LJ, Morris WL, Hedley PE et al (2005) Metabolic engineering of high carotenoid potato tubers containing enhanced levels of beta-carotene and lutein. J Exp Bot 56:81–89

Dugo P, Herrero M, Giuffrida D et al (2008) Analysis of native carotenoid composition in orange juice using C30 columns in tandem. J Sep Sci 31:2151–2160

Dutt S, Manjul AS, Chauhan M et al (2019) Biotechnology for nutritional and associated processing quality improvement in potato. In: Nutritional quality improvement in plants. Springer, Cham, pp 429–483

Eggersdorfer M, Wyss A (2018) Carotenoids in human nutrition and health. Arch Biochem Biophys 652:18–26

Emenhiser C, Sander LC, Schwartz SJ (1995) Capability of a polymeric C30 stationary phase to resolve cis-trans carotenoid isomers in reversed-phase liquid chromatography. J Chromatogr A 707(2):205–216

Ezekiel R, Singh N, Sharma S et al (2013) Beneficial phytochemicals in potato—a review. Food Res Int 50(2):487–496

Fernandez-Orozco R, Gallardo-Guerrero L, Hoirnero-Méndez D (2013) Carotenoid profiling in tubers of different potato (Solanum sp) cultivars: accumulation of carotenoids mediated by xanthophylls esterification. Food Chem 141:2864–2872

Fogelman E, Oren-Shamir M, Hirschberg J et al (2019) Nutritional value of potato (*Solanum tuberosum*) in hot climates: anthocyanins, carotenoids, and steroidal glycoalkaloids. Planta 249(4):1143–1155

Fraser PD, Bramley PM (2004) The biosynthesis and nutritional uses of carotenoids. Prog Lipid Res 43(3):228–265

Fraser PD, Romer S, Shipton CA, Mills PB, Kiano JW, Misawa N, Drake RG, Schuch W, Bramley PM (2002) Evaluation of transgenic tomato plants expressing an additional phytoene synthase in a fruit-specific manner. Proc Natl Acad Sci 99(2):1092–1097

Frohlich, Conrad, Schmid et al (2007) Isolation and structural elucidation of different geometrical isomers of lycopene. Int J Vitam Nutr Res 77(6):369–375

Gann PH, Khachik F (2003) Tomatoes or lycopene versus prostate cancer: is evolution anti-reductionist? CancerSpectrum Knowl Environ 95:1563–1565

Gerjets T, Sandmann G (2006) Ketocarotenoid formation in transgenic potato. J Exp Bot 57 (14):3639–3645

Griffiths DW, Dale MF, Morris WL et al (2007) Effects of season and postharvest storage on the carotenoid content of *Solanum phureja* potato tubers. J Agric Food Chem 55(2):379–385

Grodstein F, Kang JH, Glynn RJ et al (2007) A randomized trial of beta carotene supplementation and cognitive function in men: the Physicians' Health Study II. Arch Intern Med 167 (20):2184–2190

Gupta P, Sreelakshmi Y, Sharma R (2015) A rapid and sensitive method for determination of carotenoids in plant tissues by high performance liquid chromatography. Plant Methods 11(1):5

Hamouz K, Pazderu K, Lachman J et al (2016) Effect of cultivar, flesh colour, locality and year on carotenoid content in potato tubers. Plant Soil Environ 62:86–91

Hart DJ, Scott KJ (1995) Development and evaluation of an HPLC method for the analysis of carotenoids in foods, and the measurement of the carotenoid content of vegetables and fruits commonly consumed in the UK. Food Chem 54(1):101–111

Hauptmann R, Eschenfeldt WH, English J, Inventors, BP Corporation North America Inc, assignee, et al (1997) Enhanced carotenoid accumulation in storage organs of genetically engineered plants. United States patent US 5,618,988

Haynes KG (2010) Genotype-environment interactions for potato tuber carotenoid content. J Am Soc Hortic Sci 135(3):250–258

Haynes KG, Clevidence BA, Rao D et al (2011) Inheritance of carotenoid content in tetraploid x diploid potato crosses. J Am Soc Hortic Sci 136(4):265–272

Heinrich U, Gärtner C, Wiebusch M et al (2003) Supplementation with β-carotene or a similar amount of mixed carotenoids protects humans from UV-induced erythema. J Nutr 133 (1):98–101

Hejtmánková K, Kotíková Z, Hamouz K et al (2013) Influence of flesh colour, year and growing area on carotenoid and anthocyanin content in potato tubers. J Food Compos Anal 32(1):20–27

Herbers K (2003) Vitamin production in transgenic plants. J Plant Physiol 160(7):821–829

Hirschberg J (2001) Carotenoid biosynthesis in flowering plants. Curr Opin Plant Biol 4 (3):210–218

Iwanzik W, Tevini M, Stute R, Hilbert R (1983) Carotinoidgehalt undzusammensetzung verschiedener deutscherKartoffelsorten und deren Bedeutung für die Fleischfarbe derKnolle. Potato Res 26:149–162

Johnson EJ, McDonald K, Caldarella SM et al (2008) Cognitive findings of an exploratory trial of docosahexaenoic acid and lutein supplementation in older women. Nutr Neurosci 11:75–83

Jonasson L, Wikby A, Olsson AG (2003) Low serum β-carotene reflects immune activation in patients with coronary artery disease. Nutr Metab Cardiovasc Dis 13(3):120–125

Khachik F, Spangler CJ, Smith JC et al (1997) Identification, quantification, and relative concentrations of carotenoids and their metabolites in human milk and serum. Anal Chem 69:1873–1881

Kloosterman B, Oortwijn M, uit de Willigen J et al (2010) From QTL to candidate gene: genetical genomics of simple and complex traits in potato using a pooling strategy. BMC Genomics 11:158

Knekt P, Heliövaara M, Rissanen A et al (1992) Serum antioxidant vitamins and risk of cataract. BMJ 305(6866):1392–1394

Kobayashi A, Ohara-Takada A, Tsuda S et al (2008) Breeding of potato variety 'Inca-no-hitomi' with a very high carotenoid content. Breed Sci 58:77–82

Köpcke W, Krutmann J (2008) Protection from sunburn with β-carotene-a meta-analysis. Photochem Photobiol 84(2):284–288

Kopec RE, Cooperstone JL, Cichon MJ et al (2012) Analysis methods of carotenoids. In: Analysis of antioxidant-rich phytochemicals. Wiley-Blackwell, Hoboken, pp 105–149

Kotikova Z, Hejtmankova A, Lachman J et al (2007) Effect of selected factors on total carotenoid content in potato tubers (*Solanum tuberosum* L.). Plant Soil Environ 53(8):355

Kotíková Z, Šulc M, Lachman J, Pivec V, Orsák M, Hamouz K (2016) Carotenoid profile and retention in yellow-, purple- and red-fleshed potatoes after thermal processing. Food Chem 197:992–1001

Krinsky NI, Landrum JT, Bone RA (2003) Biologic mechanisms of the protective role of lutein and zeaxanthin in the eye. Annu Rev Nutr 23(1):171–201

Lachman J, Hamouz K, Musilová J et al (2013) Effect of peeling and three cooking methods on the content of selected phytochemicals in potato tubers with various colour of flesh. Food Chem 138 (2-3):1189–1197

Lachman J, Hamouz K, Orsák M et al (2016) Carotenoids in potatoes–a short overview. Plant Soil Environ 62(10):474–481

LaFountain AM, Pacheco C, Prum RO et al (2013) Nuclear magnetic resonance analysis of carotenoids from the burgundy plumage of the Pompadour Cotinga (*Xipholena punicea*). Arch Biochem Biophys 539(2):133–141

Li H, Deng Z, Wu T et al (2012) Microwave-assisted extraction of phenolics with maximal antioxidant activities in tomatoes. Food Chem 130:928–936

Lindgren LO, Stålberg KG, Höglund AS (2003) Seed-specific overexpression of an endogenous Arabidopsis phytoene synthase gene results in delayed germination and increased levels of carotenoids, chlorophyll, and abscisic acid. Plant Physiol 132(2):779–785

Lindley MG (1998) The impact of food processing on antioxidants in vegetable oils, fruits and vegetables. Trends Food Sci Technol 9(8-9):336–340

Lu W, Haynes K, Wiley E et al (2001) Carotenoid content and color in diploid potatoes. J Am Soc Hortic Sci 126:722–726

Luterotti S, Kljak K (2010) Spectrophotometric estimation of total carotenoids in cereal grain products. Acta Chim Slov 57(4):781–787

Mares J (2016) Lutein and zeaxanthin isomers in eye health and disease. Annu Rev Nutr 36:571–602

Mares-Perlman JA, Brady WE, Klein R et al (1995) Serum antioxidants and age-related macular degeneration in a population-based case-control study. Arch Ophthalmol 113(12):1518–1523

Maurer MM, Mein JR, Chaudhuri SK et al (2014) An improved UHPLC-UV method for separation and quantification of carotenoids in vegetable crops. Food Chem 165:475–482

Mínguez-Alarcón L, Mendiola J, López-Espín JJ et al (2012) Dietary intake of antioxidant nutrients is associated with semen quality in young university students. Hum Reprod 27(9):2807–2814

Morris WL, Ducreux L, Griffiths DW et al (2004) Carotenogenesis during tuber development and storage in potato. J Exp Bot 55(399):975–982

Morris WL, Ducreux LJ, Fraser PD et al (2006) Engineering ketocarotenoid biosynthesis in potato tubers. Metab Eng 8(3):253–263
Müller L, Caris-Veyrat C, Lowe G et al (2016) Lycopene and its antioxidant role in the prevention of cardiovascular diseases-a critical review. Crit Rev Food Sci Nutr 56(11):1868–1879
Othman R (2009) Biochemistry and genetics of carotenoid composition in potato tubers. Doctoral dissertation, Lincoln University
Palermo M, Pellegrini N, Fogliano V (2014) The effect of cooking on the phytochemical content of vegetables. J Sci Food Agric 94(6):1057–1070
Palmer AC, West KP, Dalmiya N et al (2012) The use and interpretation of serum retinol distributions in evaluating the public health impact of vitamin A programmes. Public Health Nutr 15:1201–1215
Palozza P, Krinsky NI (1992) Antioxidant effects of carotenoids in vivo and in vitro: an overview. Methods Enzymol 213:403–420
Payyavula RS, Navarre DA, Kuhl JC et al (2012) Differential effects of environment on potato phenylpropanoid and carotenoid expression. BMC Plant Biol 12(1):39
Rao AV, Rao LG (2007) Carotenoids and human health. Pharmacol Res 55(3):207–216
Rivera S, Vilaró F, Canela R (2011) Determination of carotenoids by liquid chromatography/mass spectrometry: effect of several dopants. Anal Bioanal Chem 400(5):1339–1346
Römer S, Lubeck J, Kauder F et al (2002) Genetic engineering of a zeaxanthin-rich potato by antisense inactivation and co-suppression of carotenoid epoxidation. Metab Eng 4:263–272
Rothwell JA, Medina-Remón A, Pérez-Jiménez J et al (2015) Effects of food processing on polyphenol contents: a systematic analysis using Phenol-Explorer data. Mol Nutr Food Res 59(1):160–170
Sander LC, Sharpless KE, Craft NE et al (1994) Development of engineered stationary phases for the separation of carotenoid isomers. Anal Chem 666:1667–1674
Sander LC, Sharpless KE, Pursch M (2000) C30 stationary phases for the analysis of food by liquid chromatography. J Chromatogr A 880:189–202
Seddon JM, Ajani UA, Sperduto RD, Hiller R, Blair N, Burton TC, Farber MD, Gragoudas ES, Haller J, Miller DT, Yannuzzi LA (1994) Dietary carotenoids, vitamins A, C, and E, and advanced age-related macular degeneration. JAMA 272(18):1413–1420
Smallcombe SH, Patt SL, Keifer PA (1995) WET solvent suppression and its applications to LC NMR and high-resolution NMR spectroscopy. J Magn Reson 117(2):295–303
Stahl W, Sies H (2005) Bioactivity and protective effects of natural carotenoids. BBA-Mol Basis Dis 1740(2):101–107
Stahl W, Sies H (2012) β-Carotene and other carotenoids in protection from sunlight. Am J Clin Nutr 96(5):1179S–1184S
Stahl W, Heinrich U, Jungmann H et al (2000) Carotenoids and carotenoids plus vitamin E protect against ultraviolet light–induced erythema in humans. Am J Clin Nutr 71(3):795–798
Sulli M, Mandolino G, Sturaro M, Onofri C, Diretto G, Parisi B, Giuliano G (2017) Molecular and biochemical characterization of a potato collection with contrasting tuber carotenoid content. PLoS One 12(9):e0184143
Sun T, Yuan H, Cao H et al (2018) Carotenoid metabolism in plants: the role of plastids. Mol Plant 11(1):58–74
Taleuzzaman M, Ali S, Gilani SJ et al (2015) Ultra performance liquid chromatography (UPLC) - a review. Austin J Anal Pharm Chem 2(6):1056
Tatarowska B, Milczarek D, Wszelaczyńska E et al (2019) Carotenoids variability of potato tubers in relation to genotype, growing location and year. Am J Potato Res 96:493–504
Thorup TA, Tanyolac B, Livingstone KD et al (2000) Candidate gene analysis of organ pigmentation loci in the Solanaceae. Proc Natl Acad Sci U S A 97:11192–11197
Tian J, Chen J, Lv F et al (2016) Domestic cooking methods affect the phytochemical composition and antioxidant activity of purple-fleshed potatoes. Food Chem 197:1264–1270
Tierno R, Hornero-Méndez D, Gallardo-Guerrero L et al (2015) Effect of boiling on the total phenolic, anthocyanin and carotenoid concentrations of potato tubers from selected cultivars and introgressed breeding lines from native potato species. J Food Compos Anal 41:58–65

Tode C, Maoka T, Sugiura M (2009) Application of LC-NMR to analysis of carotenoids in foods. J Sep Sci 32(21):3659–3663
Updike AA, Schwartz SJ (2003) Thermal processing of vegetables increases cis isomers of lutein and zeaxanthin. J Agric Food Chem 51(21):6184–6190
Valcarcel J, Reilly K, Gaffney M, O'Brien N (2015) Total carotenoids and l-ascorbic acid content in 60 varieties of potato (*Solanum tuberosum* L.) grown in Ireland. Potato Res 58(1):29–41
Van den Berg H, Faulks R, Granado HF et al (2000) The potential for the improvement of carotenoid levels in foods and the likely systemic effects. J Sci Food Agr 80(7):880–912
Van Eck J, Conlin B, Garvin DF et al (2007) Enhancing beta-carotene content in potato by RNAi-mediated silencing of the beta-carotene hydroxylase gene. Am J Potato Res 84(4):331
Van Vliet T, van Vlissingen MF, van Schaik F et al (1996) β-Carotene absorption and cleavage in rats is affected by the vitamin A concentration of the diet. J Nutr 126(2):499–508
VanBoekel M, Fogliano V, Pellegrini N et al (2010) A review on the beneficial aspects of food processing. Mol Nutr Food Res 54(9):1215–1247
Varela J, Pereira H, Vila M et al (2015a) Production of carotenoids by microalgae: achievements and challenges. Photosynth Res 127(2):285–286
Varela JC, Pereira H, Vila M et al (2015b) Production of carotenoids by microalgae: achievements and challenges. Photosynth Res 125(3):423–436
Walk AM, Khan NA, Barnett SM et al (2017) From neuro-pigments to neural efficiency: the relationship between retinal carotenoids and behavioral and neuroelectric indices of cognitive control in childhood. Int J Psychophysiol 118:1–8
Wan L, Tan HL, Thomas-Ahner JM et al (2014) Dietary tomato and lycopene impact androgen signaling-and carcinogenesis-related gene expression during early TRAMP prostate carcinogenesis. Cancer Prev Res 7(12):1228–1239
Weller P, Breithaupt DE (2003) Identification and quantification of zeaxanthin esters in plants using liquid chromatography-mass spectrometry. J Agric Food Chem 51:7044
Williams DJ, Edwards D, Hamernig I et al (2013) Vegetables containing phytochemicals with potential anti-obesity properties: a review. Food Res Int 52(1):323–333
Wolters AMA, Uitdewilligen JGAML, Kloosterman BA et al (2010) Identification of alleles of carotenoid pathway genes important for zeaxanthin accumulation in potato tubers. Plant Mol Biol 73:659–671
Yuan JP, Peng J, Yin K et al (2011) Potential health-promoting effects of astaxanthin: a high-value carotenoid mostly from microalgae. Mol Nutr Food Res 55(1):150–165

Anthocyanins: Coloured Bioactive Compounds in Potatoes

10

Tanuja Mishra, Satish Kumar Luthra, Pinky Raigond, and Bandana

Abstract

In potatoes, anthocyanins are present in the flesh and skin of several purple and red fleshed landraces found in the Andes. They are also present in varieties bred in several countries for their health promoting abilities as they have been characterized for strong anti-oxidative activity, anti-influenza virus activity and anti-stomach cancer activity. Several authors have reported the antioxidant potential of potato and the phytochemicals responsible for the same have free-radical scavenging activity, which not only delays ageing process by preventing neuronal degeneration but also reduces the risk of various chronic diseases. Coloured varieties have a denser texture and slightly nuttier, earthier flavour than white potatoes. Purple- or red-fleshed potatoes are a tasty way to add a pop of colour to meal while enjoying serving of health benefits.

Keywords

Anthocyanins · Structure · Cooking · Storage · Health benefits · Genetic modifications

10.1 Introduction

Anthocyanins are found in variety of species of plant kingdom as one of the most important and largest group of water soluble pigments. They contribute towards different kinds of pigmentation, viz. purple, red, orange and blue in many

T. Mishra · P. Raigond
ICAR - Central Potato Research Institute, Shimla, Himachal Pradesh, India

S. K. Luthra (✉) · Bandana
ICAR - Central Potato Research Institute, Meerut, Uttar Pradesh, India
e-mail: skluthra@hotmail.com

P. Raigond et al. (eds.), *Potato*, https://doi.org/10.1007/978-981-15-7662-1_10

vegetables, flowers and fruits due to presence of chromophores in them (Smeriglio et al. 2016). Fruits containing anthocyanins include blackberry, blueberry, raspberry, cherry and bilberry, while vegetables include red coloured radish, onion, lettuce, cabbage and potato as well as eggplant and purple or orange sweet potato.

Potato (*Solanum tuberosum*) is known to be one of the most important staple foods in the world and is a rich source of starch, vitamins C and B6. Starchy tubers of potato cultivars comprise of a huge biodiversity including red and purple coloured types. Different colours of peels or flesh are found in potatoes owing to the presence of pigment called 'anthocyanin' that accumulates in vacuoles and impart red or purple colouration to different varieties of potato (Fossen and Andersen 2000). Potatoes rich in anthocyanins (purple-fleshed or red-fleshed potatoes) are known to be associated with health benefits. Anthocyanins from potato are regarded as health promoters as they have been characterized for strong anti-oxidative activity (Ishii et al. 1996), anti-stomach cancer and anti-influenza virus activity (Hayashi et al. 2006). Several authors have reported the antioxidant potential of potato and the phytochemicals responsible for the same have free-radical scavenging activity, which not only delays ageing process by preventing neuronal degeneration but also reduces the risk of various chronic diseases (Teow et al. 2007; Zarzecka and Gugala 2011). Coloured potatoes are more beneficial to human health and hence are bred for consumption and processing industry (Lachman et al. 2013).

10.2 Structure of Anthocyanins

Around 600 molecular structures have been identified in different anthocyanins that impart red, purple or blue colour to various plant parts like flowers, fruits or vegetables (Kita et al. 2013). The basic structure of anthocyanin includes two benzyl rings, i.e. 2-phenylbenzopyrilium containing glycosylation with multiple hydroxyl or methoxyl groups (He and Giusti 2010; Zhao et al. 2017). Different types of anthocyanins occur due to differences in number of sugar groups attached with the basic glucoside unit (Andersen and Markham 2005) and the colours are the measure of type of functional group attached. Usually reddish shades are imparted by more methoxyl groups, while blue colour is indicative of presence of more hydroxyl groups (Heredia et al. 1998; He and Giusti 2010). Major anthocyanins present in the red-fleshed and purple-fleshed potatoes are acylated with hydroxycinnamic acid (Fossen and Andersen 2000). Some studies suggest that the anthocyanins found in dark-fleshed potatoes are glycosylated and acylated with ferulic acid and p-coumaric acid. Glucoside derivatives of anthocyanidins (petunidin, pelargonidin, cyanidin, peonidin, malvidin and delphinidin) have been observed in coloured potato varieties, viz. Vitelotte, Highland Burgundy Red, Shetland Black and Hermanns Blaue (Eichhorn and Winterhalter 2005). The anthocyanidins are basically anthocyanins in de-glycosylated forms (aglycones) and are specifically distributed among different pigmentation like malvidin, petunidin, delphinidin and peonidin in purple-fleshed potatoes, while pelargonidin is found in tubers with red-flesh colour (Kita et al. 2013). Although the type of anthocyanidin alters with respect to the flesh colour but acylation process with ferulic acid and p-coumaric acid remains same like, in potatoes with purple-flesh, petunidin and malvidin-3-rutinoside-5-glycosides are acylated, whereas for red-fleshed ones, pelargonidin and peonidin-3-rutinoside-5-glycosides get acylated (Lewis et al. 1998). Mori et al. (2010) reported that total anthocyanin content in red potato cultivar, Kintoki-imo, was in the range of 2–816 mg/100 g FW. Broadly the major anthocyanins present were pelargonidin 3-(6-O-(4-O-(E)-p-coumaroyl-α-L-rhamnopyranosyl)-β-D-glucopyranoside)-5-β-D-glucopyranoside (pelanin), and pelargonidin 3-(6-O-(4-O-(E)-feruloyl-α-L-rhamnopyranosyl)-β-D-glucopyranoside)-5-β-D-glucopyranoside. Following five types were postulated as per the pattern of anthocyanin in varieties-

1. Type A: having pelanin and Pg 3-Fr.Rut-5-Glc in major quantity
2. Type B: containing pelanin and Pg 3-Fr.Rut-5-Glc in major quantity like type A, but also having purple pigments, viz. peonanin and Pn 3-Rut-5-Glc
3. Type C: similar to type B, but with higher peonanin content
4. Type D: containing pelanin as the major pigment and no peonanin and
5. Type E: containing petanin as the major pigment, but no pelanin.

Hillebrand et al. (2009) reported that pigment in potato varieties with blue flesh, viz. Hermanns Blaue, Vitelotte, Shetland Black and Valfi is mainly attributed by 3-p-coumaroylrutinoside-5-glucosides of peonidin, malvidin and petunidin. In red fleshed varieties major anthocyanins identified were cyanidin, peonidin and pelargonidin, whereas in purple fleshed varieties, petunidin and malvidin were also present besides these three predominant anthocyanins (Giusti et al. 2014).

Anthocyanin profile of 12 purple-fleshed and 6 red-fleshed accession from International Potato Centre's genebank showed that purple-fleshed accessions have petunidin-3-(coumaroyl) rutinoside-5-glucoside (petanin) as the predominant anthocyanin, that represents from 37 to 78% of the total anthocyanins (ex CIP 705534 and CIP 702363). Pelargonidin- 3(coumaryl) rutinoside-5-glucoside is the predominant anthocyanin in red-fleshed accessions, representing 41–75% of the total anthocyanins (CIP 703625 and CIP 702453) (Burgos 2014).

10.3 Most Preferred Methods for Estimation of Anthocyanins

Anthocyanin pigments can be extracted to their maximum amount by employing acetone, an organic solvent mixed with alcohol (ethanol or methanol). The amount or quality of anthocyanin following extraction may be determined by spectrophotometric method or chromatography techniques. In acidic medium ($pH \leq 2$), anthocyanin pigments show maximum absorption between 260 and 280 nm (in UV region), 415 nm and 490–540 nm (in visible range) because absorption spectra of anthocyanins is pH dependent. The wavelength at which maximum absorption is observed varies slightly with respect to the differences in structure of individual anthocyanin. However, in general, 520–540 nm is the most preferred wavelength for spectrophotometric method of total anthocyanins' estimation (Harborne 1958).

As spectrophotometry is rapid and easy method, it is often used in routine lab analysis to determine anthocyanin content in different plant materials as it does not require tissue homogenization or any other complex processing of raw material. But this method is not suitable for precise characterization or molecular identification of particular anthocyanins in the given plant material. Among sophisticated techniques, most widely employed methods include diode array detection (DAD), MS or tandem mass spectrometry (MS/MS) and HPLC (Barnes et al. 2009; Jin et al. 2015). Also, recent advances in techniques like NMR, HPLC coupled with MS, MS/MS are also useful in structure elucidation and individual pigment identification (Jin et al. 2015; Li et al. 2016).

10.4 Anthocyanin Content in Potatoes

Anthocyanins are water soluble plant pigments that belong to a class of flavonoids and constitute largest group of phenolic pigments (Gould et al. 2008; Fang 2015). High concentration of anthocyanins is present in flesh and peel of several purple- and red-fleshed potatoes found in Andes. Cultivar-dependent variations in anthocyanin structures and concentrations have been observed (Brown et al. 2003). The highest anthocyanin concentration reported in a dark purple-fleshed potato (above 400 mg/ 100 g FW, Andre et al. 2007) is lower than in blueberries, cranberries, eggplant and purple corn. However, due to high intake of potatoes, i.e. upto 500 g/day in area like Andean highlands, contribution of purple potatoes' anthocyanin to diet is quite high than other sources of anthocyanins.

Various reports are available on anthocyanin concentration in white/cream and coloured potatoes (Table 10.1). Total anthocyanin concentrations in Andean potatoes ranged from 14–16,330 μg/g DW (Andre et al. 2007). The investigation on 38 cultivars selected for high pigment expression at CIP, Lima (Brown et al. 2008) revealed significantly higher mean levels of total anthocyanins in tetraploids (9.2 mg/100 g FTW) as compared to diploids (2.6) and triploids (5.5). The higher anthocyanin is known to prevalent in red and purple flesh potatoes. Amount of total anthocyanins varies according to flesh colour as in purple-fleshed potatoes nearly double amount (20–38 mg/100 g) was observed as compared to 17–20 mg/100 g in those with red flesh colour (Brown et al. 2005). Raw and cooked purple-fleshed potatoes contain 63 to 588 mg/100 g FW and from 71 to 453 mg/100 g FW, of total anthocyanins, respectively (Burgos et al. 2013). Whereas cooked red-fleshed potatoes contain 8.2 to 55.3 mg/100 g FW total anthocyanins (Burgos 2014). Cultivars with solidly pigmented flesh may have as much as 40 mg/100 g FW of total anthocyanins compared to a red or purple skinned white fleshed cultivar with 1.5 mg per 100 g FW.

The amount of anthocyanin present can be reflected by the richness of flesh colour in potato as red or purple flesh colour shows four to five fold better antioxidant capacity due to higher anthocyanin content (61.5–573.5 cyanidin mg/kg) when compared to tubers with yellow or white flesh colour (Hamouz et al. 2011). In red fleshed varieties, the amount of total anthocyanins ranged from 8.2 to 90.9 pg-3-glu

Table 10.1 Range of total anthocyanins reported in different potato varieties

Potato type	Reported range	Reference
Diploid	2.6 mg/100 g FTW	Brown et al. (2008)
Triploid	5.5 mg/100 g FTW	Brown et al. (2008)
Tetraploid	9.2 mg/100 g FTW	Brown et al. (2008)
White flesh	1.5 mg/100 g FTW	Brown et al. (2008)
Red fleshed	22 mg/100 g FTW	Lewis et al. (1998)
	17–20 mg/100 g FTW	Brown et al. (2005)
	158.78 cyanidin mg/kg	Hamouz et al. (2011)
	8.2–90.9 pg-3-glu quivalents/100 g DW	Giusti et al. (2014)
Purple flesh	368 mg/100 g FTW	Lewis et al. (1998)
	20–38 mg/100 g FTW	Brown et al. (2005)
	40 mg/100 g FTW	Brown et al. (2008)
	235.4 cyanidin mg/kg	Hamouz et al. (2011)
	16.8–152.7 mg cy-3-glu equivalents/100 g DW	Giusti et al. (2014)
	588 mg/100 g FW	Burgos (2014)
Red/purple flesh	61.5 to 573.5 cyanidin mg/kg	Hamouz et al. (2011)
Dark coloured varieties	1.5–48 mg/100 g FW	Tierno et al. (2016)
	80.9–269.67 mg/ 100 g	Dalamu et al. (2015)
Outer skin of tuber	0.65 g/kg FW	Jansen and Flamme (2006)
Flesh of tuber	0.22 g/kg FW	Jansen and Flamme (2006)
Processing varieties	18.6–22.9 mg/100 g DW	Furrer et al. (2017)

equivalents/100 g DW. Similarly, in purple fleshed varieties, range was 16.8–152.7 mg of cy-3-glu equivalents/100 g DW (Giusti et al. 2014). Recent study by Dalamu et al. (2015) in a few Indian varieties, revealed that genotypes with darker flesh colour (CP4242, Lal Mitti-2, Barielly Red) had maximum anthocyanin content ranging from 21 to 27 mg/100 g FW on whole tuber basis, which indicates that potato colour is directly proportional to the amount of anthocyanins present.

In dark fleshed breeding lines, total anthocyanin content (1.5–48.0 mg/100 g FW) is usually three to fourfolds higher than white fleshed lines that attract consumers due to high antioxidant potential (Tierno et al. 2016). Similarly, in processing varieties total anthocyanin content has been reported in the range of 18.6–22.9 mg/100 g DW (Furrer et al. 2017). Luthra et al. (2018) identified red purple skin/flesh coloured clone MSP/15–51, which possessed two time higher anthocyanin content than its female parent Bareilly Red (cream flesh colour with red broad vascular bundle) and four time higher anthocyanin content in male parent CP3770 (yellow flesh colour).

Anthocyanins are distributed unevenly throughout the potato tubers. Lewis et al. (1998) reported that concentration of flavonoids was almost double in purple or red-fleshed cultivars compared to white-fleshed cultivars and the concentration was considerably high in peel that was almost 900 mg/100 g FW in purple-fleshed and 500 mg/100 g FW in red-fleshed potatoes. It has been reported that they are denser in the outer skin (0.65 g/kg FW) than that in flesh (0.22 g/kg FW) or whole potato tuber (Jansen and Flamme 2006). Reyes et al. (2005) also reported that total anthocyanin content in red- or purple-fleshed potatoes to be 0.9 folds higher in peel than flesh thereby accounting for about 20% of the total content.

10.5 Effect of Growing Conditions, Storage As Well As Cooking on Anthocyanins

The amount of various bioactives present in any food item mainly depends upon enzyme activity, pH, storage conditions (like light, oxygen), cooking or processing methods (Palermo et al. 2014). Longer days and cooler temperatures favoured about 2.5–1.5× higher anthocyanin and phenolic content (Reyes et al. 2004). Reyes et al. 2004 investigated the changes in the content and yield of anthocyanins and total phenolics during development of purple- and red-flesh potato and reported that concentration of anthocyanins and phenolics decreases with the tuber growth and maturity. Anthocyanin concentrations increase with increase in height above sea level, higher annual sum of precipitation and lower annual average temperatures (Lachman et al. 2009). Jansen and Flamme (2006) after 2 years' study on anthocyanin contents of 23 cultivars reported that the anthocyanin content varied non-significantly even though the weather conditions during plant growth were different during the 2 years. Kotikova et al. (2007) reported that fertilizers affected the anthocyanin content of potatoes non-significantly. Dark storage of wounded cultivar All-Blue showed more anthocyanin concentration compared to storage in light (Reyes and Cisneros-Zevallos 2003). Foliar application of 2,4-D at low dose intensified skin color, improved appearance, and reduced oversized tubers in several red potato cultivars (Nelson and Bristol 1975).

Jansen and Flamme (2006) suggested that during storage upto 135 days, the anthocyanin content remains unchanged without any significant variation. Hamouz et al. (2018) studied the effect of long-term cold storage and different location on the total anthocyanin content in 12 coloured potato cultivars and observed that the anthocyanin content is greatly enhanced (around 1.25 fold) in cooler climatic location when compared with the warmer ones. Also, the period of water stress at both the locations enhanced the content. Similarly, 6 months storage at low temperature of 4 °C affected different varieties variably, i.e. total anthocyanin content increased in some cultivars while decreased in others may be due to genetic makeup of cultivar or differences in sugar accumulating behaviour during cold storage. The anthocyanin concentration of purple potatoes decreases at high temperatures (100–150 °C) (Nayak et al. 2011). In contrast, compared to raw boiling, steaming, baking or microwaving showed higher anthocyanin contents in some red-fleshed and purple-fleshed potatoes (Lachman et al. 2013).

The concentration of total anthocyanins in potato has been observed to increase when exposed to different cooking treatments (Lachman et al. 2013). On deep frying, as in case of French fries, potatoes have been found to show higher anthocyanin content that may be due to thermal transfer of components between oil and fries (Lemos and Sivaramareddy 2013). The anthocyanin content in potato varieties with red flesh has been observed to get reduced to only 25% during processing (peeling and pre-drying process) of dehydrated potato cubes. But low temperature processing as well as storage may reduce these losses (Delgado-Vargas and Paredes-López 2003). It is noteworthy that increase in temperature results in gradual degradation of anthocyanin pigments to chalcones and other brown products (Delgado-Vargas et al. 2000). Thermal exposure of potatoes during processing is well known to decrease bioactives' concentration, deactivation of various enzymes and results in loss of anthocyanins that in turn results in brown texture of product (Patras et al. 2010; Tian et al. 2016). Perla et al. (2012) reported that although cooking decreases the level of anthocyanins in 14 potato varieties, but those containing coloured flesh (red or purple) retained good concentration even after cooking. On the contrary, Lachman et al. (2013) reported that concentration of total anthocyanin remained constant with no significant decrease with cooking through any method such as microwaving, boiling and baking. Similarly, no significant losses in the anthocyanin content have been observed in Andean varieties when exposed to various cooking methods (Navarre et al. 2010; Andre et al. 2015). Cooking has been reported to release the anthocyanins from the food matrix hence these antioxidants show high recovery after cooking (Tian et al. 2016).

Burgos et al. (2013) compared anthocyanin content between different purple fleshed Andean varieties of *Solanum andigenum* (both raw and boiled in freeze dried form). It was observed that maximum anthocyanin content (418 mg/100 g FW) was observed in deep purple fleshed boiled sample (Guincho). Frying process has been reported to decrease total anthocyanin content from 38% to 70% in potato varieties with coloured (red or purple) flesh. Further, malvidin and pelargonidin derivatives were observed to be comparatively more stable than petunidin derivatives during frying (Kita et al. 2013). Effect of cooking on anthocyanin content reported in literature has been compiled in Fig. 10.1.

10.6 Benefits of Anthocyanins

Anthocyanins confer numerous advantages not only to the human consumers but also to the plant in which they are present (Camire et al. 2009). As anthocyanins are present in variety of fruits and vegetables, they are widely ingested by humans, either directly or in the form of red wines (Galvano et al. 2004). With the advent in food industry, juices and extracts with maximum anthocyanin content are commercially available that are gaining popularity thereby increasing daily intake of anthocyanins (Rodríguez-Werner et al. 2019). In fresh and processed fruits and vegetables, anthocyanins improve the overall appearance of food, and simultaneously also contribute to consumers' health and well-being (Stintzing and Carle 2004). Anthocyanins' daily intake has been estimated around 180 mg (Galvano et al. 2004).

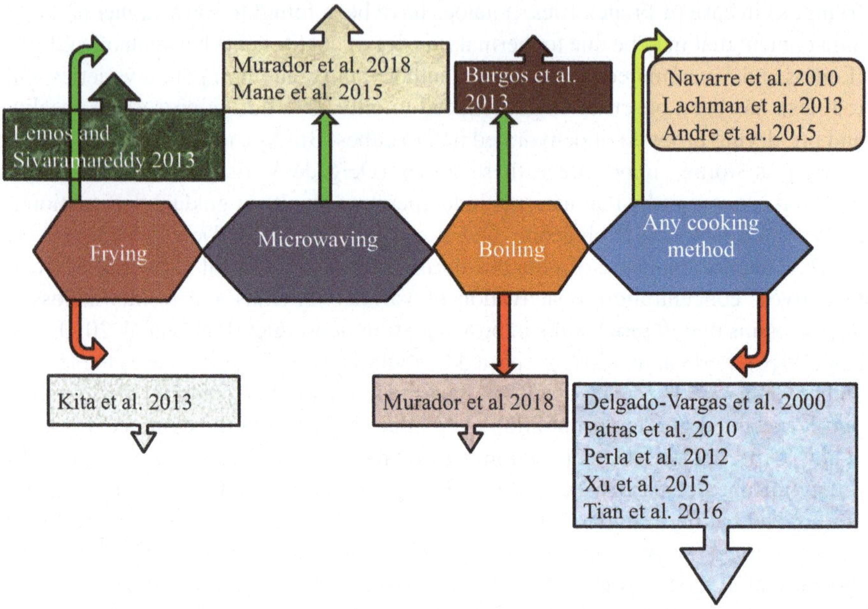

Fig. 10.1 Effect of cooking methods on the concentration of total anthocyanins in potato. Green arrows and up arrow boxes indicate increase in anthocyanin concentration after cooking, while red arrows and down arrow boxes indicate decrease in concentration. Yellow arrow or simple box indicates no significant loss or gain in anthocyanin

10.6.1 Appealing and Better Prices in the Market

Tuber skin colour and appearance are the most important quality factors considered by consumers in purchasing specialty potatoes (Jemison Jr et al. 2008). Specialty potatoes usually command a higher price, subject to size, quality and packaging as per consumer expectations. Potatoes cultivated and consumed in most parts of South America, i.e. the native home of potatoes and Europe exhibits more diversity. In recent years, consumers are showing interest in dark fleshed potato varieties thereby leading to release of new cultivars with gradual improvement in their quality (Oertel et al. 2017). Purple potatoes are eye-catching and some common varieties of the region include Purple Peruvian, All Blue, Congo, Purple Majesty, Purple Fiesta, Adirondack Blue and Vitelotte. In India, Kufri Neelkanth is the purple skinned and yellow-fleshed specialty potato variety having high antioxidants known to be beneficial for human health (Fig. 10.2). Due to the functional edible pigments, coloured potato cultivars 'Hongyoung' and 'Jayoung' have been recommended nationwide in Korea. These potatoes have a denser texture and are slightly nuttier with earthy flavour than white potatoes. Purple potatoes are a tasty way to add a pop of colour to meal while enjoying a serving of health benefits.

Fig. 10.2 Coloured potatoes developed in India

10.6.2 Improving Human Health

Anthocyanins provide numerous health benefits as they possess antioxidant, antimicrobial (Bontempo et al. 2013), anti-carcinogenic (Madiwale et al. 2012; Bontempo et al. 2013) and anti-inflammatory (Zafra-Stone et al. 2007; DeFuria et al. 2009) properties. Antioxidant potential of these pigments is contributed by the hydroxyl groups in the B-ring of their structure (Cooke et al. 2005). Han et al. (2006) suggested that flakes of purple-fleshed potato varieties Hokkai no. 92 and Kitamurasaki rich in anthocyanin (petanin) show higher antioxidant potential in terms of DPPH radical scavenging property, inhibition of linoleic acid oxidation, increase of hepatic superoxide dismutase, and glutathione peroxidase mRNA expression both in vivo and in vitro.

Moser et al. (2015) reported content of anthocyanins decline in order of purple, red and white coloured potato peel extracts and these extracts were observed to be the modulators in glycemic response. Isolated anthocyanin pigments show wide therapeutic potential that has been proven by various controlled in vitro and in vivo studies (Lila 2004). Anthocyanins extend a three line protection system for connective tissue; neutralization of enzyme that are detrimental to connective tissue, antioxidation of damaged tissue and repair of damaged proteins in walls of blood vessels (Stintzing et al. 2002).

Anthocyanins have been reported to be effective against different diseases including colorectal cancer studied in mouse models (Lippert et al. 2017) and dementia by enhancing cognition (Kent et al. 2017). Another report of cancer prevention is documented in case of anthocyanin rich potatoes through Wnt/β-catenin signaling inhibition and thus increased apoptosis or reduced cancer stem cells (Charepalli et al. 2015). Similarly, Reddivari et al. (2007) reported that potato extracts of four popular cultivars (ATTX98462–3, ATTX98491–3, CO112F2–2 and PATX99P32–2) were effective against LNCaP and PC-3 prostate cancer cells by up-regulating levels of cyclin-dependent kinase inhibitor p27. As evident through laboratory experiments, their mode of action is inhibition of epidermal growth factor receptor in cancer cells where cyanidin and delphinidin were observed to be more effective than malvidin (Meiers et al. 2001). Kubow et al. (2017) suggested that purple-fleshed potato cultivar Leona can behave as potent inducer of cytotoxicity of colonic Caco-2 cancer cells and decrease their viability. Besides, these pigments play important role in combating alcohol induced liver problems by attenuating cytochrome P450 2E1 expression (Jiang et al. 2016).

Moser et al. (2018) confirmed the role of purple potato chips in blood glucose reduction therefore acting as glycemic response modulators through animal model studies. They have been reported to be effective against visual acuity or night vision defects (Nakaishi et al. 2000; Lila 2004) as well as platelet aggregation thereby reducing the risk of coronary heart disease (Rechner and Kroner 2005). Vinson et al. (2012) reported in an intervention study that including anthocyanin rich potatoes in diet reduced diastolic blood pressure by around 4.3% in case of hypertensive patients thereby decreasing the risk of heart diseases and stroke without any sign of weight gain.

In a study on different Romanian potatoes, it was observed that the purple fleshed ones when added to bread dough (15% supplementation) not only retained anthocyanin content but also improved bread quality in terms of enhancing product shelf life, elasticity and porosity in crumb structure, and raised volume of bread (Liliana et al. 2016). Among popular potato cultivars, there is a diversified range of cultivars with red or purple coloured flesh that account for anthocyanins present in them responsible for rich antioxidant behaviour (Oertel et al. 2017). Potatoes with coloured flesh play a very important role in promoting public health as they are rich source of anthocyanins (Gutiérrez-Quequezana et al. 2018).

10.6.3 Food and Nutraceutical Industry

Red-flesh and purple-flesh potatoes are used as a source of anthocyanin natural colourants for food or non-food industry (De Vogel et al. 2005). Coloured potatoes may have economic value as natural food colourants and as food products such as novelty potato crisps and coloured potato salads. Acylated anthocyanins are known to be stable and hence can be considered as promising natural colourants for the food industry. Anthocyanins with multiple aromatic acyl-groups have potential to replace synthetic blue food colourants (Appelhagen et al. 2018).

10.6.4 Protection Against Biotic and Abiotic Factors

Coloured (red or purple) potato varieties are known to have a high level of durable resistance to late blight and also other several abiotic stresses, including wounding, light, temperature, etc. Wegener and Jansen (2007) reported average better resistance to soft rot in coloured fleshed potato varieties than white/yellow-fleshed potato cultivars. The resistance of coloured potatoes positively correlated with the higher concentrations of anthocyanins, soluble phenols and polyphenol oxidase activity in potatoes.

10.7 Genetic Modifications to Improve Anthocyanin Content in Potatoes

10.7.1 Conventional Breeding

Potato crop justifies 'nutrition transition' as it is wholesome food that provides the dietary requirements of all nutrients when taken in adequate quantity. Biofortification is the process for development of nutrient dense varieties either through plant breeding techniques, agronomic practices or biotechnological interventions. Worldwide, potato breeding, in general, and Indian potato breeding in particular have till now focused on yield improvement and biotic and abiotic stress resistance. Increasing potato consumption necessitates development of genetically superior varieties for yield with added advantage of high nutritional values. The potato breeding programme now at ICAR-CPRI, Shimla focuses on development of biofortified potato varieties for β-carotene, total carotenoids, anthocyanins, ascorbic acid, resistant starch, phenols and micronutrients like iron and zinc (Luthra et al. 2020).

Anthocyanins are beneficial for human health due to presence of anti-oxidative properties and impart colour variations in potato skin and/or flesh. Antioxidant capacity of red or blue coloured potatoes is 2–3 time higher than white/yellow-fleshed potatoes and breeders focus on the breeding of these phenotypes of potatoes which involve different variants of skin and flesh colour. Studies showed that the presence or absence of anthocyanin is controlled by single gene, whereas the distribution pattern is multigenic in nature (Brown et al. 2003). Synthesis of anthocyanidin pigments in potatoes is based on the dihydroflavonol-4-reductase activity, which catalyses reduction of dihydrokaempferol to leucopelargonidin. The potato R-locus encodes a basic factor required for the production of red pelargonidin-based anthocyanin pigments, which encodes dihydroflavonol 4-reductase, is present in all red coloured potato tubers (De Jong et al. 2003), whereas is absent in the tubers with white flesh. In present time, many coloured varieties are known, e.g. Norland, Red Norland, Congo, Dark Red Norland, Blaue Hindelbank, Red Pearl, All Blue, Purple Peruvian, Russet Norkotah, Cranberry Red, etc. (Groza et al. 2004). Numerous genes control presence and absence of these pigments. Genes controlling synthesis of red pigment are located on chromosome

2, while those imparting blue and purple colour are present on chromosome 11 and 4, respectively (Brown et al. 2005). The distribution pattern of potato skin and flesh colour is under complex genetic control. Zhang et al. (2019) identified quantitative trait loci (QTLs) influencing extent of flesh pigmentation on chromosomes 5, 8 and 9; while Parra-Galindo et al. 2019 identified seven QTLs located on chromosomes 1, 2, 10 and 11 regulating anthocyanin content in potato tubers.

Germplasm screening is the basic step to know the extent of variations available in the target gene pool and identification of promising parents for creation of genetic variability. Globally, screening of potato germplasm suggested presence of substantial variations for nutrient traits in cultivated and wild species. Genotype × environment interactions were significant for total carotenoids, anthocyanin, iron and zinc content. However, environmental effect does not lead to any significant changes in ranking of genotypes based on nutrient content. High estimates of heritability support the major role of genetics governing nutrient content. Heritable genetic variability for nutrient content, minimal inhibitors, viz. phytates and high ascorbic acid content makes potato ideal for biofortification. Presence of significant genetic variations facilitates exploitation of transgressive segregation, additive gene effect and heterosis. The breeding scheme involves identification of promising parents, crossing of parents to create genetic variability, evaluation of segregating population in clonal stages for agronomical, nutritional and quality attributes.

10.7.2 Genetic Transformation

Application of genetic engineering techniques so as to improve the levels of anthocyanin in different vegetable crops are well known. Some of the studies in potato perspective have been observed to be successful to increase anthocyanin content. Lukaszewicz et al. (2004) reported that overexpression of either single gene or simultaneously two genes encoding chalcone synthase (CHS), chalcone isomerase (CHI) and dihydroflavonol reductase (DFR) enhances the levels of phenolic acids and anthocyanins. This transgenic approach by using vector carrying multigene construct was used to improve the antioxidant quality of potato cultivars in Poland. Lorenc-Kukuła et al. (2005) reported that ectopic expression of anthocyanin 5-O-glucosyltransferase (5-UGT) in potato tubers not only significantly enhanced levels of pelargonidin 3-rutinoside-5-glucoside acylated with p-coumaric acid and peonidin 3-rutinoside-5-glucoside acylated with p-coumaric acid but also conferred pathogen resistance against *Erwinia carotovora* subsp. *carotovora* in transgenic lines. Rommens et al. (2008) introduced transgenic potato tubers enriched in flavonols and anthocyanin. Wei et al. (2012) identified six transgenic lines of potato produced by *Agrobacterium*-mediated transformation to upregulate UDP-glucose: flavonoid-3-*O*-glucosyltransferase (*3GT*) gene which resulted in enhanced levels of anthocyanin content. The said gene was inserted behind GBSSI promoter of pBin19 and the resulting construct was introduced in potato. Jin et al. (2016) successfully introduced potato breeding clones KPG16 and KPG5 with enhanced levels of total anthocyanins that can be considered as a functional food.

10.8 Future Directions

The potatoes with coloured flesh are quite common and are gaining much interest and attention at the consumer end due to attractive colour and nutritional superiority. Although the pigment responsible for their colour, i.e. anthocyanin is well known, but less is known about methods to enhance their level in potato through manipulations in genotype before sowing, climatic conditions during cultivation and further storage after harvest. Lachman et al. (2012) revealed that these factors play important role in anthocyanin accumulation. The standards for anthocyanin quantification are not readily available commercially which make it difficult to estimate low quantities of this pigment with sophisticated techniques. This challenge needs to be addressed in the near future so as to bring out new techniques for their estimation.

References

Andersen OM, Markham KR (eds) (2005) Flavonoids: chemistry, biochemistry and applications. CRC Press, Boca Raton, FL

Andre CM, Evers D, Ziebel J et al (2015) In vitro bioaccessibility and bioavailability of iron from potatoes with varying vitamin C, carotenoid, and phenolic concentrations. J Agric Food Chem 63(41):9012–9021

Andre CM, Oufir M, Guignard C et al (2007) Antioxidant profiling of native Andean potato tubers (*Solanum tuberosum* L.) reveals cultivars with high levels of β-carotene, alpha-tocopherol, chlorogenic acid, and petanin. Agric Food Chem 55:10839–10849

Appelhagen I, Wulff-Vester AK, Wendell M et al (2018) Colour bio-factories: towards scale-up production of anthocyanins in plant cell cultures. Metab Eng 48:218–232

Barnes JS, Nguyen HP, Shen S et al (2009) General method for extraction of blueberry anthocyanins and identification using high performance liquid chromatography–electrospray ionization-ion trap-time of flight-mass spectrometry. J Chromatogr A 1216(23):4728–4735

Bontempo P, Carafa V, Grassi R et al (2013) Antioxidant, antimicrobial and anti-proliferative activities of *Solanum tuberosum* L. var. Vitelotte. Food Chem Toxicol 55:304–312

Brown CR, Culley D, Yang CP et al (2005) Variation of anthocyanin and carotenoid contents and associated antioxidant values in potato breeding lines. J Am Soc Hort Sci 130(2):174–180

Brown CR, Durst RW, Wrolstad R et al (2008) Variability of phytonutrient content of potato in relation to growing location and cooking method. Potato Res 51(3–4):259

Brown CR, Wrolstad R, Durst R et al (2003) Breeding studies in potatoes containing high concentrations of anthocyanins. Amer J Pot Res 80:241–250

Burgos G (2014) Concentración y bioaccesibilidad de carotenoides y compuestos fenólicos en papas cocidas. Doctoral dissertation, Universidad de La Laguna

Burgos G, Amoros W, Muñoa L et al (2013) Total phenolic, total anthocyanin and phenolic acid concentrations and antioxidant activity of purple-fleshed potatoes as affected by boiling. J Food Comp Anal 30(1):6–12

Camire ME, Kubow S, Donnelly DJ (2009) Potatoes and human health. Crit Rev Food Sci Nutr 49 (10):823–840

Charepalli V, Reddivari L, Radhakrishnan S et al (2015) Anthocyanin-containing purple-fleshed potatoes suppress colon tumorigenesis via elimination of colon cancer stem cells. J Nutr Biochem 26(12):1641–1649

Cooke D, Steward WP, Gescher AJ et al (2005) Anthocyans from fruits and vegetables–does bright colour signal cancer chemopreventive activity? Eur J Cancer 41(13):1931–1940

Dalamu SB, Barwal S et al (2015) Assessment of phytochemical diversity in Indian potato cultivars. Ind J Hort 72(3):447–450

De Jong WS, De Jong DM, De Jong H et al (2003) An allele of dihydroflavonol 4-reductase associated with the ability to produce red anthocyanin pigments in potato (*Solanum tuberosum* L.). Theor Appl Genet 107(8):1375–1383

De Vogel J, Jonker-Termont DS, Van Lieshout EM et al (2005) Green vegetables, red meat and colon cancer: chlorophyll prevents the cytotoxic and hyperproliferative effects of haem in rat colon. Carcinogenesis 26(2):387–393

DeFuria J, Bennett G, Strissel KJ et al (2009) Dietary blueberry attenuates whole-body insulin resistance in high fat-fed mice by reducing adipocyte death and its inflammatory sequelae. J Nutr 139(8):1510–1516

Delgado-Vargas F, Jiménez AR, Paredes-López O (2000) Natural pigments: carotenoids, anthocyanins, and betalains—characteristics, biosynthesis, processing, and stability. Crit Rev Food Sci Nutr 40(3):173–289

Delgado-Vargas F, Paredes-López O (2003) Anthocyanins and betalains. In: Natural colorants for food and nutraceutical uses. CRC Press, Boca Raton, pp 167–219

Eichhorn S, Winterhalter P (2005) Anthocyanins from pigmented potato (*Solanum tuberosum* L.) varieties. Food Res Int 38(8–9):943–948

Fang J (2015) Classification of fruits based on anthocyanin types and relevance to their health effects. Nutrition 31(11–12):1301–1306

Fossen T, Andersen ØM (2000) Anthocyanins from tubers and shoots of the purple potato, *Solanum tuberosum*. J Hort Sci Biotechnol 75(3):360–363

Furrer A, Cladis DP, Kurilich A et al (2017) Changes in phenolic content of commercial potato varieties through industrial processing and fresh preparation. Food Chem 218:47–55

Galvano F, La Fauci L, Lazzarino G et al (2004) Cyanidins: metabolism and biological properties. J Nutr Biochem 15(1):2–11

Giusti MM, Polit MF, Ayvaz H et al (2014) Characterization and quantitation of anthocyanins and other phenolics in native Andean potatoes. J Agric Food Chem 62(19):4408–4416

Gould K, Davies KM, Winefield C (eds) (2008) Anthocyanins: biosynthesis, functions, and applications. Springer Science & Business Media, New York

Groza HI, Bowen BD, Kichefski D et al (2004) Red pearl: a new gourmet red potato variety. Am J Pot Res 81(3):209–213

Gutiérrez-Quequezana L, Vuorinen AL, Kallio H et al (2018) Improved analysis of anthocyanins and vitamin C in blue-purple potato cultivars. Food Chem 242:217–224

Hamouz K, Lachman J, Bečka D et al (2018) Effect of growing conditions and storage on the total anthocyanin content in potatoes with coloured flesh. Plant Soil Environ 64(9):435–440

Hamouz K, Lachman J, Pazderů K et al (2011) Differences in anthocyanin content and antioxidant activity of potato tubers with different flesh colour. Plant Soil Environ 57(10):478–485

Han KH, Sekikawa M, Shimada KI et al (2006) Anthocyanin-rich purple potato flake extract has antioxidant capacity and improves antioxidant potential in rats. Br J Nutr 96(6):1125–1134

Harborne JB (1958) The chromatographic identification of anthocyanin pigments. J Chromatogr A 1:473–488

Hayashi K, Hibasami H, Murakami T et al (2006) Induction of apoptosis in cultured human stomach cancer cells by potato anthocyanins and its inhibitory effects on growth of stomach cancer in mice. Food Sci Technol Res 12(1):22–26

He J, Giusti MM (2010) Anthocyanins: natural colorants with health-promoting properties. Annu Rev Food Sci Technol 1:163–187

Heredia FJ, Francia-Aricha EM, Rivas-Gonzalo JC et al (1998) Chromatic characterization of anthocyanins from red grapes—I. pH effect. Food Chem 63(4):491–498

Hillebrand S, Naumann H, Kitzinski N et al (2009) Isolation and characterization of anthocyanins from blue-fleshed potatoes (*Solanum tuberosum* L.). FoodReview 3(1):96–101

Ishii G, Mori M, Umemura Y (1996) Antioxidative activity and food chemical properties of anthocyanins from the colored tuber flesh of potatoes. Nippon Shokuhin Kogyo Gakkai-Shi 43(8):962–966

Jansen G, Flamme W (2006) Coloured potatoes (Solanum tuberosum L.)–anthocyanin content and tuber quality. Genet Resour Crop Ev 53(7):1321

Jemison JM Jr, Sexton P, Camire ME (2008) Factors influencing consumer preference of fresh potato varieties in Maine. Am J Pot Res 85(2):140–149

Jiang Z, Chen C, Wang J et al (2016) Purple potato (Solanum tuberosum L.) anthocyanins attenuate alcohol-induced hepatic injury by enhancing antioxidant defense. J Nat Med 70(1):45–53

Jin C, Lee W, Choi H et al (2016) Selection of the excellent potato clones based on total polyphenol, anthocyanin and vitamin C contents. Korean J Hort Sci Technol 34(3):488–494

Jin H, Liu Y, Yang F et al (2015) Characterization of anthocyanins in wild Lycium ruthenicum Murray by HPLC-DAD/QTOF-MS/MS. Anal Methods 7(12):4947–4956

Kent K, Charlton K, Roodenrys S et al (2017) Consumption of anthocyanin-rich cherry juice for 12 weeks improves memory and cognition in older adults with mild-to-moderate dementia. Eur J Nutr 56(1):333–341

Kita A, Bąkowska-Barczak A, Hamouz K et al (2013) The effect of frying on anthocyanin stability and antioxidant activity of crisps from red-and purple-fleshed potatoes (*Solanum tuberosum* L.). J Food Comp Anal 32(2):169–175

Kotikova Z, Hejtmankova A, Lachman J et al (2007) Effect of selected factors on total carotenoid content in potato tubers (*Solanum tuberosum* L.). Plant Soil Environ 53:355–360

Kubow S, Iskandar MM, Melgar-Bermudez E et al (2017) Effects of simulated human gastrointestinal digestion of two purple-fleshed potato cultivars on anthocyanin composition and cytotoxicity in colonic cancer and non-tumorigenic cells. Nutrients 9(9):953

Lachman J, Hamouz K, Musilová J et al (2013) Effect of peeling and three cooking methods on the content of selected phytochemicals in potato tubers with various colour of flesh. Food Chem 138 (2–3):1189–1197

Lachman J, Hamouz K, Orsák M et al (2012) Impact of selected factors–cultivar, storage, cooking and baking on the content of anthocyanins in coloured-flesh potatoes. Food Chem 133 (4):1107–1116

Lachman J, Hamouz K, Šulc M et al (2009) Cultivar differences of total anthocyanins and anthocyanidins in red and purple-fleshed potatoes and their relation to antioxidant activity. Food Chem 114:836–843

Lemos MA, Sivaramareddy A (2013) Effect of oil quality on the levels of total phenolics, total anthocyanins and antioxidant activity of French fries. In: Inside Food Symposium. Leuven, Belgium, pp 1–6

Lewis CE, Walker JR, Lancaster JE et al (1998) Light regulation of anthocyanin, flavonoid and phenolic acid biosynthesis in potato minitubers in vitro. Funct Plant Biol 25(8):915–922

Li SY, He F, Zhu BQ et al (2016) A systematic analysis strategy for accurate detection of anthocyanin pigments in red wines. Rapid Commun Mass Spectrom 30(13):1619–1626

Lila MA (2004) Anthocyanins and human health: an in vitro investigative approach. Biomed Res Int 2004(5):306–313

Liliana BC, Cristina CM, Damşa Florentina D et al (2016) Effects of several purple potato additions on bread quality. Int J Eng Res Appl 6(4):1–10

Lippert E, Ruemmele P, Obermeier F et al (2017) Anthocyanins prevent colorectal cancer development in a mouse model. Digestion 95(4):275–280

Lorenc-Kukuła K, Jafra S, Oszmiański J et al (2005) Ectopic expression of anthocyanin 5-O-glucosyltransferase in potato tuber causes increased resistance to bacteria. J Agric Food Chem 53(2):272–281

Lukaszewicz M, Matysiak-Kata I, Skala J et al (2004) Antioxidant capacity manipulation in transgenic potato tuber by changes in phenolic compounds content. J Agric Food Chem 52 (6):1526–1533

Luthra SK, Gupta VK, Tiwari JK, et al (2020). Potato Breeding in India. *CPRI, Shimla Technical Bulletin No 74* (Revised), 214p
Luthra SK, Tiwari JK, Dalamu et al (2018) Breeding for coloured flesh potatoes: molecular, agronomical and nutritional profiling. Potato J 45(2):81–92
Madiwale GP, Reddivari L, Stone M et al (2012) Combined effects of storage and processing on the bioactive compounds and pro-apoptotic properties of color-fleshed potatoes in human colon cancer cells. J Agric Food Chem 60(44):11088–11096
Mane S, Bremner DH, Tziboula-Clarke A, Lemos A (2015) Effect of ultrasound on the extraction of total anthocyanins from purple majesty potato. Ultrason Sonochem 27:509–514
Meiers S, Keměny M, Weyand U et al (2001) The anthocyanidins cyanidin and delphinidin are potent inhibitors of the epidermal growth-factor receptor. J Agric Food Chem 49(2):958–962
Mori M, Hayashi K, Ohara-Takada A et al (2010) Anthocyanins from skins and fleshes of potato varieties. Food Sci Technol Res 16(2):115–122
Moser S, Aragon I, Furrer A et al (2015) Potato phenolics modulate rate of glucose transport in a Caco-2 human intestinal cell model. The. FASEB J 29(1_supplement):606–606
Moser S, Aragon I, Furrer A et al (2018) Potato phenolics impact starch digestion and glucose transport in model systems but translation to phenolic rich potato chips results in only modest modification of glycemic response in humans. Nutr Res 52:57–70
Murador D, Braga AR, Da Cunha D, De Rosso V (2018) Alterations in phenolic compound levels and antioxidant activity in response to cooking technique effects: a meta-analytic investigation. Crit Rev Food Sci Nutr 58(2):169–177
Nakaishi H, Matsumoto H, Tominaga S et al (2000) Effects of black currant anthocyanoside intake on dark adaptation and VDT work-induced transient refractive alteration in healthy humans. Altern Med Rev 5(6):553–562
Navarre DA, Shakya R, Holden J et al (2010) The effect of different cooking methods on phenolics and vitamin C in developmentally young potato tubers. Am J Pot Res 87(4):350–359
Nayak B, Berrios JDJ, Powers JR et al (2011) Thermal degradation of anthocyanins from purple potato (Cv. Purple Majesty) and impact on antioxidant capacity. J Agric Food Chem 59 (20):11040–11049
Nelson DC, Bristol DW (1975) 2,4-D on potatoes. Agricultural Extension Service. University of Minnesota and North Dakota State Univ. Red River Valley Potato Facts number 5
Oertel A, Matros A, Hartmann A et al (2017) Metabolite profiling of red and blue potatoes revealed cultivar and tissue specific patterns for anthocyanins and other polyphenols. Planta 246 (2):281–297
Palermo M, Pellegrini N, Fogliano V (2014) The effect of cooking on the phytochemical content of vegetables. J Sci Food Agric 94(6):1057–1070
Parra-Galindo MA, Piñeros-Niño C, Soto-Sedano JC et al (2019) Chromosomes I and X Harbor consistent genetic factors associated with the anthocyanin variation in potato. Agronomy 9 (7):366
Patras A, Brunton NP, O'Donnell C et al (2010) Effect of thermal processing on anthocyanin stability in foods; mechanisms and kinetics of degradation. Trends Food Sci Tech 21(1):3–11
Perla V, Holm DG, Jayanty SS (2012) Effects of cooking methods on polyphenols, pigments and antioxidant activity in potato tubers. LWT-Food Sci Technol 45(2):161–171
Rechner AR, Kroner C (2005) Anthocyanins and colonic metabolites of dietary polyphenols inhibit platelet function. Thrombosis Res 116(4):327–334
Reddivari L, Vanamala J, Chintharlapalli S et al (2007) Anthocyanin fraction from potato extracts is cytotoxic to prostate cancer cells through activation of caspase-dependent and caspase-independent pathways. Carcinogenesis 28(10):2227–2235
Reyes LF, Cisneros-Zevallos L (2003) Wounding stress increases the phenolic content and antioxidant capacity of purple-fresh potatoes (*Solanum tuberosum* L.). J Agric Food Chem 51:5296–5300

Reyes LF, Miller JC, Cisneros-Zevallos L (2004) Environmental conditions influence the content and yield of anthocyanins and total phenolics in purple- and red-flesh potatoes during tuber development. Am J Potato Res 81(3):187–193

Reyes LF, Miller JC, Cisneros-Zevallos L (2005) Antioxidant capacity, anthocyanins and total phenolics in purple-and red-fleshed potato (*Solanum tuberosum* L.) genotypes. Am J Pot Res 82 (4):271

Rodríguez-Werner M, Winterhalter P, Esatbeyoglu T (2019) Phenolic composition, radical scavenging activity and an approach for authentication of Aronia melanocarpa berries, juice, and Pomace. J Food Sci 84(7):1791–1798

Rommens CM, Richael CM, Yan H et al (2008) Engineered native pathways for high kaempferol and caffeoylquinate production in potato. Plant Biotechnol J 6(9):870–886

Smeriglio A, Barreca D, Bellocco E et al (2016) Chemistry, pharmacology and health benefits of anthocyanins. Phytother Res 30(8):1265–1286

Stintzing FC, Carle R (2004) Functional properties of anthocyanins and betalains in plants, food, and in human nutrition. Trends Food Sci Technol 15(1):19–38

Stintzing FC, Stintzing AS, Carle R et al (2002) Color and antioxidant properties of cyanidin-based anthocyanin pigments. J Agric Food Chem 50(21):6172–6181

Teow CC, Truong VD, McFeeters RF et al (2007) Antioxidant activities, phenolic and β-carotene contents of sweet potato genotypes with varying flesh colours. Food Chem 103(3):829–838

Tian J, Chen J, Lv F et al (2016) Domestic cooking methods affect the phytochemical composition and antioxidant activity of purple-fleshed potatoes. Food Chem 197:1264–1270

Tierno R, López A, Riga P et al (2016) Phytochemicals determination and classification in purple and red fleshed potato tubers by analytical methods and near infrared spectroscopy. J Sci Food Agric 96(6):1888–1899

Vinson JA, Demkosky CA, Navarre DA et al (2012) High-antioxidant potatoes: acute in vivo antioxidant source and hypotensive agent in humans after supplementation to hypertensive subjects. J Agric Food Chem 60(27):6749–6754

Wegener CB, Jansen G (2007) Soft-rot resistance of coloured potato cultivars (*Solanum tuberosum* L.): the role of anthocyanins. Potato Res 50:31–44

Wei Q, Wang QY, Feng ZH et al (2012) Increased accumulation of anthocyanins in transgenic potato tubers by overexpressing the 3GT gene. Plant Biotechnol Rep 6(1):69–75

Xu J, Su X, Lim S et al (2015) Characterisation and stability of anthocyanins in purple-fleshed sweet potato P40. Food Chem 186:90–96

Zafra-Stone S, Bagchi M, Bagchi D (2007) Health benefits of edible berry anthocyanins: novel antioxidant and anti-angiogenic properties. In: Antioxidant measurement and applications. American Chemical Society, Chapter 22, Washington, pp 337–351. https://doi.org/10.1021/bk-2007-0956.ch022

Zarzecka K, Gugala M (2011) The effect of herbicides and soil tillage systems on the content of polyphenols in potato tubers. Pol J Environ Stud 20(2):513–517

Zhang T, Zhang F, Cao Y et al (2019) Effect of whole purple potato flour on dough properties and quality of steamed bread. J Food Res 8(4):122–130

Zhao CL, Yu YQ, Chen ZJ, Wen GS, Wei FG, Zheng Q, Wang CD, Xiao XL (2017) Stability-increasing effects of anthocyanin glycosyl acylation. Food Chem 214:119–128

Potato Glycoalkaloids

11

Brajesh Singh, Som Dutt, and Pinky Raigond

Abstract

Potato contains a range of essential nutrients such as carbohydrates, proteins, vitamins, dietary fibres, and minerals and several other health promoting metabolites. In addition to these, potatoes do contain few molecules with harmful effects if consumed in higher amounts. Steroidal glycoalkaloids are group of such molecules, and α-chaconine and α-solanine are the two most predominant glycosides present in potato tubers. Concentrations of the glycoalkaloids in all the commercial cultivars of potato are much lower than the recommended safe limit for human consumption. However, exposure to light, various biotic factors lead to increase in the glycoalkaloids levels. Genetic factors have also been shown to regulate the glycoalkaloids levels in potato. Globally, efforts have been/are being made to develop potato cultivars with lower/minimum glycoalkaloids and also to constantly improving the sensitivity, accuracy, and throughput of the methods employed for estimation of the glycoalkaloids.

Keywords

Solanine · Chaconine · HPLC · MALDI-TOF · Aglycone

11.1 Introduction

Potatoes are important vegetables containing key nutrients such as vitamins, carbohydrates, proteins, minerals, and several other health promoting molecules. Apart from these positively health impacting constituents, potatoes also contain few potentially harmful compounds. One such group of compounds are glycoalkaloids.

B. Singh (✉) · S. Dutt · P. Raigond
ICAR - Central Potato Research Institute, Shimla, Himachal Pradesh, India
e-mail: birju16@gmail.com

P. Raigond et al. (eds.), *Potato*, https://doi.org/10.1007/978-981-15-7662-1_11

Glycoalkaloids are naturally occurring compounds in plants of Solanaceae family. When these glycoalkaloids are present in very high levels much beyond the safe limit these alkaloids affect nervous system, resulting in general disruption of membranes. Other symptoms of potato glycoalkaloids toxicity include nausea, diarrhoea, abdominal pain, vomiting, and headache.

Glycoalkaloids in potatoes contribute flavour; however, their higher concentrations impart bitter taste. Their natural function in potato plants has been suggested to be stress metabolites and to protect plants against from insect pests, pathogens, and unfavourable environmental factors. For human consumption upper safe limit of glycoalkaloids content in potato has been recommended as 20 mg/100 g of potato (fresh weight). In most of the countries, glycoalkaloids content has been included as one of the important criteria for releasing any new variety. Commercially grown varieties of potatoes though differ in their glycoalkaloids level, tubers of all of them contain glycoalkaloids much below the upper fixed limit (i.e. 20 mg/100 g potato) considered safe for human consumption. Peel of potato contains comparatively higher concentration of glycoalkaloids as compared to that of flesh.

The average content in the peel ranges from 3 to 100 mg/100 g of peel. For peeled potatoes, the average content varies from 0.10 to 4.50 mg/100 g. Being present in minor quantities and having health concern, glycoalkaloids detection needs sensitive methodologies for their accurate estimation. For this reason various methods based on liquid chromatography and gas chromatography linked with mass spectrometers or diode array detectors, biosensors, immunoassays, etc. have been employed for identification and/or estimation of glycoalkaloids. Glycoalkaloids content have been found to be affected by various environmental factors (light, drought), mechanical injury, storage conditions, genetic factors, and various biotic factors. In this chapter, all these aspects of potato glycoalkaloids are discussed.

11.2 Content of Glycoalkaloids in Potato

Upper limit for glycoalkaloids content has been fixed as 20 mg/100 g of potato. Several glycoalkaloids have been reported in potato; however, α-chaconine and α-solanine (Fig. 11.1) constitute up to 95% of the total glycoalkaloids present in it (Friedman et al. 2003). In one study, Shakya and Navarre (2008) identified in potato over 50 glycoalkaloids with solanidane or solanidane-like aglycones. Several studies have reported the estimation of glycoalkaloids content in different potato cultivars and in different tissue of the potato tubers. The glycoalkaloids content have been reported on per kg fresh weight/dry weight or per 100 g fresh weight/dry weight basis. Panovská et al. (1997) estimated glycoalkaloids level in different potato cultivars grown in the Czech Republic. The potato cultivars were collected from three regions and monitored during 2 years. The average of total glycoalkaloids content for tested cultivars ranged from 31 to 166 mg/kg fresh weight. No significant difference between glycoalkaloid levels in tubers from different regions was observed. Friedman et al. (2003) estimated total glycoalkaloids in whole potato, peel, and flesh, both in dry as well as wet state. In dried tissue 40–883, 5–592,

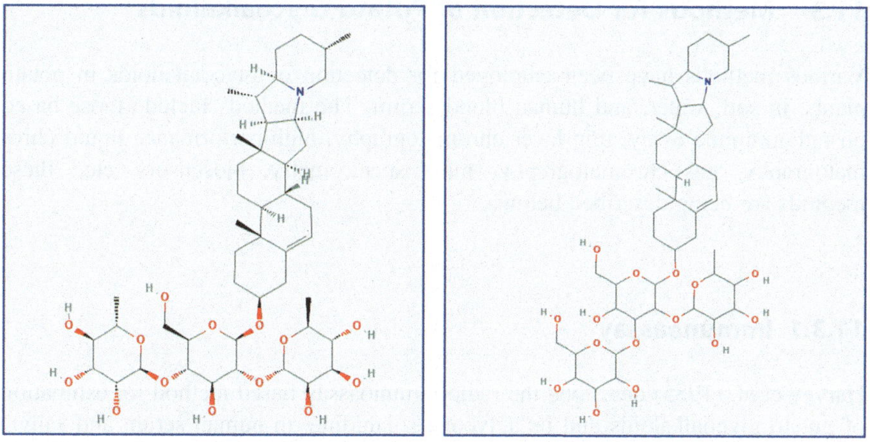

alpha-Chaconine
(MW: 852.1 g/mol)

alpha-Solanine
(MW: 868.1 g/mol)

Fig. 11.1 Structure of major glycoalkaloids of potato

84–2226 mg/kg total glycoalkaloids were found in whole potato, flesh, and peel, respectively. In wet state 12–429, 1–148, and 12–429 mg/kg total glycoalkaloids were found in whole potato, flesh, and peel, respectively. Shimoi et al. (2007) estimated the glycoalkaloids content in 27 cultivars of potato and found the content ranging from 21 to 230 mg/kg potato on fresh weight basis.

Aziz et al. (2012) estimated total glycoalkaloids, α-chaconine, and α-solanine content in selected potato cultivars grown in Pakistan and expressed on per 100 g dry weight basis. The total glycoalkaloids concentration was recorded to be ranging from 177–5450 and 3–15 mg in peel and fresh, respectively, of all the potato cultivars tested. The α-solanine content ranged from 46–2743 and 4–2466 mg, in peel and fresh, respectively. Similarly, α-chaconine content ranged from 4–6818 and 4–475 mg in peel and fresh, respectively. Deusser et al. (2012) estimated glycoalkaloids content in potato cultivars grown in Luxembourg. They reported that glycoalkaloids content were highest in the peel and lowest in the inner flesh. Total glycoalkaloids content ranged from 150–8133 and 4–957 mg/kg dry weight in flesh and peel, respectively, when analysed in 60 varieties of potato by Valcarcel et al. (2014). In peel, α-chaconine accumulated more as compared to that of α-solanine. Singh et al. (2016) estimated glycoalkaloids in peels of 41 Indian potato cultivars. The total glycoalkaloids content was found to be in the range of 0.24–3.62 mg/100 g tuber fresh weight (FW). α-Solanine and α-chaconine were found to be in the range of 0.14–0.76 mg/100 g FW and 0.11–2.00 mg/100 g FW, respectively. Wayumba et al. (2019) evaluated 21 potato clones for cold chip processing and observed the levels of α-solanine glycoalkaloid in the range of 0.15–15.54 mg/100 g fresh weight.

11.3 Methods for Detection of Potato Glycoalkaloids

Various methods have been employed for detection of glycoalkaloids in potato plants, in soil, water, and human blood serum. The methods include those based on radioimmune assay, thin-layer chromatography, high-performance liquid chromatography, gas chromatography, mass spectrometry, biosensors, etc.; these methods are being described below.

11.3.1 Immunoassay

Harvey et al. (1985) described the radioimmunoassay based method for estimation of potato glycoalkaloids and its aglycone solanidine, in human serum and saliva. Based on the data collected it was concluded that alkaloids concentration in the serum and saliva were significantly higher in a group of persons (subjects) eating potatoes containing unusually high concentrations of alkaloids when compared with those subjects from a group eating their normal diets. Glorio-Paulet and Durst (2000) used liposome immunomigration, liquid-phase competition immunoassay for quantification of glycoalkaloids in potato. They raised polyclonal anti-solanine antibodies and used in the assembly of liposome immunomigration, liquid-phase competition strip immunoassay. However, they observed a similar cross-reactivity between the glycoalkaloids α-solanine and α-chaconine. Therefore, the developed strip assay was recommended for estimation of total glycoalkaloids from potatoes. Driedger et al. (2000a) developed a solution-phase for the immunoassay of potato glycoalkaloids using capillary electrophoresis along with laser-induced fluorescence. Solanidine was coupled to 4′-(aminomethyl)fluorescein and a Polyclonal antibody solution was used as the immunoreagent. For detection of unbound fluorescent solanidine capillary electrophoresis along with laser-induced fluorescence using 488/520 nm excitation/emission was used. For improving resolution 50 mM phosphate containing 10% methanol and 1.5 mM SDS (pH 7.5) buffer was used. Using this developed method, analysis of extracts of sprouted and unsprouted *Yukon Gold* potato tubers showed total glycoalkaloids of 98 and 55 μg/g, respectively, in two parts. Subsequently, the developed method was used for analysis of potato glycoalkaloids (Driedger et al. 2000b). The basis of immunoassay was the competition between fluorescently labelled solanidine and glycoalkaloids from potato. Reaction products were separated in the capillary zone electrophoresis mode. The developed method was found suitable for glycoalkaloids concentration in the range of 40–200 μg/g fresh weight. Further, it was recommended that immune-capillary electrophoresis: laser-induced fluorescence method might be a rapid alternative to commonly used HPLC and ELISA methods.

11.3.2 Thin-Layer Chromatography

Thin-layer chromatography (TLC) has been employed by various researches for detection of glycoalkaloids in potato. Cadle et al. (1978) reported use of high-performance thin-layer chromatography (HPTLC) for identification and estimation of glycoalkaloids from potatoes. For extraction of α-chaconine and α-solanine from dehydrated potatoes, boiling methanol-acetic acid (95 and 5% v/v) was used as extraction solvent. The analytes were separated on HPTLC plate by a saturated mixture of dichloromethane–methanol–water-concentrated ammonium hydroxide. For visualization, a modified Carr-Price reagent, 20% (w/v) antimony(III) chloride in acetic acid was used. For quantification, densitometry by reflectance scanning was recorded at 507 nm. Simonovska and Vovk (2000) utilized HPTLC for the quantification of glycoalkaloids (α-chaconine and α-solanine) in different parts of the potato plant. For extraction of samples, acetic acid having ion-pairing reagent, for purification, solid-phase extraction was used and applied to the HPTLC silica gel plates. The separation was performed using mobile phase as methonal:chloroform:2% NH_4OH (30:65:5). They tested a range of detection reagents for densitometric quantification. The fluorescence enabled the detection up to lower limit of 10 ng of α-chaconine as well as α-solanine. Mäder et al. (2009) used HPTLC method for estimating the composition of glycoalkaloids: α-solanine and α-chaconine during processing of potato at commercial scale. Processing of potatoes to potato flakes resulted in remarkable decrease in α-solanine and α-chaconine content. The decrease was mainly attributed to peeling and leaching. Thermal exposure effect appeared to be not very significant. About 10% of the initial glycoalkaloids remained after processing. As expected, byproducts developed from peel had highest content of glycoalkaloids.

11.3.3 High Performance Liquid Chromatography

High-performance liquid chromatography (HPLC) based methods in combination with diode array detectors and/or mass spectrometers have been used in various studies pertaining to separation and detection of potato glycoalkaloids. Chen et al. (1994) analysed potato shoots for glycoalkaloids content by four-sector tandem mass spectrometry with scanning-array detection. Positive and negative ion modes were used in the analysis. In positive ion mode, collision-induced dissociation tandem mass spectra of the [M + H] + ions induced three major fragmentation processes. The analysis revealed that the positive ion method had the advantage of a lower detection limit than in conventional mass spectrometry, numerous and intense fragment ions which were structurally informative. In 1997, in a collaborative study, 12 labs evaluated parameters pertaining to liquid chromatographic method for detection and quantification of the glycoalkaloids α-solanine and α-chaconine in potato tubers (Hellenäs and Branzell 1997). Frozen potato tuber homogenates distributed as three blind duplicates and three split-level pairs were used as samples. The analytical methods included aqueous extraction, workup on disposable solid-phase extraction cartridges, and reversed-phase chromatography with photometric detection at 202 nm. Results received from all the participating labs when compared showed similar relative standard deviations for reproducibility for α-solanine and α-chaconine ranging from 8% to 13% in the applied concentration range of 12–260 mg/kg FW. Kuronen et al. (1999) developed a reversed-phase liquid chromatography based method for separation and profiling of potato steroidal glycoalkaloids and their aglycones simultaneously. Using acetonitrile in combination with triethylammonium phosphate buffer at pH 3.0 as elution solvent for isocratic and gradient elution conditions exhibited most reproducible retention behaviour for the glycoalkaloids as well as their aglycones. Friedman et al. (2003) reported improvement in the HPLC method, for estimation of glycoalkaloids in potato. They studied the effects of (1) mobile phase pH and composition (phosphate buffer and acetonitrile) (2) of phosphate buffer concentration, (3) column packing capacity values of HPLC amino columns, (4) column temperature, on the retention times of glycoalkaloids. Effect of all variables, except for pH, was observed to be significant on the retention times of the glycoalkaloids. Limit of detection for this improved HPLC method was found to be 150 ng. The improved method was evaluated using extracts from the cortex of one potato variety grown in Japan, peels and flesh from the eight varieties from the USA. Total glycoalkaloids content in different tissues were found to be ranging as 5–592, 84–2226, 40–883, 1–148, 12–429, 7–187 mg/kg in dry flesh, dry peel, dry whole potatoes, wet flesh, wet peel, wet whole potatoes, respectively. Turakainen et al. (2004) analysed the effect of selenate supplementation on glycoalkaloids (alpha-solanine and alpha-chaconine) content of potato. They used reverse-phase high-performance liquid chromatography (RP-HPLC) coupled with diode array detector. For extraction of glycoalkaloids from freeze-dried tubers of two potato cultivars (Satu and Sini), 5% acetic acid was used. They concluded that acceptable application levels of selenate did not have an effect on the glycoalkaloids concentration in immature potato tubers. Matsuda et al. (2004) developed HPLC-electrospray ionization/mass spectrometry ESI/MS method for estimation of glycoalkaloids, α-solanine, and α-chaconine, in potato tubers. Extraction of potato samples was carried out using 5% aqueous acetic acid, and subjected directly to HPLC-ESI/MS after filtration. The developed method was found to have the detection limits of 38 and 14 ppb for α-solanine (m/z 868) and α-chaconine (m/z 852), respectively. The authors suggested that the high sensitivity and selectivity of MS detection effectively reduced the time of analysis, thus could be employed for a high throughput estimation of glycoalkaloids in potato tubers.

Shakya and Navarre (2008) used LC-MS for analysis of solanidane glycoalkaloid diversity among tubers of four wild potato species and three cultivars. Through this study they tentatively identified over 50 glycoalkaloids with solanidane or solanidane-like aglycones, many of which were indicated to be novel glycoalkaloids, and thus suggesting that number of glycoalkaloids present in potato might remarkably exceed the numbers previously thought. Jensen et al. (2008) employed liquid chromatography–time-of-flight mass spectrometry for quantification of potato glycoalkaloids in soil. The lower detection limits of the employed method, for glycoalkaloids was found to be in the range of 2.2–4.7 μg/L, and had linearities up to 2000 μg/L for alpha-solanine and alpha-chaconine and up to 760 μg/L for solanidine. Carputo et al. (2010) estimated glycoalkaloids in potato haploids derived from *Solanum tuberosum* x *S. bulbocastanum* somatic hybrids. Glycoalkaloids were analysed in haploid plants obtained from tetraploid *Solanum bulbocastanum x S. tuberosum* hybrids through in vitro anther culture. Glycoalkaloids were extracted from tubers and analysed by HPLC. Haploids generally showed the occurrence of parental glycoalkaloids. However, in several cases loss of parental glycoalkaloids and gain of new glycoalkaloids lacking in the parents were observed. Deusser et al. (2012) estimated glycoalkaloids (α-chaconine and α-solanine) contents of peels, outer flesh, and inner flesh of potato tubers grown in Luxembourg, using UPLC-DAD and HPLC-MS/MS. They observed the glycoalkaloids contents to be lowest in the inner flesh and highest in the peel. Glycoalkaloids contents were found below guideline limits in all the tested cultivars. Lachman et al. (2013) studied the effect of peeling and three cooking methods on glycoalkaloids content in potato tubers with various colour of flesh. They used HPLC-ESI/MS/MS for estimation of glycoalkaloids, α-chaconine, and α-solanine. Significant reduction in total glycoalkaloids, α-solanine and α-chaconine content were observed in case of all the three cooking methods; however, highest decrease in total glycoalkaloids was recorded in case of boiling of peeled tubers.

11.3.4 Solid–Liquid–Liquid Chromatography

Nikolic and Stankovic (2003) reported using solid–liquid–liquid systems for solanidine hydrolytic extraction and separation from the potato vines. The developed single-step method comprised of (1) glycoalkaloids extraction from potato vines, (2) hydrolysis of glycoalkaloids to corresponding aglycone i.e. solanidine, and (3) solanidine extraction. They developed a procedure by combining three different processes: extraction of glycoalkaloids from potato vines, their hydrolysis to solanidine, and the extraction of solanidine, in a single step. Use of 10% HCl in 50% methanol as the first liquid phase and chloroform as the second liquid phase resulted in more than 98% solanidine hydrolytic extraction and yielded 0.24 g solanidine per100 g of potato vines.

Nielsen et al. (2020) developed a method for quantification of glycoalkaloids (α-solanine, α-chaconine) and their aglycone form; solanidine, in potato protein isolates, using liquid chromatography-mass spectrometry single quadruple in single ion monitoring mode. Validation of the methods was performed with food-grade potato protein powder and glycoalkaloids standards. The developed method was used for glycoalkaloids quantification in a range of potato protein isolates. Estimation of total glycoalkaloids using the developed method revealed increase in total glycoalkaloids during protein isolates storage. Significant reduction in glycoalkaloids content was observed upon washing of the potato protein isolates with water.

11.3.5 Fourier Transform Based Methods

Cahill et al. (2010) used Orbitrap Fourier transform (FT) mass spectrometry to determine the mass fragmentation pattern of α-solanine, α-chaconine and aglycones (demissidine and solasodine). They used the linear ion trap mass spectrometry, multistage collision-induced dissociation (CID), and higher energy collision-induced dissociation (HCD). Through the hybrid mass spectrometer MS(3) were produced from MS(2) product ions. The proposed ion formulae were confirmed using accurate mass data in the MS(3) spectra. Loss of up to sugars from different regions was observed from the precursor and product ions from glycoalkaloids. Lohne et al. (2015) reported the presence of potato glycoalkaloids in the dog food when analysed using single-stage Orbitrap mass spectrometer in combination with differential analysis software.

11.3.6 Gas Chromatography

In 1977, Roosen-Runge and Schneider had reported the glycoalkaloids extraction from potato using pyridine and their subsequent estimation using gas chromatography (GC) after silylation. King (1980) developed gas–liquid chromatography based method for glycoalkaloids analysis from potato. For the determination of components of the alkaloids they performed hydrolysis with 1 N HCl in ethanol. 0.25 mg/100 g was found to be the lower limit of sensitivity of the developed method. The developed method was also found to be capable of differentiating between demissidine and solanidine on the basis of formation of corresponding β-trifluoroacetates. Laurila et al. (1999) developed gas chromatography/mass spectrometry (GC-MS) based methods for estimation of solanidine- and tomatidine-type glycoalkaloid aglycones. Pentafluoropropionylation and trimethylsilylation were used in the developed method. This resulted in the production of more abundant and specific fragmentation. The developed method could suitably and simultaneously determine both solanidane- and spirosolane-type steroidal glycoalkaloid aglycones from a range of *Solanum* species and hybrids. All steps of the developed method (extraction of glycoalkaloids, their hydrolysis, derivatization, and quantification) were found to be highly reproducible with less than 6% coefficient of variation.

11.3.7 Matrix-Assisted Laser Desorption/Ionization: Time of Flight (MALDI-TOF) Mass Spectrometry

Driedger and Sporns (1999) employed MALDI-TOF mass spectrometric method for determination of glycoalkaloids concentration in potatoes. Extraction of samples were carried out using methanol-water and deposited on 2,4,6-trihydroxyacetophenone crystals. Positive ions were analysed with a MALDI-TOF mass spectrometer having 337 nm laser. The detection limit of the developed method was shown to be 2 μg/g. For purification of potato glycoalkaloids from blood serum and their subsequent detection, Driedger and Sporns (2001) developed immunoaffinity and MALDI-TOF mass spectrometry based method. C_{18} solid-phase extraction cartridges were used for extracting potato glycoalkaloids from serum and were captured on agarose beads coated with antibody and finally eluted with methanol. The extracted glycoalkaloids when detected using MALDI-TOF mass spectrometer it was observed that purification of glycoalkaloids using the immunoaffinity purification method reduced the signal suppression significantly when compared with that of unpurified sample analysis.

11.3.8 Biosensors

Korpan et al. (2002) designed enzyme based biosensor for detection of glycoalkaloids. They used pH-sensitive field effect transistors as transducers and immobilized butyrylcholinesterase as a biorecognition element. The developed biosensor was capable of measuring the total glycoalkaloids of potatoes in 0.5–100 μM concentration range with the detection limits of 0.5 μM for α-chaconine and of 2.0 μM for α-solanine as well as for solanidine, respectively. Arkhypova et al. (2003) developed pH-sensitive field effect transistors and cholinesterases biosensors for detection of glycoalkaloids with improved sensitivity. The lowest limits of the detectable quantity by the developed biosensor, of alpha-chaconine and alpha-solanine and solanidine were found to be 0.2, 0.5, and 1.0 μM, respectively. Further, Soldatkin et al. (2005) reported the applicability of pH-sensitive field effect transistors based biosensor for detection of total glycoalkaloids content with high degree of reproducibility, less relative standard deviation of around 3% and with capability of operating directly on potato juice, or even directly on a suspension of potato or plant material. Also, they reported the operational and storage stability of the developed biosensor as quite good with a drift lower than 1% per day and response stability for more than 3 months. Korpan et al. (2006) investigated the use of the ion sensitive field effect transistor (ISFET)-based biosensor specificity exclusively to the glycoalkaloids. Utilization of EDTA and phosphotriesterase resulted in detection of glycoalkaloids concentration with high degree of specificity. The developed biosensor was validated for determination of glycoalkaloids in potato varieties of Ukraine. The results obtained were found to be comparable to those of HPLC and TLC. Further, Arkhypova et al. (2008) optimized biosensors based on pH-sensitive field effect transistors and butyrylcholinesterase to estimate glycoalkaloids. The biosensor was tested and used for glycoalkaloids estimation in tubers of different potato varieties. The results were found to be in line with those from other commonly used protocols. Espinoza et al. (2014) developed disposable biosensors based technology for detection of glycoalkaloids. The methodology was based on the cholinesterase-based sensors. The high sensitivity of the biosensor was attributed to the use of a genetically engineered enzyme acetylcholinesterase, in combination with a single-step detection method on the basis of inhibition slope measurement. Use of butyrylcholinesterase, an enzyme able to detect α-solanine and α-chaconine with differential inhibition kinetics ensured the selectivity. Immobilization of the enzymes was carried out directly on the working electrode surface via entrapment in PVA-AWP polymer. The developed method allowed the detection of glycoalkaloids at 50 ppb in potato matrix.

11.3.9 Other Lesser Used Methods

Bianco et al. (2002) developed nonaqueous capillary electrophoresis-electrospray ionization-ion trap mass spectrometry method for detection of glycoalkaloids and aglycones. For separation and detection of these alkaloids, a solvent comprising acetonitrile:methanol (90:10), 1.2 M acetic acid, and 50 mM ammonium acetate was used, with the voltage of 25.5 kV. Mass spectra obtained using the positive ionization mode showed the presence of protonated molecular ion of solanine, chaconine, and solanidine as m/z of 868, 852, and 398, respectively. Use of MS/MS mode further improved the quantification of the alkaloids.

Backleh et al. (2004) developed adsorptive bubble separation with a pH gradient based method for enrichment of α-solanine, α-chaconine, and solanidine aglycone from potato juice. Gas flow rate, bubble size, and pH value were shown to be influencing the enrichment of the alkaloids into the foam. Concentrations of samples in the range of 2–6 mg per L of the alkaloids, having pH 6.0, exhibited highest efficiency. Glossman-Mitnik (2007) used CHIH-DFT chemistry model within Density Functional Theory, to calculate the molecular structure, infrared and ultraviolet spectra of solanidine. The calculated values were comparable with the experimental data available for the solanidine.

Attoumbré et al. (2013) developed a preparative method of isolation of solanidine from potato using centrifugal partition chromatography (CPC). Initially, they developed a process for getting solanidine by hydrolysing the glycoalkaloids obtained from peel and sprouts of potato. Subsequently, isolation of solanidine was carried out by optimizing CPC parameters. Acetate/butanol/water (42.5:7.5:50) was used as solvent for extraction of glycoalkaloids from crude extract. Flow rate of the mobile phase was kept at 8 mL/min and rotation was maintained at 2500 rpm. One-step process resulted in 86.7% solanidine recovery from the crude extract with a yield of 98 mg of solanidine per run, having more than 98% purity. Hossain et al. (2014) compared extraction of steroidal alkaloids from potato peel waste employed by ultrasonic assisted extraction and by solid–liquid extraction. In both the methods methanol was used for extraction of steroidal alkaloids from potato peel waste. Further, optimal conditions were identified (amplitude 61 μm, extraction time 17 min) for ultrasonic assisted extraction of glycoalkaloids and aglycones (-α-chaconine, α-solanine, demissidine, and solanidine). 1102 μg steroidal alkaloids were recovered from per g dried potato peel, when ultrasonic assisted extraction was used. On the other hand solid–liquid extraction resulted in recovery of 710.51 glycoalkaloid μg/g dried potato peel. Recoveries of α-solanine, α-chaconine, solanidine, and demissidine were observed to be 273, 542.7, 231, and 55.3 μg/g dried potato peel, in case of ultrasonic assisted extraction. Solid–liquid extraction yielded 180.3,

337.6, 160.2, and 32.4 μg α-solanine, α-chaconine, solanidine, and demissidine, respectively, from 1 g of dried potato peel. This study indicated that UAE had strong potential as an extraction method for steroidal alkaloids from potato peel waste. Tata et al. (2015) used imprint imaging desorption electrospray ionization mass spectrometry (DESI-MS) for monitoring fluctuations of glycoalkaloids present in *Pythium ultimum* infected sprouted potatoes. The main advantage was suggested to be the requirement of minimal sample preparation for glycosides analysis. A decrease of the relative abundance of α-chaconine, α-solanine, and an increase in the relative abundance of solasodenone, solanidine, solasodiene, solanaviol, solaspiralidine, γ-chaconine, γ-solanine, β-chaconine, and β-solanine was observed on day 8 after the inoculation.

11.4 Factors Affecting Glycoalkaloids in Potato

Glycoalkaloids levels in potato have been reported to be affected by a range of factors associated with plants growth and development, plant interaction with surrounding environmental cues, genetic factors, etc. Tjamos and Kuć (1982) reported that arachidonic acids and eicosapentaenoic from *Phytophthora infestans* inhibited accumulation of glycoalkaloids in potato tubers. Uppal (1987) studied the environmental and varietal effect on the glycoalkaloids content of potato. He reported the glycoalkaloids content in the tubers of 14 potato varieties is within the safe limits for human consumption. Total glycoalkaloids in the peels represented 60–70% of the total glycoalkaloids in the whole tuber. A correlation was also observed in glycoalkaloids level in tubers and leaves. Exposure to sunlight led to the increase in the glycoalkaloids content in the peels @ 1.9 mg/100 g fresh weight per day.

Munshi and Mondy (1992) reported that application of selenium by banding resulted into greater decreases in glycoalkaloid content in potato tubers as compared to that of application as foliar spray. However, Turakainen et al. (2004) demonstrated that acceptable application levels of selenate had no effect on glycoalkaloid concentration in immature potato tubers. Griffiths and Dale (2001) studied the effect of light exposure on the glycoalkaloid content of potato tubers. They used 11 genotypes of *Solanum phureja* and estimated and compared the individual glycoalkaloids prior to and after the light exposure. In all the genotypes, light exposure resulted in an increase in total glycoalkaloid content. Nine genotypes also accumulated a tomatidenol-based glycoalkaloid, α-solamarine in addition to solanidine-based glycoalkaloids, α-solanine and α-chaconine. King and Calhoun (2012) had reported that although the normal levels and types of glycoalkaloids found in commercial varieties of potato presented no hazard to human health; however, domestic crosses with a wild *Solanum oplocense* accession resulted in appearance of novel (non-indigenous) glycoalkaloids in the resulting cross cultivars, thus suggested the need to monitor glycoalkaloids in all breeding programs involving wild potato species. Petersson et al. (2013) reported that light exposure and mechanical wounding raised glycoalkaloids levels in potato tubers, though the relative response was seen to be varying among cultivars. On the other hand, elevated temperature had no effect on glycoalkaloids content. Thus the study demonstrated a large variation among potato cultivars with regard to post-harvest glycoalkaloids changes. Shepherd et al. (2016) studied the impact of light exposure on the metabolites of transgenic potato tubers with manipulated synthesis of glycoalkaloids. The metabolic composition of transgenic potato with reduced glycoalkaloids content via the down-regulation of the SGT1 gene revealed that transgenic tubers exhibited an almost complete knock-out of α-solanine production and light had less impact on its accumulation. Exposure to light led to significant increase in levels of α-chaconine in the peel of both the control and transgenic lines, though more particularly in the transgenic line. Zhang et al. (2019) reported that steroidal glycoalkaloids content in potato tubers increased when exposed to light, along with the accumulation of steroidal glycoalkaloids, expression of abiotic as well as biotic stress-responsive genes was also induced.

11.5 Effect of Post-Harvest Storage and Processing of Potato on Glycoalkaloids

Studies on the effect of post-harvest storage parameters and processing of potato tubers have been undertaken by various researchers. Wilson et al. (1983) studied the effect of growth-location and length of storage on glycoalkaloids content of roadside-stand potatoes as stored by consumers. In a simple experiment, they purchased potatoes from roadside stands at 25 locations in the State of Maine and were stored from 1 to 3 months under home storage conditions at 12.2 °C. Alpha-chaconine, α-solanine, and total glycoalkaloids content were estimated in the stored tubers, after 1 and 3 months of storage. Mean α-chaconine, α-solanine, and total glycoalkaloids content of the tubers ranged from 0.41 to 3.45, 0.35 to 1.51, and 0.75 to 6.16 mg/100 g wet weight of tuber, respectively. The interaction of location and storage time was found to be significant on concentration of the individual as well as total glycoalkaloids in the tubers. The results also demonstrated that storage under those suboptimum conditions did not cause an increase of glycoalkaloids to a toxic level. Grunenfelder et al. (2006) developed the colour photographic indices for grading fresh market potatoes for estimating extent of greening under lighting conditions prevailing in retail markets. The scale was calibrated for total glycoalkaloids content by measuring total glycoalkaloids and chlorophyll accumulation in four potato cultivars, over the respective greening scales. Total glycoalkaloids concentration in tubers of two of the studied varieties exhibited high degree of correlation with greening level and chlorophyll concentrations. Using the developed scale the peels were found to contain 3.4–6.8-fold higher total glycoalkaloids concentration than the flesh. The total glycoalkaloids concentration in periderm ranged from 37 to 160 mg/100 g dry weight.

Mäder et al. (2009) estimated the composition of glycoalkaloids: alpha-solanine and alpha-chaconine during commercial potato processing. Processing of potatoes to potato flakes resulted in remarkable decrease in alpha-solanine and alpha-chaconine content. The decrease was mainly attributed to peeling and leaching. The influence of thermal exposure was found to be less significant. About 10% of the initial glycoalkaloids remained unchanged after processing. Lachman et al. (2013) studied the effect of peeling and three cooking methods on glycoalkaloids content in potato tubers with various colour of flesh. They used HPLC-ESI/MS/MS for estimation of glycoalkaloid, α-chaconine, and α-solanine. They observed that all cooking treatments reduced total glycoalkaloids, α-chaconine, and α-solanine. Boiling of peeled tubers resulted in the highest decrease in total glycoalkaloids content among the three tested cooking methods. Rytel et al. (2013) studied the effect of different temperatures of blanching and pre-drying used in the production of dried potato dice on the content of glycoalkaloids in red and blue fleshed potato varieties. They reported that peeling and blanching removed 70% and 29% of the glycoalkaloids, respectively. Further, blanching of potato dice at 85 °C and pre-drying at 120 °C showed the lowest glycoalkaloids concentration. On the other hand the highest glycoalkaloids content was found in blanched potato dice at the 65 °C and pre-dried at 120 °C.

Decayed potatoes are reported to contain harmful metabolites including glycoalkaloids. Thus, for safety consideration, the decayed potatoes are discarded from the lots aimed for consumption or processing. However, it leads to loss in net yield and thus economic loss. Keeping this in view Koffi et al. (2017) investigated the possibility of using these discarded potatoes to generate value added compounds from the toxic glycoalkaloids. They isolated glycoalkaloids from green, sprouting, and rotting potato tubers and subjected them to chemoenzymatic treatment for solanidine recovery. The glycoalkaloids in the flesh of different types of decayed potatoes were extracted, estimated, and compared. The results showed that turned green, sprouting, and rotting potato flesh contained high amounts of solanine and chaconine, exceeding by two to fivefold of the recommended limit, and ranging from 2578–5063 mg/kg of dry weight potato flesh. Further, they developed a simple chemoenzymatic protocol comprising a partial acidic hydrolysis followed by an enzymatic treatment with the β-glycosidase enzyme which led to the efficient conversion of α-chaconine to solanidine. Liu et al. (2020) studied the effect of acid soaking as a pre-treatment on the glycoalkaloids content in French fries. It was observed that soaking raw potato cuts in acetic acid solutions for at least 8 h reduced 90% of α-solanine and α-chaconine in potato samples. Further, processing of pre-acid soaked potato cuts into French fries led to an additional more than 50% decrease in the glycoalkaloids content. Thus, the findings demonstrated that acid soaking was an effective way to reduce/minimize glycoalkaloids level in potatoes.

11.6 Degradation or Modification of Glycoalkaloids

Oda et al. (2002) isolated three strains of filamentous fungi from potato sprouts to obtain glycoalkaloids degrading enzyme. All of the strains were found to be capable of hydrolysing α-chaconine and not α-solanine. For the *Plectosphaerella cucumerina* strain α-chaconine hydrolysing enzyme was purified whose kinetic studies suggested the purified enzyme to be rhamnosidase specific for the hydrolysis of the rhamnose–glucose linkage in α-chaconine. For this study the authors hypothesized that conversion of α-chaconine to β1-chaconine might be the first step of detoxification for filamentous fungi to grow on potato sprouts that accumulate antifungal α-chaconine. Jensen et al. (2009a) analysed the fate of potato glycoalkaloids in potato field. They surveyed the glycoalkaloids in groundwater, plant, and soil in a potato field to estimate their distribution and fate during the season. In plants, glycoalkaloids content was up to 25 kg/ha during maturity and decreased to below 0.63 kg/ha during plant senescence. The upper soil had only up to 0.6 kg/ha of glycoalkaloids, whereas in the groundwater, no traces of glycoalkaloids were detected. From the study it was concluded that possibility/potential of leaching of the glycoalkaloids was very less. Further, Jensen et al. (2009b) studied the degradation of the potato α-solanine in three agricultural soils. They followed the degradation of potato α-solanine glycoalkaloid for 42 days in three agricultural soils with common texture and carbon contents. A similar degradation pattern was found in all soils. Overall, degradation rates for the initial first reaction were found to be in the range 0.22–1.64/day. And the estimated half-lives were shown to be in the range 1.8–4.1 days for the three top soils at 15 °C; the fastest degradation was observed in the sandy soil. From this study the authors concluded that overall, fast degradation of α-solanine was found in both top and subsoil even at low temperatures, and thus the risk for α-solanine leaching to the groundwater appeared to be low. Also, Jensen et al. (2009c) studied the degradation of the potato glycoalkaloids in groundwater. They followed abiotic and microbial degradation of alpha-solanine and alpha-chaconine in groundwater sampled from below a potato field and spiked with the glycoalkaloids (115 nmol/L). Degradation of the spiked glycoalkaloids was found to be primarily microbial and the glycoalkaloids were degraded within 21–42 days. The metabolites β(1)-solanine, γ-solanine, and solanidine were formed from α-solanine, while β-chaconine, γ-chaconine, and solanidine were detected from α-chaconine. The findings suggested that indigenous groundwater microorganisms are capable of degrading the glycoalkaloids.

Hennessy et al. (2019) sequenced two Glycoalkaloid-degrading *Arthrobacter* strains isolated from green potato peel. The genome sequence based information might be vital for deciphering the genetic mechanism of glycoalkaloids degradation by these *Arthrobacter* strains. Hennessy et al. (2020) have reported a bacterial gene cluster for deglycosylation of potato steroidal glycoalkaloids (α-chaconine and α-solanine). The first step in the detoxification of both α-chaconine and α-solanine is the removal of the trisaccharide. They performed whole-genome sequencing of a bacterial isolate, *Arthrobacter sp.* S41, which is capable of completely degrading α-chaconine and α-solanine and identified a gene cluster probably involved in the deglycosylation of glycoalkaloids. Functional characterization confirmed the enzymatic activity of the gene cluster involved in the complete deglycosylation of both α-chaconine and α-solanine. The novel enzymes capable of deglycosylation of the potato glycoalkaloids might be quite useful in the bioconversion of feed proteins (from potato) to food proteins suitable for human nutrition.

11.7 Key Genes for Manipulating Glycoalkaloids in Potato

A range of studies have been conducted globally in order to identify and use the genes and the corresponding enzymes involved in the synthesis of glycoalkaloids, for manipulating glycoalkaloids level in potato. Two main glycoalkaloids of potato i.e. α- solanine and α-chaconine are synthesized from geranyl pyrophosphate (GPP). GPP in turn is synthesized via mevalonate and/or DOXP pathway (Fig. 11.2). McCue et al. (2007) proved that glycosterol rhamnosyltransferase catalyses the terminal step in triose side-chain (of steroidal glycoalkaloids) biosynthesis in potato. Krits et al. (2007) reported that potato genotypes exhibiting different levels of steroid glycoalkaloid content showed an association between high steroid glycoalkaloid levels in their leaves and tubers and high expression of 3-hydroxy-3-methylglutaryl coenzyme A reductase 1 (hmg1) and squalene synthase 1 (pss1), genes of the mevalonic/isoprenoid pathway. Sørensen et al. (2008) mapped a highly significant QTL to chromosome I in potato tubers from *Solanum tuberosum* x *S. sparsipilum*. Ginzberg et al. (2012) overexpressed HMG-CoA reductase and squalene synthase genes in potato and observed higher glycoalkaloids level in the developed transgenics.

Itkin et al. (2013) performed comparative co-expression analysis between tomato and potato coupled with chemical profiling and observed that an array of ten genes is

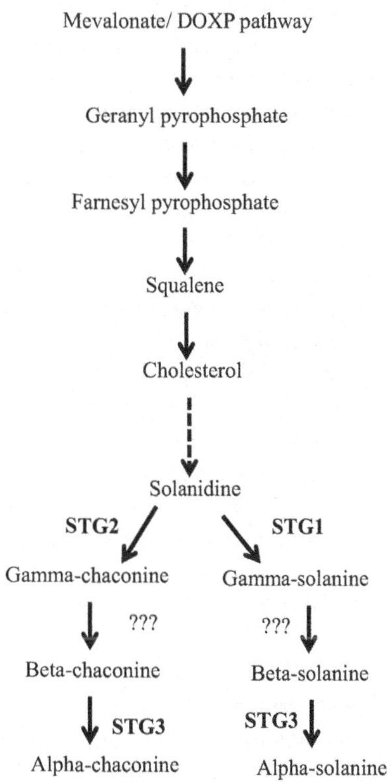

Fig. 11.2 Glycosides biosynthesis pathway in potato. STG1: Solanine UDP-galactose galactosyltransferase; STG2: Solanidine glucosyltransferase; STG3: Rhamnosyltransferase

involved in steroidal glycoalkaloids biosynthesis in potato. Six of the ten genes were mapped as a cluster on chromosome 7, and an additional two were found to be adjacent in a duplicated genomic region on chromosome 12. Manrique-Carpintero et al. (2014) reported allelic variation in genes of glycoalkaloid biosynthesis pathway in a diploid interspecific population of potato. Allelic sequences spanning coding regions of four candidate genes (3-hydroxy-3-methylglutaryl coenzyme A reductase, squalene epoxidase, solanidine galactosyltransferase, and solanidine glucosyltransferase) were cloned from two potato species differing in glycoalkaloid composition (*Solanum chacoense* and *Solanum tuberosum* group Phureja). A whole-genome SNP genotyping analysis of F2 subsample verified the importance of alleles at two of the four genes for glycoalkaloids synthesis and accumulation. Shepherd et al. (2015) used transgenic approach for manipulating glycoalkaloids content in potato. They further performed metabolite profiling to assess the potential composition changes in potato genetically modified by independent down-regulation of three genes (SGT1, SGT2, and SGT3) to reduce glycoalkaloids content (Fig. 11.2). Metabolic profiling revealed expected changes in specific glycoalkaloids level in the transgenic lines. Mariot et al. (2016) characterized the genes involved in glycoalkaloids biosynthesis in new varieties of potato. Using HPLC-MS for total glycoalkaloids quantification and RT-PCR for expression analysis of GAME, SGT1, and SGT3 genes relationship between total glycoalkaloid content and the relative expression of GAME, SGT1, and SGT3 genes in potato tubers was revealed. Umemoto et al. (2016) identified two cytochrome P450 monooxygenases (PGA1 and PGA2) associated with biosynthesis of steroid glycoalkaloids in potato. Using transgenic approach knockdown plants for the two identified cytochrome P450 monooxygenase genes were generated which contained very little glycoalkaloids without any effect on vegetative growth and tuber production. From the results of in vitro enzyme assays it was suggested that PGA1 and PGA2 catalysed the 26- and 22-hydroxylation steps, respectively, in the glycoalkaloids biosynthetic pathway.

Kumar et al. (2017) identified a potato lanosterol synthase (LAS)-related gene and used it for developing potato transgenic lines. Expression analysis of the transgene and the estimation of steroidal glycoalkaloids in the transgenic lines indicated the involvement of lanosterol synthase spatial variation in steroidal glycoalkaloids in potato. Nakayasu et al. (2017) established that a dioxygenase enzyme catalyses steroid 16α-hydroxylation in steroidal glycoalkaloid biosynthesis in potato. Nahar et al. (2017) investigated two potato cultivars differing in their glycoalkaloids accumulation during wounding or light exposure, to identify genes responsible for variation in tuber glycoalkaloids levels. From the data obtained from microarray analysis, sterol and glycoalkaloids quantification, and sense/antisense transgenic plants analysis the authors identified two of these genes, encoding distinct types of sterol Δ24-reductases which work co-ordinately to regulate glycoalkaloids synthesis in potato in response to wounding and light exposure. Zhang et al. (2019) reported that steroidal glycoalkaloids content in potato tubers increased when exposed to light. Along with the accumulation of steroidal glycoalkaloids, expression of abiotic as well as biotic stress-responsive genes was also induced. Yasumoto et al. (2019) developed a highly active Platinum TALEN expression vector construction system, and applied to reduce the steroidal glycoalkaloids contents in potato. They targeted SSR2 gene, which encoded a key enzyme for steroidal glycoalkaloids biosynthesis and generated knock-out mutants having reduced levels of steroidal glycoalkaloids and having no effects on potato plant phenotype or yield.

11.8 Conclusion

Potato cultivars grown worldwide contain low amount of glycoalkaloids (mainly α-solanine and α-chaconine) in amounts much lower than the acceptable limit for human consumption. However, exposure of potato tubers during developmental and/or post-harvest storage stages to environmental factors (especially light and drought), mechanical wounding, or insect pest/pathogen encounters may raise glycoalkaloids to toxic levels, thus making them unfit for human consumption. Although, during cooking or other post-harvest processing glycoalkaloids are destroyed, yet potato cultivars having less glycoalkaloids are preferred over those with comparatively higher glycoalkaloids. Because of their low concentrations in potatoes and potential toxic effects, instrumentations and methodologies for their detection and quantification need to be highly sensitive, precise, and reproducible. Shorter half life of glycosides in soil and less possibility of them leaching into the groundwater, make glycosides as low potential threat to the environment.

References

Arkhypova VN, Dzyadevych SV, Soldatkin AP et al (2003) Development and optimisation of biosensors based on pH-sensitive field effect transistors and cholinesterases for sensitive detection of solanaceous glycoalkaloids. Biosens Bioelectron. https://doi.org/10.1016/S0956-5663(02)00222-1

Arkhypova VN, Dzyadevych SV, Jaffrezic-Renault N et al (2008) Biosensors for assay of glycoalkaloids in potato tubers. Appl Biochem Microbiol. https://doi.org/10.1134/S0003683808030162d

Attoumbré J, Giordanengo P, Baltora-Rosset S (2013) Solanidine isolation from Solanum tuberosum by centrifugal partition chromatography. J Sep Sci. https://doi.org/10.1002/jssc.201300188

Aziz A, Randhawa MA, Butt MS et al (2012) Glycoalkaloids (α-Chaconine and α-Solanine) contents of selected Pakistani potato cultivars and their dietary intake assessment. J Food Sci. https://doi.org/10.1111/j.1750-3841.2011.02582.x

Backleh M, Ekici P, Leupold G et al (2004) Enrichment of the glycoalkaloids α-solanine and α-chaconine from potato juice by adsorptive bubble separation using a pH gradient. J Sep Sci. https://doi.org/10.1002/jssc.200301685

Bianco G, Schmitt-Kopplin P, De Benedetto G et al (2002) Determination of glycoalkaloids and relative aglycones by nonaqueous capillary electrophoresis coupled with electrospray ionization-ion trap mass spectrometry. Electrophoresis. https://doi.org/10.1002/1522-2683(200209)23:17<2904::AID-ELPS2904>3.0.CO;2-1

Cadle LS, Stelzig DA, Harper KL, Young RJ (1978) Thin-layer chromatographic system for identification and quantitation of potato tuber glycoalkaloids. J Agric Food Chem. https://doi.org/10.1021/jf60220a033

Cahill MG, Caprioli G, Vittori S, James KJ (2010) Elucidation of the mass fragmentation pathways of potato glycoalkaloids and aglycons using Orbitrap mass spectrometry. J Mass Spectrom 45 (9):1019–1025. https://doi.org/10.1002/jms.1785

Carputo D, Savarese S, Andolfi A et al (2010) Glycoalkaloid profile in potato haploids derived from *Solanum tuberosum - S. bulbocastanum* somatic hybrids. Chem Biodivers. https://doi.org/10.1002/cbdv.200900391

Chen S, Derrick PJ, Mellon FA, Price KR (1994) Analysis of Glycoalkaloids from potato shoots and tomatoes by four-sector tandem mass spectrometry with scanning-Array detection: comparison of positive ion and negative ion methods. Anal Biochem. https://doi.org/10.1006/abio.1994.1155

Deusser H, Guignard C, Hoffmann L, Evers D (2012) Polyphenol and glycoalkaloid contents in potato cultivars grown in Luxembourg. Food Chem. https://doi.org/10.1016/j.foodchem.2012.07.028

Driedger DR, Sporns P (1999) Glycoalkaloid concentration in by-products of potato starch extraction as measured by matrix-assisted laser desorption/ionization mass spectrometry. J Food Process Preserv. https://doi.org/10.1111/j.1745-4549.1999.tb00392.x

Driedger DR, Sporns P (2001) Immunoaffinity sample purification and MALDI-TOF MS analysis of α-solanine and α-chaconine in serum. J Agric Food Chem. https://doi.org/10.1021/jf0004513

Driedger DR, Leblanc RJ, Leblanc EL, Sporns P (2000a) A capillary electrophoresis laser-induced fluorescence method for analysis of potato glycoalkaloids based on a solution-phase immunoassay. 2. Performance evaluation. J Agric Food Chem. https://doi.org/10.1021/jf000157a

Driedger DR, LeBlanc RJ, LeBlanc EL, Sporns P (2000b) A capillary electrophoresis laser-induced fluorescence method for analysis of potato glycoalkaloids based on a solution-phase immunoassay. 1. Separation and quantification of immunoassay products. J Agric Food Chem. https://doi.org/10.1021/jf990680t

Espinoza MA, Istamboulie G, Chira A et al (2014) Detection of glycoalkaloids using disposable biosensors based on genetically modified enzymes. Anal Biochem. https://doi.org/10.1016/j.ab.2014.04.005

Friedman M, Roitman JN, Kozukue N (2003) Glycoalkaloid and calystegine contents of eight potato cultivars. J Agric Food Chem. https://doi.org/10.1021/jf021146f

Ginzberg I, Thippeswamy M, Fogelman E et al (2012) Induction of potato steroidal glycoalkaloid biosynthetic pathway by overexpression of cDNA encoding primary metabolism HMG-CoA reductase and squalene synthase. Planta. https://doi.org/10.1007/s00425-011-1578-6

Glorio-Paulet P, Durst RA (2000) Determination of potato glycoalkaloids using a liposome immunomigration, liquid-phase competition immunoassay. J Agric Food Chem. https://doi.org/10.1021/jf990349+

Glossman-Mitnik D (2007) CHIH-DFT determination of the molecular structure and IR and UV spectra of solanidine. J Mol Model 13(1):43–46. https://doi.org/10.1007/s00894-006-0123-1

Griffiths DW, Dale MFB (2001) Effect of light exposure on the glycoalkaloid content of *Solanum phureja* tubers. J Agric Food Chem. https://doi.org/10.1021/jf010656r

Grunenfelder LA, Knowles LO, Hiller LK, Knowles NR (2006) Glycoalkaloid development during greening of fresh market potatoes (*Solanum tuberosum* L.). J Agric Food Chem. https://doi.org/10.1021/jf0607359

Harvey MH, Morris BA, Mcmillan M, Marks V (1985) Measurement of potato steroidal alkaloids in human serum and saliva by radioimmunoassay. Hum Exp Toxicol. https://doi.org/10.1177/096032718500400506

Hellenäs KE, Branzell C (1997) Liquid chromatographic determination of the glycoalkaloids alpha-solanine and alpha-chaconine in potato tubers: NMKL Interlaboratory study. Nordic committee on food analysis. J AOAC Int 80(3):549–554

Hennessy RC, Park B, Pathiraja D et al (2019) Draft genome sequences of two glycoalkaloid-degrading arthrobacter strains isolated from green potato peel. Microbiol Resour Announc. https://doi.org/10.1128/mra.00226-19

Hennessy RC, Nielsen SD, Greve-Poulsen M et al (2020) Discovery of a bacterial gene cluster for Deglycosylation of toxic potato steroidal Glycoalkaloids α-Chaconine and α-Solanine. J Agric Food Chem. https://doi.org/10.1021/acs.jafc.9b07632

Hossain MB, Tiwari BK, Gangopadhyay N et al (2014) Ultrasonic extraction of steroidal alkaloids from potato peel waste. Ultrason Sonochem. https://doi.org/10.1016/j.ultsonch.2014.01.023

Itkin M, Heinig U, Tzfadia O et al (2013) Biosynthesis of antinutritional alkaloids in solanaceous crops is mediated by clustered genes. Science 80. https://doi.org/10.1126/science.1240230

Jensen PH, Juhler RK, Nielsen NJ et al (2008) Potato glycoalkaloids in soil-optimising liquid chromatography-time-of-flight mass spectrometry for quantitative studies. J Chromatogr A. https://doi.org/10.1016/j.chroma.2007.12.049

Jensen PH, Jacobsen OS, Henriksen T et al (2009a) Degradation of the potato glycoalkaloids - α-solanine and α-chaconine in groundwater. Bull Environ Contam Toxicol. https://doi.org/10.1007/s00128-009-9698-4

Jensen PH, Pedersen RB, Svensmark B et al (2009b) Degradation of the potato glycoalkaloid α-solanine in three agricultural soils. Chemosphere. https://doi.org/10.1016/j.chemosphere.2009.04.008

Jensen PH, Strobel BW, Hansen HCB, Jacobsen OS (2009c) Fate of toxic potato Glycoalkaloids in a potato field. J Agric Food Chem. https://doi.org/10.1021/jf803564v

King RR (1980) Analysis of potato glycoalkaloids by gas-liquid chromatography of alkaloid components. J Assoc Off Anal Chem. https://doi.org/10.1093/jaoac/63.6.1226

King RR, Calhoun LA (2012) Complete 1H and 13C NMR spectral assignments for the glycoalkaloid dehydrocommersonine. Magn Reson Chem. https://doi.org/10.1002/mrc.3844

Koffi GY, Remaud-Simeon M, Due AE, Combes D (2017) Isolation and chemoenzymatic treatment of glycoalkaloids from green, sprouting and rotting Solanum tuberosum potatoes for solanidine recovery. Food Chem. https://doi.org/10.1016/j.foodchem.2016.10.014

Korpan YI, Volotovsky VV, Martelet C et al (2002) A novel enzyme biosensor for steroidal glycoalkaloids detection based on pH-sensitive field effect transistors. Bioelectrochemistry. https://doi.org/10.1016/s1567-5394(01)00150-5

Korpan YI, Raushel FM, Nazarenko EA et al (2006) Sensitivity and specificity improvement of an ion sensitive field effect transistors-based biosensor for potato glycoalkaloids detection. J Agric Food Chem. https://doi.org/10.1021/jf0529316

Krits P, Fogelman E, Ginzberg I (2007) Potato steroidal glycoalkaloid levels and the expression of key isoprenoid metabolic genes. Planta. https://doi.org/10.1007/s00425-007-0602-3

Kumar A, Fogelman E, Weissberg M et al (2017) Lanosterol synthase-like is involved with differential accumulation of steroidal glycoalkaloids in potato. Planta. https://doi.org/10.1007/s00425-017-2763-z

Kuronen P, Väänänen T, Pehu E (1999) Reversed-phase liquid chromatographic separation and simultaneous profiling of steroidal glycoalkaloids and their aglycones. J Chromatogr A. https://doi.org/10.1016/S0021-9673(99)00954-1

Lachman J, Hamouz K, Musilová J et al (2013) Effect of peeling and three cooking methods on the content of selected phytochemicals in potato tubers with various colour of flesh. Food Chem. https://doi.org/10.1016/j.foodchem.2012.11.114

Laurila J, Laakso I, Väänänen T et al (1999) Determination of solanidine- and tomatidine-type glycoalkaloid aglycons by gas chromatography/mass spectrometry. J Agric Food Chem. https://doi.org/10.1021/jf981009b

Liu H, Roasa J, Mats L et al (2020) Effect of acid on glycoalkaloids and acrylamide in French fries. Food Addit Contam Part A Chem Anal Control Expo Risk Assess. https://doi.org/10.1080/19440049.2020.1743883

Lohne JJ, Turnipseed SB, Andersen WC et al (2015) Application of single-stage orbitrap mass spectrometry and differential analysis software to nontargeted analysis of contaminants in dog food: detection, identification, and quantification of glycoalkaloids. J Agric Food Chem. https://doi.org/10.1021/acs.jafc.5b00959

Mäder J, Rawel H, Kroh LW (2009) Composition of phenolic compounds and glycoalkaloids α-solanine and α-chaconine during commercial potato processing. J Agric Food Chem. https://doi.org/10.1021/jf901066k

Manrique-Carpintero NC, Tokuhisa JG, Ginzberg I, Veilleux RE (2014) Allelic variation in genes contributing to glycoalkaloid biosynthesis in a diploid interspecific population of potato. Theor Appl Genet. https://doi.org/10.1007/s00122-013-2226-2

Mariot RF, De Oliveira LA, Voorhuijzen MM et al (2016) Characterization and transcriptional profile of genes involved in glycoalkaloid biosynthesis in new varieties of *Solanum tuberosum* L. J Agric Food Chem. https://doi.org/10.1021/acs.jafc.5b05519

Matsuda F, Morino K, Miyazawa H et al (2004) Determination of potato glycoalkaloids using high-pressure liquid chromatography-electrospray ionisation/mass spectrometry. Phytochem Anal. https://doi.org/10.1002/pca.755

McCue KF, Allen PV, Shepherd LVT et al (2007) Potato glycosterol rhamnosyltransferase, the terminal step in triose side-chain biosynthesis. Phytochemistry. https://doi.org/10.1016/j.phytochem.2006.10.025

Munshi CB, Mondy NI (1992) Glycoalkaloid and nitrate content of potatoes as affected by method of selenium application. Biol Trace Elem Res. https://doi.org/10.1007/BF02784000

Nahar N, Westerberg E, Arif U et al (2017) Transcript profiling of two potato cultivars during glycoalkaloid-inducing treatments shows differential expression of genes in sterol and glycoalkaloid metabolism. Sci Rep. https://doi.org/10.1038/srep43268

Nakayasu M, Umemoto N, Ohyama K et al (2017) A dioxygenase catalyzes steroid 16-α-hydroxylation in steroidal glycoalkaloid biosynthesis. Plant Physiol. https://doi.org/10.1104/pp.17.00501

Nielsen SD, Schmidt JM, Kristiansen GH et al (2020) Liquid chromatography mass spectrometry quantification of α-solanine, α-chaconine, and solanidine in potato protein isolates. Foods. https://doi.org/10.3390/foods9040416

Nikolic NC, Stankovic MZ (2003) Solanidine hydrolytic extraction and separation from the potato (*Solanum tuberosum* L.) vines by using solid-liquid-liquid systems. J Agric Food Chem. https://doi.org/10.1021/jf020426s

Oda Y, Saito K, Ohara-Takada A, Mori M (2002) Hydrolysis of the potato glycoalkaloid α-chaconine by filamentous fungi. J Biosci Bioeng. https://doi.org/10.1263/jbb.94.321

Panovská Z, Hajšlová J, Kosinková P, Čepl J (1997) Glycoalkaloid content of potatoes sold in Czechia. Nahrung – Food. https://doi.org/10.1002/food.19970410306

Petersson EV, Arif U, Schulzova V et al (2013) Glycoalkaloid and calystegine levels in table potato cultivars subjected to wounding, light, and heat treatments. J Agric Food Chem. https://doi.org/10.1021/jf400318p

Roosen-Runge C, Schneider E (1977) Zur Bestimmung der Solanum-Alkaloide Solanin und Chaconin on the determination of the solanum alkaloids solanine and chaconine. ZeitschriftfrLeb und -forsch. https://doi.org/10.1007/bf01354310

Rytel E, Tajner-Czopek A, Aniolowska M, Hamouz K (2013) The influence of dehydrated potatoes processing on the glycoalkaloids content in coloured-fleshed potato. Food Chem. https://doi.org/10.1016/j.foodchem.2013.04.131

Shakya R, Navarre DA (2008) LC-MS analysis of solanidane glycoalkaloid diversity among tubers of four wild potato species and three cultivars (*Solanum tuberosum*). J Agric Food Chem. https://doi.org/10.1021/jf8006618

Shepherd LVT, Hackett CA, Alexander CJ et al (2015) Modifying glycoalkaloid content in transgenic potato - metabolome impacts. Food Chem. https://doi.org/10.1016/j.foodchem.2015.04.111

Shepherd LVT, Hackett CA, Alexander CJ et al (2016) Impact of light-exposure on the metabolite balance of transgenic potato tubers with modified glycoalkaloid biosynthesis. Food Chem. https://doi.org/10.1016/j.foodchem.2015.12.095

Shimoi T, Ushiyama H, Kan K et al (2007) Survey of glycoalkaloids content in the various potatoes. J Food Hyg Soc Japan. https://doi.org/10.3358/shokueishi.48.77

Simonovska B, Vovk I (2000) High-performance thin-layer chromatographic determination of potato glycoalkaloids. J Chromatogr A. https://doi.org/10.1016/S0021-9673(00)00900-6

Singh B, Raigond P, Barwal S et al (2016) Glycoalkaloids in peels of Indian potatoes. Potato J 43:86–92

Soldatkin AP, Arkhypova VN, Dzyadevych SV et al (2005) Analysis of the potato glycoalkaloids by using of enzyme biosensor based on pH-ISFETs. Talanta. https://doi.org/10.1016/j.talanta.2004.11.006

Sørensen KK, Kirk HG, Olsson K et al (2008) A major QTL and an SSR marker associated with glycoalkaloid content in potato tubers from Solanum tuberosum x S. sparsipilum located on chromosome I. Theor Appl Genet. https://doi.org/10.1007/s00122-008-0745-z

Tata A, Perez CJ, Hamid TS et al (2015) Analysis of metabolic changes in plant pathosystems by imprint imaging DESI-MS. J Am Soc Mass Spectrom. https://doi.org/10.1007/s13361-014-1039-0

Tjamos EC, Kuć JA (1982) Inhibition of steroid glycoalkaloid accumulation by arachidonic and eicosapentaenoic acids in potato. Science 80. https://doi.org/10.1126/science.217.4559.542

Turakainen M, Väänänen T, Anttila K et al (2004) Effect of selenate supplementation on glycoalkaloid content of potato (*Solanum tuberosum* L.). J Agric Food Chem. https://doi.org/10.1021/jf049132s

Umemoto N, Nakayasu M, Ohyama K et al (2016) Two cytochrome P450 monooxygenases catalyze early hydroxylation steps in the potato steroid glycoalkaloid biosynthetic pathway. Plant Physiol. https://doi.org/10.1104/pp.16.00137

Uppal DS (1987) Varietal and environmental effect on the glycoalkaloid content of potato (*Solanum tuberosum* L.). Plant Foods Hum Nutr. https://doi.org/10.1007/BF01092209

Valcarcel J, Reilly K, Gaffney M, O'Brien N (2014) Effect of genotype and environment on the glycoalkaloid content of rare, heritage, and commercial potato varieties. J Food Sci. https://doi.org/10.1111/1750-3841.12443

Wayumba BO, Choi HS, Seok LY (2019) Selection and evaluation of 21 potato (*Solanum tuberosum*) breeding clones for cold chip processing. Foods. https://doi.org/10.3390/foods8030098

Wilson AM, Mcgann DF, Bushway RJ (1983) Effect of growth-location and length of storage on glycoalkaloid content of roadside-stand potatoes as stored by consumers. J Food Prot. https://doi.org/10.4315/0362-028x-46.2.119

Yasumoto S, Umemoto N, Lee HJ et al (2019) Efficient genome engineering using platinum talein potato. Plant Biotechnol. https://doi.org/10.5511/plantbiotechnology.19.0805a

Zhang W, Zuo C, Chen Z et al (2019) RNA sequencing reveals that both abiotic and biotic stress-responsive genes are induced during expression of steroidal glycoalkaloid in potato tuber subjected to light exposure. Genes (Basel). https://doi.org/10.3390/genes10110920

New Health-Promoting Compounds in Potatoes

12

Pinky Raigond, Sastry S. Jayanty, and Som Dutt

Abstract

Potato (*Solanum tuberosum* L.) is a nutrient-dense vegetable consumed worldwide due to its long shelf life. There is a significant contribution of potato in meeting dietary requirements in both the developing and developed world. Colored-flesh potatoes are gaining popularity among consumers due to the higher concentration of phytochemicals. The antioxidant activity measured in colored-flesh potatoes is comparable to blueberries and pomegranates which are considered as superfoods. Potatoes are rich in starch but also contain a diverse set of nutrients such as amino acids, minerals, vitamins, lectins, proteinase inhibitors, terpenes polyamines, polyphenolic compounds, along with antinutritive compounds, i.e., glycoalkaloids. Chlorogenic acid and its derivatives are the predominant phenolic compounds in potato tubers that provide protection against chronic diseases. Based on the current knowledge, potato-derived nutrients have various potential health-promoting properties such as antioxidant, anticancer, anti-inflammatory, antidiabetic, anti-obesity, and hypocholesterolemic potential as reported by several researchers. Bioaccessibility and bioavailability of nutrients present in potatoes are higher compared with other staple crops. Further studies are needed to establish the possible in vivo efficacy of these components as well as their long-term health benefits. In this chapter, we review the recent relevant literature about potato metabolites and their potential health-promoting properties.

P. Raigond (✉) · S. Dutt
Crop Physiology, Biochemistry & Post-Harvest Technology Division, ICAR-Central Potato Research Institute, Shimla, Himachal Pradesh, India
e-mail: Pinky.Raigond@icar.gov.in; jariapink@gmail.com

S. S. Jayanty
San Luis Valley Research Center, Department of Horticulture and LA, Colorado State University, Fort Collins, CO, USA

P. Raigond et al. (eds.), *Potato*, https://doi.org/10.1007/978-981-15-7662-1_12

Keywords

Antidiabetic compounds · Anticancerous compounds · Antihypertensive compounds

12.1 Introduction

Importance of nutrition has been emphasized through various studies to combat diet and lifestyle related disorders such as diabetes, cancer, and cardiovascular diseases. Low intake of fruits and vegetables is considered as the sixth main risk factor for mortality. It has been reported that worldwide consumption of fruits and vegetables is less than the daily recommended serving (Raigond et al. 2018). Though potatoes are consumed in large quantities compared to other vegetables in most of the countries, they are always being unappreciated compared to other vegetables due to misconception related to their role as contributor to obesity and diabetes development. Potatoes are labeled as high carbohydrate food and people believe it to be a high calorie and high-fat food compared to other carbohydrate rich foods such as rice. Presence of relatively high concentrations of phytonutrients with bioactivities to combat chronic disease development is vastly underestimated in case of potatoes. Various studies have indicated potential of potatoes to confer health benefits in human cell culture, experimental animals, and human clinical studies and their anticancer, anti-inflammatory, hypocholesterolemic, antidiabetic, and anti-obesity potential has been reported by several researchers. Although potatoes contain antinutritional compounds such as glycoalkaloids, lectins, and proteinase inhibitors, these compounds when present in permissible limits have shown various health benefits (Burgos et al. 2019).

In recent years, researchers have mainly focused on quantifications and bioaccessibility/bioavailability studies of phenolics, carotenoids, anthocyanins, proteins, minerals, and vitamins (Mishra et al. 2019). Potato contain myriad of health-promoting compounds and are not only a good source of energy but also exhibit secondary metabolites that are essential to prevent many diseases (Fig. 12.1). Presence of high concentration of carbohydrates along with antidiabetic compounds such as antioxidants, biguanide and related compounds, and α-glucosidase inhibitors complicates the evaluation of potatoes for management/prevention of diabetes.

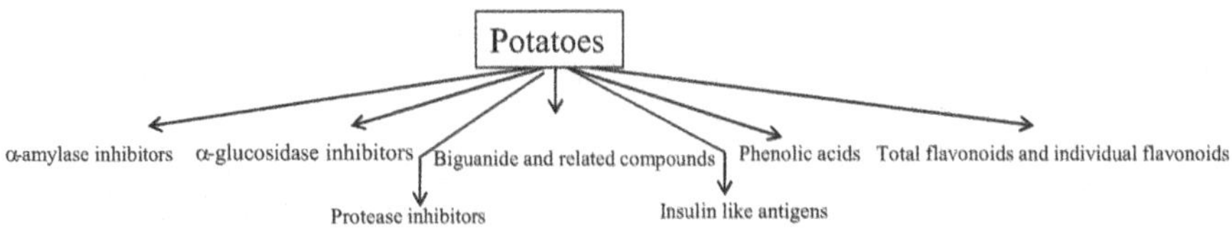

Fig. 12.1 Different types of health-promoting compounds in potatoes

This chapter will focus on the health-promoting compounds that are either reported recently from potatoes or their role in health improvement is established in recent past.

12.2 Type of Health Promoting Compounds Reported in Recent Past

12.2.1 Antidiabetic Compounds

Since long the antioxidants present in potatoes are thought to be solely responsible for their antidiabetic potential. Since the last decade, many researchers have reported some other compounds also that contribute towards diabetes management. Oral anti-hyperglycemic agents have been compiled in Fig. 12.2.

12.2.1.1 Antioxidants

Among the antioxidants, phenols and anthocyanin have shown high impact in diabetes management. Administration of purple potato extract extracted from Blue Congo potato variety to Streptozotocin (STZ)-induced diabetic rats decreased blood glucose, improved glucose tolerance, and decreased the amount of glycated hemoglobin. Also suppression of malondialdehyde (stress indicator) levels, inhibition of oxidative modified proteins, advanced glycation end product and advanced oxidation protein products and hence, restoration of antioxidant enzyme activities in STZ-induced diabetic rats were reported with administration of this extract. The study revealed presence of five anthocyanins and six phenolic acids, among which acylated anthocyanin as petunidin-3-p-coumaroyl-rutinoside-5-glucoside was dominant anthocyanin in purple potato extract that may be responsible for antidiabetic and antioxidative properties of the extract (Strugała et al. 2019). In vitro study has shown that dried peels of potatoes processed by extrusion cooking bound bile acids and carcinogen benzo(a)pyrene (Camire et al. 1993, 1995). Potato peels have shown amelioration of oxidative stress on rat and human erythrocytes (Singh and Rajini

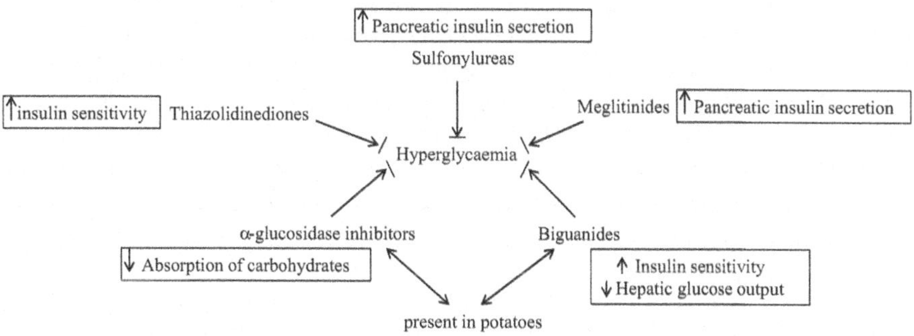

Fig. 12.2 Oral anti-hyperglycemic agents and their mode of action (modified from Lankatillake et al. 2019)

2008). Incorporation of potato peels (10%) in diet of diabetic rats managed some aspects of oxidative stress. Diabetic rats fed with 5% and 10% potato peel powder had 16% and 33% lower plasma glucose level compared to control diabetic rats, respectively (Singh et al. 2005). Potatoes are a good source of polyphenols with chlorogenic and caffeic acids being the most abundant phenolic acids in potatoes. Potato extracts rich in these acids have been reported to prevent type II diabetes and cardiovascular diseases (Paynter et al. 2006; Morton et al. 2000). Administration of chlorogenic acid to obese, hyperlipidemic, and insulin resistant rats showed slow absorption of glucose in the gut and increased insulin sensitivity without affecting insulin release (Rodriguez de Sotillo and Hadley 2002). Chlorogenic acid is reported to selectively inhibit hepatic glucose-6-phosphatase that is known as a rate limiting enzyme in gluconeogenesis process (Arion et al. 1997).

12.2.1.2 Biguanide and Related Compounds (BRCs)

Defects in insulin secretion or insulin action are major cause of type II diabetes. Type II diabetes is linked to hyperglycemia and pharmacological agents such as sulphonylureas, biguanides, thiazolidinediones, α-glucosidase inhibitors, and meglitinide are used for treatment of diabetes. These agents have different mode of action (Raigond et al. 2018). Biguanide and related compounds that include guanidine, metformin, galegine, phenformin, urea, L-arginine, and biuret are known to increase insulin sensitivity, decrease glucose absorption, and hence reduce hyperglycemia. Among all biguanide and related compounds, metformin is the most popular and first line drug for the treatment of type II diabetes (Perla and Jayanty 2013). In medieval Europe, *Galega officinalis* L. was used to treat the symptom similar to type II diabetes. *Galega officinalis* L. is known to exhibit high concentration of guanidine that is known for its hypoglycemic activity in animals. This compound is present in the basic structure of metformin (Bailey and Day 2004). Perla and Jayanty (2013) evaluated biguanide and related compounds from many plants and plant parts such as fresh green curry leaves (*Murraya koenigii* L.), dry fenugreek seeds (*Trigonella foenum-graecum* L.), fresh green bitter gourd (*Momordica charantia* L.), orange fleshed sweet potato (*Ipomoea batatas* L.), garlic (*Allium sativum* L.), and white fleshed potatoes (*Solanum tuberosum* L.). This was the first report on presence of BRCs in potatoes. The concentration of these compounds was high in curry leaves (25.57 μg/g) followed by fenugreek seeds (18.98 μg/g) and bitter gourd (11.14 μg/g) and low in potato (7.05 μg/g). Later Raigond et al. (2018) evaluated BRCs from vegetables and fruits, viz. potato (*Solanum tuberosum* L.), turnip (*Brassica rapa* L.), mushroom (*Agaricus bisporus*), radish (*Raphanus sativus* L.), carrot (*Daucus carota* L.), garlic (*Allium sativum* L.), tomato (*Solanum lycopersicum* L.), ginger (*Zingiber officinale*), green chili (*Capsicum annuum* L.), coriander (*Coriandrum sativum* L.), grapes (*Vitis vinifera* L.), banana (*Musa paradisiaca L.*), papaya (*Carica papaya* L.), and orange (*Citrus sinensis* L.), using HPLC. They reported BRCs in range of 0.20–3.28 mg/g FW in vegetables and 0.17–1.91 mg/g FW in fruits. Potato contained 0.56 mg/g FW BRCs, and consumption of BRCs through potato is almost 29.36 mg/day when calculated on the basis of per capita potato consumption in India (19 kg/person/year).

Concentration of these compounds in potato peel is reported to be almost double than concentration present in flesh (Raigond et al. 2016). These studies have shown that potatoes contain compounds that can help to control/manage type II diabetes.

12.2.1.3 Alpha-Amylase Inhibitors

Synthetic and plant-based inhibitors of carbohydrate degrading enzyme have the ability to reduce the risk of type II diabetes and related complications. Kalita et al. (2018) evaluated colored (purple, red, yellow) and white flesh potatoes for their polyphenolic compounds and anthocyanin. In red and purple potatoes, total polyphenolic compounds were high, with chlorogenic acid being dominating phenolic acid. Among anthocyanins, the derivatives of petunidin, peonidin, malvidin, and pelargonidin are mainly present. The potato extracts were used for the evaluation of inhibition of α-amylase, α-glucosidase, and aldose reductase. Purple potato extract was most effective for inhibition of these three enzymes with IC50 values 25, 42, and 32 μg/mL, respectively. This study showed potential of colored potatoes in management of diabetes by inhibiting carbohydrate degrading enzyme. On contrary, Saleem (2010) could not find α-amylase inhibitory activity in any of the potato cultivars out of 54 tested cultivars. Raigond et al. (2017) tested 46 Indian potato varieties for presence of α-amylase and α-glucosidase inhibitory activities. They reported α-amylase inhibitory activity in only one variety, i.e., Kufri Frysona with 20.5% inhibitory activity.

12.2.1.4 Alpha-Glucosidase Inhibitors

α-glucosidase, also known as α-d-glucoside glucohydrolase, is present in microorganisms, plants, and animal tissues. It catalyzes the release of α-glucose from the non-reducing end of the substrate. This enzyme is membrane bound and present in the epithelium of small intestine, where it helps in glucose absorption by catalyzing hydrolytic cleavage of oligosaccharides into absorbable monosaccharides. Rate of hydrolytic cleavage of oligosaccharide can be decreased by inhibiting the α-glucosidase activity in the intestine. Due to this inhibition, process of carbohydrate digestion shifts to lower part of small intestine that results in delayed absorption rate of glucose into blood and hence decreases the postprandial increase in blood glucose and thereby avoids the onset of late diabetic complications. Plants have been reported to contain α-glucosidase inhibitors such as acarbose, voglibose, nojirimycin, and 1-deoxynojirimycin. Compounds such as alkaloids, phenolics, curcuminiods, terpenoids, and anthocyanin also have shown α-glucosidase inhibitory activity (Kumar et al. 2011). Saleem (2010) observed α-glucosidase inhibitory activity to be ranged from 0 to 59% in 54 varieties of Chilean potato. Raigond et al. (2017) reported the α-glucosidase inhibitory activity in 14 Indian potato cultivars out of 46 Indian potato cultivars. The inhibitory activity ranged from 0 to 52.8% in varieties, viz. Kufri Arun, Kufri Anand, Kufri Kuber, Kufri Khasigaro, Kufri Kundan, Kufri Muthu, Kufri Neela, Kufri Naveen, Kufri Pushkar, Kufri Red, Kufri Sutlej, Kufri Sadabahar, Kufri Safed and Kufri Swarna. The activity was the maximum in Kufri Kuber variety.

12.2.1.5 Chlorogenic Acid

Chlorogenic is the most abundant phenolic acid present in diet of human. Potatoes' phenols contain almost 90% of chlorogenic acid. Chlorogenic acid has been reported to have various health beneficial properties that include antibacterial, anticarcinogenic, and antioxidant activities. Though antioxidant potential of potato has been evaluated in vitro since long, its role in diabetes management has been established in the last decade. Chlorogenic acid is considered as a novel insulin sensitizer that has similar effect on insulin as that of drug metformin that is widely used for treatment of type II diabetes (Fig. 12.3). Polyphenols have been reported to have inhibitory effect on enzymes involved in carbohydrate metabolism. Narita and Inouye (2009) observed mixed type inhibitory effect of 5-caffeoylquinic acid, caffeic acid, and quinic acid on α-amylase isomers I and II, where 5-caffeoylquinic acid was observed to be the most potent inhibitor. Karim et al. (2017) studied the inhibitory effect of chlorogenic acid isomer 5-O-caffeoylquinic acid on potato starch digestion by porcine pancreatic alpha-amylase. They reported dose-dependent decrease in porcine pancreatic α-amylase activity by co-addition or pre-addition of this enzyme with 5-O-caffeoylquinic acid. The decrease in activity was variety dependent, hence they concluded that multi-factors such as phenolic content, dry matter, and starch content of variety play important role in reducing the enzymatic breakdown of starch. In vivo animal studies also supported the reports on inhibitory effect of 5-O-caffeoylquinic acid on starch hydrolytic enzymes. Chlorogenic acid is reported to significantly lower the plasma glucose absorption hence can be considered as glycemic index lowering agent. Oral administration of 5-O-caffeoylquinic acid at 3.5 mg/kg body weight decreased the glycemic peaks by 22 and 17% after 10 and 15 min of administration, respectively (Bassoli et al. 2008).

12.2.1.6 Insulin-Like Antigens

Since the discovery of insulin in dog pancreas, presence of insulin-like proteins has been investigated in plants as well. The existence of insulin-like hormones in plants was first reported by J.B Collip in 1923. Till date these antigens are reported in plants such as green tops of onions, lettuce leaves, green bean leaves, barley roots, beet roots, moongbean sprouts, rice, etc. Though Collip (1923) and Best (1924) reported glucose lowering effect of germinating potatoes in rabbit, but the compounds responsible for blood glucose control were not studied. Presence of insulin-like

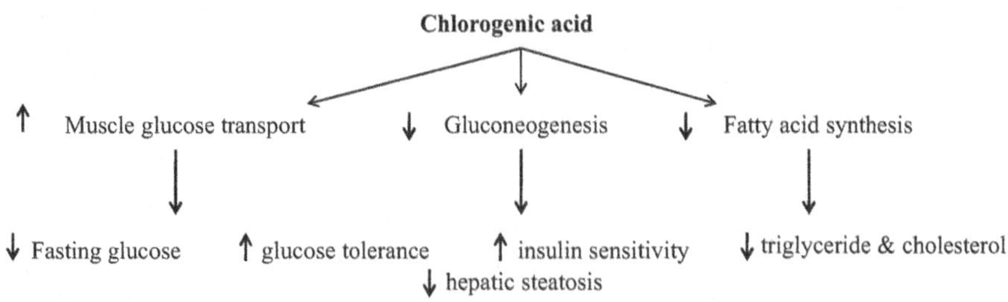

Fig. 12.3 Mode of action of chlorogenic acid to control/manage diabetes

antigens in unsprouted as well as sprouted tubers was first reported by Raigond et al. (2020). They confirmed the presence of these antigens in potatoes through western blotting and later quantified the concentration through ELISA. Anti-insulin antibody raised in guinea pig as primary antibody and goat anti-guinea pig IgG-goat polyclonal secondary antibody to guinea pig IgG-HRP as secondary antibody was used for both western blot and ELISA quantification. They reported concentration of insulin-like antigens to be almost four-fold higher in sprouted tubers compared to unsprouted tubers. The concentrations ranged from 34 to 67 μg/g DW in unsprouted tubers of five Indian potato cultivars, whereas sprouted tubers (with ~3 cm long sprouts) contained 252 μg/g DW (mean of two varieties) insulin-like antigens. Further, they selected three Indian potato varieties out of five based on popularity (Kufri Bahar), presence of α-glucosidase inhibitor (Kufri Pushkar), and high antioxidant content (Kufri Surya). The lyophilized powder of these varieties was orally administered to normal and streptozotocin induced diabetic rats. Metformin drug that is used widely to manage type II diabetes was used as control. Dose of potato powder and metformin was given at 300 mg/kg body weight. After 5 h of administration, blood glucose decrease was reported with Kufri Surya and Kufri Pushkar. Blood glucose decrease in diabetic rats was the maximum with Kufri Surya (30%) followed by Kufri Pushkar (18%), whereas Kufri Bahar showed no decrease. Kufri Surya performed better than drug metformin in lowering blood glucose in diabetic rats. The decrease in blood sugar with same dose of metformin as that of potato powder was 24%. These in vivo results correlated well with the nutritional quality of varieties. Varieties Kufri Surya and Kufri Pushkar, performed better than Kufri Bahar for all nutritional parameters. They reported insulin-like antigens, biguanide and related compounds, chlorogenic acid, ascorbic acid, and carotenoids to be the highest in Kufri Surya followed by Kufri Pushkar and least in Kufri Bahar. They concluded that presence of these compounds (antidiabetic and nutritional compounds) in higher concentration can be considered responsible for reduction in blood glucose level in diabetic rats. Their study has shown potential of potatoes to combat diabetes in human also.

12.2.2 Antihypertensive Compounds

Being naturally high in potassium and low in sodium content, potatoes have ability to counter hypertension-associated diseases.

12.2.2.1 Flavonoids

Flavonoids are known to influence flavor and color of fruits and vegetables. Catechin is the most abundant flavonoid in potatoes and its concentration varies from 0 to 204 mg/100 g dry weight. Other flavonoids in potatoes are quercetin, kaempferol rutinose, and rutin. Colored potatoes contain almost double concentration of flavonoids compared to white fleshed potatoes (Akyol et al. 2016).

Human randomized controlled trials showed that flavonoids specifically quercetin, kaempferol, and catechins exert cardiovascular protection and may reduce the chances of many cardiovascular diseases, particularly hypertension. Hypertension is a major contributing factor to mortality. Most of these studies to determine flavonoids' role in cardiovascular diseases were conducted using pure extracts of flavonoids in animal and human subjects. Flavonoids' antihypertensive action is mediated through increasing nitric oxide (NO) bioavailability. NO is important for blood pressure control and vascular homeostasis due to its vasorelaxatory effect. NO stimulates Ca^{2+}-activated potassium channels in vascular smooth muscle cells facilitating vasodilation (Maaliki et al. 2019).

Catechins maintain blood pressure balance in the body by regulating nicotinamide adenine dinucleotide phosphate (NADPH) oxidase and promoting the expression of heme oxygenase (Gómez-Guzmán et al. 2012). Tsang et al. (2018) measured arterial stiffness as pulse wave velocity in 14 healthy male and female adults after feeding 200 g of Purple Majesty or white flesh potato. Pulse wave velocity was significantly reduced ($p = 0.001$) following Purple Majesty consumption for 14 days and no changes were recorded following white potato consumption.

12.2.2.2 Angiotensin Converting Enzyme Inhibitors

Angiotensin converting enzyme (ACE) plays an important role in blood pressure regulation along with fluid and salt balance in mammals. It is involved in maintaining vascular tension. This enzyme converts inactive decapeptide, i.e., angiotensin I into angiotensin II that is a vascoconstrictor octapeptide and inactivates bradykinin (vasodilator peptide) resulting in increased blood pressure. Mäkinen et al. (2016) reported antihypertensive properties of potato and rapeseed protein derived peptides. They used Goldblatt rat model of hypertension for measurement of blood pressure after administration of potato and rapeseed peptides. Potato peptide induced mean arterial pressure of −60 mmHg, whereas rapeseed induced −50 mmHg in comparison to vehicle treated rats.

Vinson et al. (2012) reported effect of potato as a hypotensive agent. In hypertensive subjects consuming six to eight small purple microwaved potatoes twice a day for 4 weeks, no significant effect of potato on fasting plasma glucose, lipids, or HbA1c was reported. Body weight also did not increase significantly. With potato consumption diastolic and systolic blood pressure decreased significantly and the decrease was 4.3% and 3.5%, respectively.

Pihlanto et al. (2008) isolated proteins from potato tubers of different physiological stages and also from by-product of potato industry to study their ACE inhibition and radical scavenging potential. They reported that hydrolysis of protein isolates and by products with alcalase, neutrase, and esperase increased the ACE inhibition and radical scavenging activity. A protein extracted and isolated from potato tuber vascular bundle and inner tuber tissue showed inhibition of angiotensin converting enzyme I (Makinen et al. 2008).

12.2.3 Blood Glucose Lowering Compounds

Though because of its low sodium content, potatoes are considered good for persons suffering from high blood pressure. However, recent reports have shown presence of other blood glucose lowering compounds also.

12.2.3.1 Kukoamines

Potatoes have been reported to contain blood glucose lowering compounds that are present in herbs used to lower blood pressure. Parr et al. (2005) reported these compounds during metabolic profiling of potato tubers. They identified N_1, N_{12}-Bis (dihydrocaffeoyl)spermine (kukoamine A), and N_1, N_8-bis(dihydrocaffeoyl)-spermidine at concentration of several tens of micrograms per gram of dry matter. Kukoamines are present in higher concentrations in potatoes than some other compounds which have been investigated in the past. These compounds are not properly studied yet and have been previously found only in *Lycium chinense* whose bark was used to lower blood pressure. Kukoamines are reported in other solanaceous crops such as tomato and *Nicotiana sylvestris* also. Stability of these compounds after cooking is still unknown and need to be studied.

12.2.4 Anti-Obesity Compounds

Cho et al. (2010) reported role of chlorogenic acid as anti-obesity agent. With caffeic or chlorogenic acid supplemented diets @ 0.02% (wt/wt), they reported significant decrease in body weight, visceral fat mass, and plasma leptin and insulin levels in high-fat diet induced-obese mice. Chlorogenic acid supplementation reduced triglyceride content significantly with increase in plasma adiponectin level compared to high-fat control diet group. Their results indicated role of caffeic and chlorogenic acid in body weight management, lipid metabolism and obesity-linked hormone levels in high-fat diet fed mice, where chlorogenic acid is more effective than caffeic acid.

Literature showed conflicting reports on role of potato consumption on weight and body mass index. Linde et al. (2006) reported increase in consumption of French fries to increase body mass index, however, potato cooked with other methods showed no such effect. Robertson et al. (2018) concluded that there is no association between potato consumption and obesity, only French fries consumption showed an association between French fries and obesity. The possible reason could be high-fat present in fried potatoes.

Leeman et al. (2008) compared different cooking methods of potatoes for their satiety effect. They observed boiled and mashed potatoes to be more satiating than French fries when matched for energy but not for carbohydrates. Potato protein proteinase inhibitor 2 (PI2) has been studied for its satiety effect. This protein suppresses appetite by stimulating release of the peptide cholecystokinin. Potato proteinase inhibitor 2 has been incorporated into weight loss supplement, namely Slendesta R (Hill et al. 1990).

Komarnytsky et al. (2011) studied the satiety-promoting effects of potato protease inhibitors. Acute oral administration of potato protease inhibitors promoted satiety, delayed gastric emptying along with decrease in proteolytic activity in the duodenum, whereas weight gain reduced and plasma cholecystokinin levels increased in rats with repeated oral intake of protease inhibitor. The authors concluded potato protease inhibitors to be effective in controlling food intake along with weight gain via increase in circulating cholecystokinin levels.

12.2.5 Anti-Cancerous Compounds

Several in vitro and in vivo studies using potatoes have indicated phenolic acids, anthocyanins, glycoalkaloids, fibers, and proteinase inhibitors to exhibit anticancerous effect.

12.2.5.1 Antioxidants

Potato anthocyanins have shown anticancerous properties. Reddivari et al. (2007) studied the effect of extract of colored potato cultivars, viz. CO112F2-2, PATX99P32-2, ATTX98462-3, and ATTX98491-3 on prostate cancer cells. They reported extract of cultivar CO112F2-2 as well as its anthocyanin fraction at 5 mg chlorogenic acid eq./ml to inhibit cell proliferation and to increase the cyclin-dependent kinase inhibitor p27 levels in both LNCaP and PC-3 cells. Potato extract and anthocyanin fraction activates mitogen-activated protein kinase and c-Jun N-terminal kinase. These kinases activate caspase-independent apoptosis through nuclear translocation of endonuclease G (Endo G) and apoptosis-inducing factor in both cell lines. Purple and red potatoes have shown protective role against stomach cancer. Hayashi et al. (2006) reported benzo(a)pyrene induced repression of stomach cancer growth in mouse. Role of anthocyanins in induction of apoptosis in human stomach cancer cell lines and also suppression of mouse stomach cancer proliferation showed its potential as a bioactive antitumor component. The anthocyanin rich purple potatoes showed high in vitro antioxidant, antiproliferative, and antimicrobial activities against the colon cancer cells Caco-2, SW48, breast cancer cells MCF-7 and MDA-MB-231 after simulated gastrointestinal digestion (Ombra et al. 2015). Compared to white and yellow fleshed potatoes, purple flesh potatoes rich in anthocyanins have been reported to suppress proliferation and elevated apoptosis of colon cancer cells (Madiwale et al. 2011). Extracts of purple potatoes suppressed colon tumorigenesis via elimination of colon cancer stem cells (Charepalli et al. 2015). Chlorogenic acid has been reported to protect from liver, colon, and prostate cancer cells in human; it inhibits the proliferation of colon and prostate cancer cells significantly.

Potato carotenoids mainly contain lutein and zeaxanthin that are known to be deposited in eye macula and protect eyes against macular degeneration and cataracts (Wu et al. 2015). These carotenoids also protect against developing degenerative diseases such as cancers and cardiovascular diseases (Krinsky and Johnson 2005).

As per American Optometric Association (2009), lutein supplementation at 10 mg per day and zeaxanthin supplementation at 2 mg per day show health benefits.

12.2.5.2 Potato Proteinase Inhibitors

Huang et al. (1997) demonstrated role of potato proteinase inhibitor I and II in blocking UV-induced activation of transcription factor activator protein-1. This protein is responsible for tumor growth in mammalian cells. Hence blocking this protein helps in controlling UV-induced skin carcinoma.

12.2.5.3 Glycoalkaloids

Since the last decade anticancer/antitumor properties of potato glycoalkaloids are studied by many researchers. Exposure of cancer cells both in culture and in vivo to potato glycoalkaloids, i.e., α-chaconine and α-solanine or their hydrolysis product such as mono-, di-, and trisaccharide derivatives and the aglycones solasodine, solanidine, and tomatidine showed inhibition of tumor cells. As these compounds are naturally present in all potato varieties, therefore, consumption of potato can significantly protect against multiple type of cancers (Burgos et al. 2019). α-solanine was reported to inhibit growth and induce apoptosis in multiple cancer cells including human colon (HT29) and liver (HepG2) cancer cells (Lee et al. 2004). Potato glycoalkaloids, i.e., α-solanine and α-chaconine extracted from potato varieties, viz. Dejima, Jowon, Sumi, Toya, and Vora Valley showed antiproliferative effect against human tumor cell lines such as cervical (HeLa), liver (HepG2), lymphoma (U937), stomach (AGS and KATO III) cells, and on normal liver (Chang) cells. These glycolkaloids showed dose-dependent antiproliferative effect and α-chaconine was more effective compared to α-solanine (Friedman et al. 2005). Yang et al. (2006) also observed α-chaconine to induce apoptosis of HT-29 human colon cancer cells through caspase-3 activation and inhibition of extracellular signal-regulated kinase phosphorylation. α-chaconine increases cyclin-dependent kinase inhibitor p27 levels in cancer cell line LNCap and PC3, whereas α-solanine inhibits cancer cell growth through caspase 3-dependent mitochondrial apoptosis and decreases the expression of metastasis-related proteins MMP-2 and MMP-9 (Reddivari et al. 2010; Sun et al. 2014). α-solanine is also reported to inhibit proliferation of PANC-1, sw1990, MIA PaCa-2 cells in a dose-dependent manner. It is reported to inhibit pancreatic cancer via suppressing pathways involving proliferation, angiogenesis, and metastasis in vitro and in vivo (Lv et al. 2014).

12.2.5.4 Fiber

Langner et al. (2009) studied the anticancer activity of water extract of heated potato fiber (Potex). Heated potex exhibited antiproliferative effect against tumor cell cultures, including rhabdomyosarcoma-medulloblastoma (TE671), glioma (C6), breast (T47D), colon (HT-29), and lung (A549) carcinoma. Potato fiber extract, i.e., Potex decreased cancer cell motility and induced apoptotic cell death without affecting the viability of normal human skin fibroblast, rat oligodendrocytes, and mouse neurons.

Asli et al. (2018) conducted a study for 11.4 years to find correlation between potato consumption and risk of pancreatic cancer. Their study included 114,240 men and women in prospective HELGA cohort using Cox proportional hazard models. They reported non-significantly high risk of pancreatic cancer with high potato consumption by comparing highest v/s lowest quartile of potato consumption. Their study indicated that association between high potato consumption and pancreatic cancer was significant for women and not for men. However, it was concluded that potato consumption is not consistently associated with high risk of pancreatic cancer.

12.2.6 Anti-Hyperlipidemic Compounds

In vivo animal studies have indicated that cholesterol lowering properties of potatoes are attributed to presence of protein, phenols, fibers, resistant starch, phosphorylated starch, and glycoalkaloids (Friedman 2006). Rats supplemented with diet of cooked unpeeled potatoes for 3 weeks showed lower plasma cholesterol and triglyceride along with decrease in liver cholesterol content (Robert et al. 2006). Retrograded starch is reported to decrease serum total cholesterol and triglyceride concentrations. Retrograded starch is considered to increase bile acids excretion and also to inhibit the synthesis of fatty acids at the mRNA levels of fatty acid synthase (FAS) and SREBP-1c resulting in low serum cholesterol concentration (Hashimoto et al. 2006). Cholesterol free diet with 20% (w/w) potato peptides increased serum high-density lipoprotein (HDL) cholesterol and fecal steroid output along with decrease in non-HDL cholesterol concentrations compared to 20% casein peptide diet in rats (Liyanage et al. 2008). Potato peptides reduced non-high-density lipoprotein cholesterol level in rats fed a cholesterol-enriched diet and; two cholesterol enriched diets containing soy-peptides compared to two cholesterol enriched diets containing casein and decrease was 18.39% with potato peptide containing diet and 32.76% with two cholesterol enriched diets containing soy-peptide, at the end of feeding period. The study indicated role of potato peptides in increasing cecal anaerobe and *Bifidobacterium* populations and reduction in non-HDL cholesterol in cholesterol fed rats (Liyanage et al. 2009).

12.3 Methods for Estimations of Health Promoting Compounds

Analytical tools and methodologies have advanced rapidly in the last few decades for both targeted (quantification of specific analytes) and non-targeted metabolomics (global metabolite profiling). For untargeted metabolites, liquid chromatography coupled to mass spectrometry (LC-MS) has been the preferred choice. Multiple mass spectrometry techniques (UPLC- and GC-MS) were used to profile small molecules (<1.2 KDa) (Chaparro et al. 2018). For targeted analysis and quantification, the instrument of choice is high-performance liquid chromatography coupled with different detectors depending on the type of metabolite (Andre et al. 2007).

Spectrophotometric methods were used to determine the activity assays of metabolites identified in potato germplasm (Perla et al. 2012).

Potato metabolites have shown promising health-promoting effects in human cell culture, experimental animals, and human clinical studies. There are several meta-analysis papers and in cohort studies showing the effectiveness of potato related and extracted compounds in reducing chronic diseases.

12.4 Conclusion/Future Direction

The literature compiled in this chapter indicated that potatoes have potential to manage/control myriad of diseases such as diabetes, hypertension, several types of cancers and can control high blood pressure. Potato protein peptides exhibit weight controlling properties. Most of the researchers have reported colored potatoes to be nutritionally superior than white/cream flesh potatoes. These studies indicated that breeding for colored potatoes and their cultivation needs to be promoted for betterment of health. Due to high per capita consumption of potatoes worldwide, nutritionally superior potatoes can play significant role in well-being of humans.

References

Akyol H, Riciputi Y, Capanoglu E et al (2016) Phenolic compounds in the potato and its byproducts: an overview. Int J Mol Sci 17:835

American Optometric Association (2009) Diet, nutrition and eye health. http://www.aoa.org/documents/nutrition/Diet_Nutrition_Eye_Health_booklet.pdf. Accessed 28 Sept 2017

Andre CM, Oufir M, Guignard C et al (2007) Antioxidant profiling of native Andean potato tubers (*Solanum tuberosum* L.) reveals cultivars with high levels of β-carotene, alpha-tocopherol, chlorogenic acid, and petatin. J Agric Food Chem 55:10839–10849

Arion WJ, Canfield WK, Ramos FC et al (1997) Chlorogenic acid and hydroxynitrobenzaldehyde: new inhibitors of hepatic glucose 6-phosphatase. Arch Biochem Biophys 339:315–322

Asli LA, Braaten T, Olsen A et al (2018) Potato consumption and risk of pancreatic cancer in the HELGA cohort. Br J Nutr 119:1408–1415

Bailey CJ, Day C (2004) Metformin: its botanical background. Pract Diabetes Int 21(3):115–117

Bassoli BK, Cassolla P, Borba-Murad GR et al (2008) Chlorogenic acid reduces the plasma glucose peak in the oral glucose tolerance test: effects on hepatic glucose release and glycaemia. Cell Biochem Funct 26(3):320–328

Best CH (1924) Recent work on insulin. Endocrinology 8:617–629

Burgos G, Felde TZ, Andre C et al (2019) The potato and its contribution to the human diet and health. In: Campos H, Ortiz O (eds) The potato crop. Springer, Cham. https://doi.org/10.1007/978-3-030-28683-5_2, Chap. 2

Camire ME, Zhao J, Violette DA (1993) In vitro binding of bile acids by extruded potato peels. J Agic Food Chem 41:2391–2394

Camire ME, Zhao J, Dougherty MP et al (1995) In vitro binding of benzo[α]pyrene by extruded potato peel. J Agric Food Chem 43:970–973

Chaparro JM, Holm DG, Broeckling CD et al (2018) Metabolomics and ionomics of potato tuber reveals an influence of cultivar and market class on human nutrients and bioactive compounds. Front Nutr 5:36. https://doi.org/10.3389/fnut.2018.00036

Charepalli V, Reddivari L, Radhakrishnan S et al (2015) Anthocyanin-containing purple-fleshed potatoes suppress colon tumorigenesis via elimination of colon cancer stem cells. J Nutr Biochem 26(12):1641–1649

Cho A-S, Jeon S-M, Kim M-J et al (2010) Chlorogenic acid exhibits anti-obesity property and improves lipid metabolism in high-fat diet-induced-obese mice. Food Chem Toxicol 48 (3):937–943

Collip JB (1923) Glucokinin: a new hormone present in plant tissue. J Biol Chem 56:513–543

Friedman M (2006) Potato glycoalkaloids and metabolites: roles in the plant and in the diet. J Agric Food Chem 54:8655–8681

Friedman M, Lee KR, Kim HJ et al (2005) Anticarcinogenic effects of glycoalkaloids from potatoes against human cervical, liver, lymphoma, and stomach cancer cells. J Agric Food Chem 53 (15):6162–6169

Gómez-Guzmán M, Jiménez R, Sánchez M et al (2012) Epicatechin lowers blood pressure, restores endothelial function, and decreases oxidative stress and endothelin-1 and NADPH oxidase activity in DOCA-salt hypertension. Free Radic Biol Med 52:70–79

Hashimoto N, Ito Y, Han KH et al (2006) Potato pulps lowered the serum cholesterol and triglyceride levels in rats. J Nutr Sci Vitaminol 52:445–450

Hayashi K, Hibasami H, Murakami T et al (2006) Induction of apoptosis in cultured human stomach cancer cells by potato anthocyanins and its inhibitory effects on growth of stomach cancer in mice. Food Sci Technol Res 12:22–26

Hill AJ, Peikin SR, Ryan CA et al (1990) Oral administration of proteinase inhibitor II from potatoes reduces energy intake in man. Physiol Behav 48:241–246

Huang C, Ma W-Y, Ryan CA et al (1997) Proteinase inhibitors I and II from potatoes specifically block UV-induced activator protein-1 activation through a pathway that is independent of extracellular signal-regulated kinases, c-Jun N-terminal kinases, and P38 kinase. Proc Natl Acad Sci U S A 94(22):11957–11962

Kalita D, Holm DG, LaBarbera DL et al (2018) Inhibition of α-glucosidase, α-amylase, and aldose reductase by potato polyphenolic compounds. PLos One 13:e0191025. https://doi.org/10.1371/journal.pone.0191025

Karim Z, Holmes M, Orfila C (2017) Inhibitory effect of chlorogenic acid on digestion of potato starch. Food Chem 217:498–504

Komarnytsky S, Cook A, Raskin I (2011) Potato protease inhibitors inhibit food intake and increase circulating cholecystokinin levels by a trypsin-dependent mechanism. Int J Obes 35(2):236–243

Krinsky NI, Johnson EJ (2005) Carotenoid actions and their relation to health and disease. Mol Asp Med 26(6):459–516

Kumar S, Narwal S, Kumar V et al (2011) α-Glucosidase inhibitors from plants: a natural approach to treat diabetes. Pharmacogn Rev 5(9):19–29

Langner E, Rzeski W, Kaczor J et al (2009) Tumour cell growth-inhibiting properties of water extract isolated from heated potato fibre (Potex). J Pre-Clin Clin Res 3:36–41

Lankatillake C, Huynh T, Dias DA (2019) Understanding glycaemic control and current approaches for screening antidiabetic natural products from evidence-based medicinal plants. Plant Methods 15:105

Lee KR, Kozukue N, Han JS, Park JH, Chang EY, Baek EJ, Chang JS, Friedman M (2004) Glycoalkaloids and metabolites inhibit the growth of human colon (HT29) and liver (HepG2) cancer cells. J Agric Food Chem 52:2832–2839

Leeman M, Östman E, Björck I (2008) Glycaemic and satiating properties of potato products. Eur J Clin Nutr 62:87–952

Linde JA, Utter J, Jeffery RW et al (2006) Specific food intake, fat and fiber intake, and behavioral correlates of BMI among overweight and obese members of a managed care organization. Int J Behav Nutr Phys Act 3:42

Liyanage R, Han K-H, Watanabe S et al (2008) Potato and soy peptide diets modulate lipid metabolism in rats. Biosci Biotechnol Biochem 72(4):943–950

Liyanage R, Han K-H, Shimada K-I et al (2009) Potato and soy peptides alter caecal fermentation and reduce serum non-HDL cholesterol in rats fed cholesterol. Eur J Lipid Sci Technol 111:884–892

Lv C, Kong H, Dong G et al (2014) Antitumor efficacy of α-solanine against pancreatic cancer *in vitro* and *in vivo*. PLoS One 9(2):e87868

Maaliki D, Shaito AA, Pintus G et al (2019) Flavonoids in hypertension: a brief review of the underlying mechanisms. Curr Opin Pharmacol 45:57–65

Madiwale GP, Reddivari L, Holm DG et al (2011) Storage elevates phenolic content and antioxidant activity but suppresses antiproliferative and proapoptotic properties of colored-flesh potatoes against human colon cancer cell lines. J Agric Food Chem 59(11):8155–8166

Makinen S, Kelloniemi J, Pihlanto A et al (2008) Inhibition of angiotensin converting enzyme I caused by autolysis of potato proteins by enzymatic activities confined to different parts of the potato tuber. J Agric Food Chem 56:9875–9883

Mäkinen S, Streng T, Larsen LB et al (2016) Angiotensin I-converting enzyme inhibitory and antihypertensive properties of potato and rapeseed protein-derived peptides. J Funct Foods 25:160–173

Mishra T, Raigond P, Thakur N et al (2019) Recent updates on healthy phytoconstituents in potato: a nutritional depository. Potato Res. https://doi.org/10.1007/s11540-019-09442-z

Morton LW, Caccetta RA-A, Puddey IB et al (2000) Chemistry and biological effects of dietary phenolic compounds: relevance to cardiovascular disease. Clin Exp Pharmacol Physiol 27:152–159

Narita Y, Inouye K (2009) Kinetic analysis and mechanism on the inhibition of chlorogenic acid and its components against porcine pancreas alpha-amylase isozymes I and II. J Agric Food Chem 57(19):9218–9225

Ombra M, Fratianni F, Granese T et al (2015) In vitro antioxidant, antimicrobial and anti-proliferative activities of purple potato extracts (*Solanum tuberosum* cv Vitelotte noire) following simulated gastro-intestinal digestion. Nat Prod Res 29(11):1087–1091

Parr AJ, Mellon FA, Colquhoun IJ et al (2005) Dihydrocaffeoyl polyamines (kukoamine and allies) in potato (*Solanum tuberosum*) tubers detected during metabolite profiling. J Agric Food Chem 53(13):5461–5466

Paynter NP, Yeh HC, Voutilainen S et al (2006) Coffee and sweetened beverage consumption and the risk of type 2 diabetes mellitus: the atherosclerosis risk in communities study. Am J Epidemiol 164:1075–1084

Perla V, Jayanty SS (2013) Biguanide related compounds in traditional antidiabetic functional foods. Food Chem 138:1574–1580

Perla V, Holm DG, Jayanty SS (2012) Effects of cooking methods on polyphenols, pigments, and antioxidant activity in potato tubers. LWT- Food Sci Technol 45:161–171

Pihlanto A, Akkanen S, Korhonen HJ (2008) ACE-inhibitory and antioxidant properties of potato (*Solanum tuberosum*). Food Chem 109(1):104–112

Raigond P, Kaundal B, Singh B et al (2016) Biguanide related compounds (anti-diabetic) in Indian potatoes. Newsletter no. 64. ICAR-CPRI, Shimla

Raigond P, Singh B, Dutt S et al (2017) Potential of Indian potatoes for the management of hyperglycemia. Indian J Hortic 74(1):103–108

Raigond P, Kaundal B, Sood A et al (2018) Quantification of biguanide and related compounds (anti-diabetic) in vegetables and fruits. J Food Compos Anal 74:82–88

Raigond P, Bhardwaj P, Raigond B et al (2020) Insulin like antigens in potatoes: a new hope for developing insulin rich potatoes for diabetics. In: Souvenir, Global Potato Conclave, 28–-31 January 2020 at Gandhinagar, Gujarat, India, pp 117–118

Reddivari L, Vanamala J, Chintharlapalli S et al (2007) Anthocyanin fraction from potato extracts is cytotoxic to prostate cancer cells through activation of caspase-dependent and caspase-independent pathways. Carcinogenesis 28(10):2227–2235

Reddivari L, Vanamala J, Safe SH et al (2010) The bioactive compounds α-chaconine and gallic acid in potato extracts decrease survival and induce apoptosis in LNCaP and PC3 prostate cancer cells. Nutr Cancer 62:601–610

Robert L, Narcy A, Rock E et al (2006) Entire potato consumption improves lipid metabolism and antioxidant status in cholesterol-fed rat. Eur J Nutr 45(5):267–274

Robertson TM, Alzaabi AZ, Robertson MD et al (2018) Starchy carbohydrates in a healthy diet: the role of the humble potato. Nutrients 10:1764. https://doi.org/10.3390/nu10111764

Rodriguez de Sotillo D, Hadley M (2002) Chlorogenic acid modifies plasma and liver concentrations of: cholesterol, triacylglycerol, and minerals in (fa/fa) Zucker rats. J Nutr Biochem 13:717–726

Saleem F (2010) Anti-diabetic potentials of phenolic enriched Chilean potato and select herbs of Apiaceae and Lamiaceae families. (Thesis), pp 1–122

Singh N, Rajini PS (2008) Antioxidant-mediated protective effects of potato peel extract in erythrocytes against oxidative damage. Chem Biol Interact 173:97–104

Singh N, Kamath V, Rajini PS (2005) Attenuation of hyperglycemia and associated biochemical parameters in STZ-induced diabetic rats by dietary supplementation of potato peel powder. Clin Chim Acta 353:165–175

Strugała P, Dzydzan O, Brodyak I et al (2019) Antidiabetic and antioxidative potential of the blue Congo variety of purple potato extract in streptozotocin-induced diabetic rats. Molecules 24:3126

Sun H, Lv C, Yang L et al (2014) Solanine induces mitochondria-mediated apoptosis in human pancreatic cancer cells. Biomed Res Int 2014:805926

Tsang C, Smail NF, Almoosawi S et al (2018) Antioxidant rich potato improves arterial stiffness in healthy adults. Plant Foods Hum Nutr 73:203–208

Vinson JA, Demkosky CA, Navarre DA et al (2012) High-antioxidant potatoes: acute in vivo antioxidant source and hypotensive agent in humans after supplementation to hypertensive subjects. J Agric Food Chem 60(27):6749–6754

Wu J, Cho E, Willett WC et al (2015) Intakes of lutein, zeaxanthin, and other carotenoids and age-related macular degeneration during 2 decades of prospective follow-up. JAMA Ophthalmol 133(12):1415–1424

Yang SA, Paek SH, Kozukue N et al (2006) Alpha-chaconine, a potato glycoalkaloid, induces apoptosis of HT-29 human colon cancer cells through caspase-3 activation and inhibition of ERK ½ phosphorylation. Food Chem Toxicol 44(6):839–846

Potato Peel Composition and Utilization 13

Alka Joshi, Shruti Sethi, Bindvi Arora, Ahmad Farid Azizi, and B. Thippeswamy

Abstract

Globally, food industries are approaching towards achieving the goal of food safety, nutritional security and sustainability. Potato peel practically untrapped from technology interventions although a concentrated source of pharmacological indispensable compounds such as minerals, dietary fibres, phenolics and anthocyanins. Nutraceutical potential of potato peels having low glycoalkaloids content is now fetching attention to be channelized through food applications by various researchers. Recycling and utilization of waste generated from potato industries is utmost important which otherwise is termed as 'zero value'. High moisture content, adhered starch, fibrous nature, dark colour, musty flavour are the major constraints, sensory limitations for effective utilization and storability of potato peels. Therefore, provision for on site drying must be in place to provide higher stability to the perishable peels. Keeping the possible utilization dimensions of potato peels, this chapter was designed to cover composition and health benefits, various food and non-food uses of potato peel and extraction of functional ingredients from the potato peels. This envisages that use of peel can not only reduce the waste burden of environment but also can cater to the food and pharmaceutical industries significantly.

Keywords

Potato · Peel · Food use · Non-food use · Functional ingredients

A. Joshi (✉) · S. Sethi · B. Arora · A. F. Azizi · B. Thippeswamy
Division of Food Science & Postharvest Technology, ICAR–Indian Agricultural Research Institute, New Delhi, India
e-mail: alka.foodtech@gmail.com

P. Raigond et al. (eds.), *Potato*, https://doi.org/10.1007/978-981-15-7662-1_13

13.1 Introduction

Potato peel is a by-product obtained while converting potatoes into their value added products. It constitutes of about one-quarter of the input of the potato processing plant. Such huge output levels pose severe disposal problem to the potato industry, especially when the peels are wet and can undergo rapid microbial spoilage. Although it is termed as a zero value product, potato peel has an array of nutritionally important components that can be used in food and pharma industries (Charmley et al. 2019). It is an excellent source of phytonutrients such as dietary fibre, phenolics, glycoalkaloids and anthocyanins (in coloured varieties) along with vitamins and minerals especially vitamins C and B_1, B_2, B_3 and calcium, phosphorus and iron. Maldonado et al. (2014) extracted and fractionated phenolic acids and glycoalkaloids from potato peels with water/ethanol-based solvents. Additionally, secondary plant metabolites are successfully extracted from potato peels free of toxic glycoalkaloids for food preservation. Steroidal glycosides (glycoalkaloids) enriched extract is also an ingredient for pharmaceutical industry. Therefore, waste utilization is also termed as 'rest raw material' management. The growing rejection of synthetic food ingredients and additives by consumers, exploitation of natural resources such as the potato peel as functional ingredients may become a promising alternative in the times to come. The utilization of such by-products also contributes to reduced amounts of waste, generates pharmacological important ingredients at lower cost and further helps to maintain sustainable food production (Schieber et al. 2001). The high quantity of peel generated from the potato processing plants is an environment pollutant because of its high biological oxygen demand (BOD) potential. Depending on the peeling process adopted (steam, abrasion or lye), the amount of waste ranges from 15 to 40% of the amount of processed potatoes (Arapoglou et al. 2010). The utilization of this by-product can substantially reduce the waste disposal problem of the potato industries and also help in extraction of beneficial compounds and enrichment of other food items with essential nutrients. Presently, peels are mostly used as animal feed or raw material for bio-gas production.

Peel can be obtained either in its raw or boiled form from the industries depending upon the processing line. Since in its native form peel cannot be used directly in food products, therefore, technological interventions are required to modify peels for food use or to extract the valuable compounds for further utilization. This chapter includes various food and non-food uses of potato peel as well as extraction of phytonutrients (dietary fibre, phenolics, glycoalkaloids) from them.

13.2 Composition of Potato Peels

Losses caused by potato peeling range from 15% to 40%, the amount depending on the procedure applied, i.e. steam, abrasion or lye peeling (Schieber et al. 2001). Composition of the potato peel decides the techno-functionality as well as its end-use. Arapoglou et al. (2009) have documented high moisture (83.3–85.1%) and carbohydrate (8.7–12.4%) content of raw potato peels with a low overall protein

(1.2–2.3%) and lipid (0.1–0.4%) content. Hundred gram (on fwb) of potato peels contains approximately 0.021 mg of thiamine, 0.038 mg of riboflavin, 1.033 mg of niacin, 11.4 mg of ascorbic acid whereas in case of minerals, sodium, potassium, iron and calcium content were found to be 10, 413, 3.24 and 30 mg, respectively. Major carbohydrate observed was starch (7.8%). More than 50% of this carbohydrate is dietary fibre (Camire et al. 1995); therefore, addition of peels beyond a certain limit into a product imparts grittiness, hardness and musty flavour to the product. Potato peels constitute 2.5% of total dietary fibre and total ash amounting to be 0.9–1.6%. Camire and Flint (1991) and Al-Weshahy and Rao (2012) have reported dietary fibre of the potato peel (average content of 40 g/100 g) primarily constitute from the insoluble fraction such as cellulose, hemicellulose and lignin etc. The variation in the dietary fibre content is owing to the peeling method as well as extraction method. Abrasion peeling results in more starch and less dietary fibre (especially lignin) compared to the steam peeling. On the other hand, manual peeling of potato produces peels with approximately 63% of alcohol-insoluble fibres on a dry weight basis, which can be further fractionated into 3.4% pectin, 2.2% cellulose, 14.7% protein, 66.8% starch and 7.7% ash. The sugars in the alcohol-soluble fraction consist of 1.4% total soluble sugars and 0.9% reducing sugars (Mahmood et al. 1998). In addition, potato peels are reported to contain a variety of other phytochemicals including phenols, unsaturated fatty acids, amides, etc. (Schieber and Saldaña 2009). The total phenolic content ranges from 1.51 to 3.32 mg of gallic acid equivalent per gram of dry potato peel powder with higher content in the peel extracts from red-coloured potato varieties (Al-Weshahy and Rao 2009). De Sotillo et al. (1994a) reported the total phenolic compounds in potato peel waste to be 48 mg/100 g. Kähkönen et al. (1999) reported 4.3 mg of gallic acid equivalents of phenolics per g of dry potato peel. Phenolic content varied due to difference in colour of tubers and varieties used during study (Al-Saikhan et al. 1995).

Apart from the above-mentioned nutritional compounds, potatoes contain nitrogenous steroidal glycosides called glycoalkaloids that are also known as stress metabolites. They are distributed unevenly in the tuber with maximum concentration in the periderm layer (Dale et al. 1998). These compounds are toxic to humans, animals and microorganisms (Mäder et al. 2009; Schieber and Saldaña 2009). The predominant glycoalkaloids in potato peel are α-solanine, α-chaconine and solanidine. Depending on the potato cultivar, light, irradiation, conditions of storage and mechanical injury, the total concentration of α-solanine and α-chaconine in potato peel can range from 84 to 3526 ppm (with ratio of α-solanine to α-chaconine from 2.1 to 2.4) (Friedman 2004). The acceptable maximum level of total glycoalkaloid content is 20 mg/100 g of fresh tuber weight (Singh et al. 2016). Peels of Indian potatoes are relatively low in glycoalkaloids content with a mean value of 1.62 mg/100 g fresh weight. Azizi et al (2020) reported a maximum of 1.70 mg/100 g of glycoalkaloids content in peels of raw tubers of Kufri Bahar, whereas he observed lower glycoalkaloids content of 1.44, 0.86 and 0.75 mg/100 g, respectively, in peels of potato varieties Kufri Chipsona 1, Kufri Jyoti and Kufri Frysona. Thus, while utilizing the potato peel in the formulation of functional foods, care has to be taken to keep the glycoalkaloid content lower than the maximum acceptable limits.

13.3 Potato Peel—Health Benefits

Dietary fibre is a broad term that includes several carbohydrates such as cellulose, hemicelluloses, lignins, pectins, gums, etc. It is well known as a bulking agent that increases the intestinal mobility and hydration of the faeces (Forsythe et al. 1976). Potato peel comprises primarily of insoluble fibre as stated earlier, which binds to bile acids, the mechanism by which it has the ability to lower plasma cholesterol (Lazarov and Werman 1996). A study on rats regarding the potential of potato peel to give a hypocholesterolemic effect revealed 40% and 30% reduction in plasma cholesterol and hepatic fat cholesterol level, respectively, after 4 weeks of feeding with a diet enriched with potato peels. High intake of dietary fibre also affects the bowel movement and gastric emptying time which further affects the absorption of glucose in the body, thus imparting a positive influence on blood glucose levels (Onyechi et al. 1998; Chandalia et al. 2000). Dietary fibres are known to impart protective effect against mutagenesis and carcinogenesis via binding of mutagenic and carcinogenic substances, reduce transit time in intestine, increase water absorption and faecal bulk and lower faecal pH through the fermentation by intestinal microorganisms (Harris and Ferguson 1993). Earlier, Camire et al. (1993, 1995) suggested the possible role of potato peel fibre as an anticarcinogenic material. They investigated the role of extrusion, a high temperature short time pretreatment, of potato peel vis-a-vis wheat bran and cellulose as other sources of dietary fibre, in binding of benzopyrene, a carcinogen under in-vitro conditions. Unextruded steamed potato peels bound more benzopyrene than did peels extruded at 110 °C with feed moisture of 30%. Potato peels obtained by abrasion bound lower benzopyrene in comparison to peels obtained by steaming of potatoes. The higher dietary fibre and starch content in abrasion peels are believed to have maintained benzopyrene in suspension. Also, higher amount of polyphenols remaining in abrasion peel, in contrast with steam peel, play a role in lowering the levels of benzopyrene by binding (Camire et al. 1993).

Phenolic compounds present in varied crops are the first line of defence against hostile environment. In potato also, they protect against insect and pathogen attack. They are present throughout the potato tuber with maximum concentration found in the peels (50%) with concentration reducing towards the centre of the tuber. Potato peel may contain as high as 177 mg gallic acid equivalent (GAE)/100 g fresh peel total phenolics (Makris et al. 2007) with highest amount detected in red- or purple-fleshed ones. The phenolic compounds are the major contributors to impart potato peel its antioxidative nature. The peel contains both bound (1.26 mg ferulic acid equivalent/g dry weight) and free phenolics (2.66 mg ferulic acid equivalent/g dry weight), the major being ferulic acid (Nara et al. 2006). Irradiation of potato peel significantly enhances the phenolic content by 26% (Kannat et al. 2005). The presence of these active compounds in human diet prevents inflammation and degenerative diseases (Pourcel et al. 2007; Im et al. 2008).

13.4 Food Uses

13.4.1 As a Source of Dietary Fibre

Different sources of dietary fibres have been used to replace wheat flour in the preparation of bakery products. Work on inclusion of potato peel as a source of dietary fibre in baked products was initiated in the 1970s. Toma et al. (1979) reported potato peels as being superior to wheat bran in its content of total dietary fibre, water holding capacity with low quantity of starchy components. When introduced in bread, potato peel increased crumb darkening and reduced the loaf volume. Orr et al. (1982) reported a musty odour in potato peel incorporated breads that could be diminished by the extrusion of potato peel before its addition. Upon incorporation of potato peel in biscuits, Abdel-Magied (1991) observed a significant effect on the physical characteristics of the biscuits. The biscuits having 5 and 10% (w/w) of potato peel on flour replacement basis had lower stack weight and sensory score (colour and appearance) in proportion to the levels used. In addition, supplemented biscuits were harder to bite than control biscuits. Arora and Camire (1994) found that muffins with 25% potato peel were darker, lower in height and more resistant to compression. Similarly, cookies with 10 and 15% potato peel were darker, harder and smaller in diameter than control cookies. Jeddou et al. (2017) incorporated potato peel flours (PPF) as a protein and dietary fibre source in dough and cake. They observed huge colour difference between the control sample and the dough with added potato peels. The peel incorporation also significantly reduced the cake hardness along with the lightness and yellowness values. The study proved that fibre-enriched cake with good sensory quality could be produced with 5% of potato peel flour with improved dough strength and ratio of elasticity to extensibility.

Arora and Camire (1994) incorporated potato peel as a dietary fibre in baked foods @25% in wheat flour for baking cinnamon muffins. Peel incorporated muffins were darker in colour, lower in height and had higher resistance to compression. Muffins incorporated with extruded peels were less cohesive. Oatmeal cookies with peel added @10% or 15% were smaller, darker and stiffer compared to control cookies. Though sensory panelists noticed differences in colour and flavour between cookies with no peels and those with 10% peels, no variation was observed among types of peels. During 4 weeks of storage at 21 °C, all cookie formulations became harder, although differences between control cookies and those with peels decreased with time. Lower peroxide values were found for cookies with peels throughout the storage period, particularly those with 15% extruded peels. Potato peels, particularly extruded peels which have reduced glycoalkaloid levels offer bakers the ability to prolong shelf-life while increasing dietary fibre content. Curti et al. (2016) added potato peel fibre into bread @0.4 g fibre/100 g flour. Fibre addition resulted in changes in water activity and moisture content that further affected the softness of bread. It also resulted in reduced staling (non-microbial changes) of bread. Low fat fabricated potato chips with good organoleptic acceptability have been developed in Division of Food Science and Postharvest Technology, ICAR-IARI, New Delhi wherein baked potato chips were enriched with fibre using potato peel (Azizi et al 2020).

13.4.2 As a Source of Natural Antioxidants

Antioxidants inhibit oxidation of lipids in foods. Unlike the natural antioxidants, consumption of high concentration of synthetic antioxidants has carcinogenic effect on the human body (Thorat et al. 2013). Natural antioxidants have gained considerable interest in recent years for their role in preventing the auto oxidation of fats, oils and fat containing food products. Potato peels contain 10 times higher content of polyphenols as compared to the potato flesh. When extracted with mixture of 0.1% of HCl in methanol: acetone: water (60:30:10, v/v/v), potato peel was found to contain approximately 177 mg of total polyphenols/100 g of fresh weight of peel. A strong positive correlation exists between total polyphenolic compounds in potato peel extracts and their antioxidant potency (Al-Weshahy and Rao 2012). The dominant phenolic compounds of potato peel extracts are chlorogenic and gallic acids (Amado et al. 2014). In comparison with mature potato, young potato peel is an excellent source of bio-active phytochemicals with good antioxidant potential (Arun et al. 2015). The potential of potato peel extract to act as a natural antioxidant in soybean oil was evaluated by Zia-ur-Rehman et al. (2004). After 60 days of storage at 45 °C, soy bean oil, containing 1600 and 2400 ppm of petroleum ether extract of potato peels, showed lower values of free fatty acids (0.120, 0.109%) and peroxide value (10.0, 9.0 meq/kg) than the control samples. Similar antioxidant effect of potato peel extract in fish-rapeseed oil mixture was also observed by Koduvayur Habeebullah et al. (2010).

Potato peels have been used as antioxidants in different food matrices. Rowayshed et al. (2015) used potato peels as sources of natural antioxidants to retard lipid oxidation in biscuits. After 6 months of storage, biscuits containing 0.5, 1, 2 and 3% of potato peel extract showed the order of oxidative stability in the manner as potato peel extract at 3% > 2% > 1% > 0.5% > BHA (200 ppm) > control. These results illustrate that potato peel extract at various concentrations exhibited very strong antioxidant activity comparable with synthetic antioxidants (BHA) and control. Positive antioxidant results were obtained for freeze-dried potato peel extracts added in raw beef patties during cold storage at 5 °C for 12 days (Mansour and Khalil 2000). Potato peel extract has also been found to curb the ill effects of irradiation on meat. During irradiation, adverse oxidative changes in meat can be successfully retarded by using the potato peel extract (Formanek et al. 2003). The antioxidant activity of peel extract was found to be comparable to the synthetic antioxidant, BHT (Kanatt et al. 2005). Similar observations on prevention of lipid and protein oxidation were also reported through application of 2.4 and 4.8 g/kg ethanolic extracts of potato peel in mackerel mince (*Trachurus trachurus*) during storage at 5 °C for 96 h (Sabeena Farvin et al. 2012). These studies illustrate that potato peel extract exhibits very strong antioxidant activity which is almost equal to synthetic antioxidants although when applied at a higher amount. Therefore, potato peel extract in oils, fats and other food products can safely be used as natural antioxidant to suppress lipid oxidation. In comparison with synthetic preservatives potato peel extract as antioxidant is non-carcinogenic and non-mutagenic (Sepelev and Galoburda 2015).

13.4.3 As an Antimicrobial Agent

Natural antimicrobials such as the sulphur compounds (allylsulphide and allicin) found in onion and garlic are widely researched upon for their use in processed foods. Similar to those, potato peel extracts have demonstrated antimicrobial properties owing to flavonoids and terpenes (Nostro et al. 2000). Amanpour (2015) also concluded that potato peel has a non-mutagenic bacteriostatic-behaviour and is safe to use in food processing industries.

13.4.4 As Nutra and Pharma Ingredient

Dudek et al. (2013) observed wound healing and anti-ulcer activity of potato peel. The potato peel is a natural wound healer having ability to heal quickly by aiding in production of high tensile skin at the site of wound (Panda et al. 2011). Glycoalkaloids found in potato peel have also been studied for their application as a steroid hormone precursor (Schieber and Saldaña 2009). These examples corroborate that research on potato peel as a source of health improving bio-active is important and its utilization will serve to improve the nutritional and health status of individuals.

13.4.5 In Edible Film

Edible films can also be produced from potato peel waste. Kang and Min (2010) developed an edible film by using glycerol and soy lecithin. Film prepared by high pressure homogenization up to 138 MPa showed good moisture barrier, tensile and colour properties.

13.4.6 Potato Peel Biscuits

Dietary fibre extracted from potato peels has been utilized for preparing dietary fibre rich biscuits (Singh 2013).

13.5 Extraction of Functional Ingredients from Potato Peels

13.5.1 Dietary Fibre

As discussed earlier, potato peel is chief source of water insoluble dietary fibre. Generally, dietary fibre can be extracted from different plant materials using any of the five methods alone or in combination with chemical, enzymatic, enzyme-chemical combination, membrane separation and microbial fermentation. Li et al. (2019) have developed an acid–alkali method to recover up to 12.6% dietary fibre (on wet basis) from potato peel. In this method, potato peel is refluxed with 1.25% dilute sulphuric acid for 35 min followed by hot water filtration and alkaline extraction with 1.6% sodium hydroxide solution. The residue is then routed through thermal extraction or cold extraction method. In the thermal extraction procedure, residue is exposed to high thermal resistant alpha-amylase with the addition of neutral detergent and defoamer (decalin). After 1 h of micro-boiling the peel residue is filtered through hot water to get the dietary fibre for further quantification. While under cold extraction, the acid–alkali digested residue is washed with acetone, dried and converted into ash to extract and quantify the dietary fibre.

13.5.2 Phenolics

As discussed in previous sections potato peel is a rich source of phenolics that are potent antioxidants (Lisinska and Leszczynski 1987) and are known to have anticarcinogenic effects. Although the potency (effectiveness and stability) of antioxidants from potato peel are still under research, they appear to be a promising alternative as clean label antioxidants compared to those of synthetic origin. Extracts of potato peel have demonstrated strong radical scavenging activity in food systems (de Sotillo et al. 1994a, b). As stated earlier, the predominant phenolics in the potato peel are chlorogenic acid, gallic acid, caffeic acid and protocatechuic acid (Lisinska and Leszczynski 1987). Extraction of phenolics from peel has been tried using different solvent systems in various proportions to change their polarity. Moreover, in any standardized solvent, yield of captured phenolics can be increased by creating more permeation, maceration, tissue exposure to the solvent. Some generally used phenolic extraction techniques are mentioned below.

13.5.2.1 Solvent Extraction

Methanol has been reported as a superior solvent in comparison with water for extraction of phenolics from any food matrix. Besides this, hexane, acetone and ethanol in various concentrations have also been reported for the extraction of these compounds. In case of potato peels, methanol has been found the best solvent followed by water, ethanol, acetone and hexane (Samarin et al. 2012). However, Schieber and Saldaña (2009) successfully tried water, 95% ethanol, methanol (4 °C) and water (25 °C, 100 °C) for extraction of phenolics from potato peels.

Any process or techno-intervention that enhances the porosity of matrix can be used for maximum extraction of phenolics. Sahin and Sumnu (2006) have described various methods of porosity measurement such as (1) Direct Method (2) Optical Method (3) Density method (4) Gas Pycnometer method (volume fraction of air) (5) Porosimeter (liquid extrusion from the pores or liquid intrusion into the pores), etc. Few well explored techniques of cell porosity enhancement have been briefly described below.

13.5.2.2 Ultrasonic Extraction

Ultrasonication is based on cavitation theory. A high intensity ultrasound (10–1000 W cm^{-2}) at higher frequency (up to 2.5 MHz) causes physical disruption of tissues. Ultrasounds produce very rapid localized changes in pressure and temperature which causes shear disruption, cavitation and thinning of cell membrane (Fellows 2005). De Sotillo et al. (1994a) extracted phenolics from ground potato peels using different solvent system. Methanol was adjudged the best solvent; however, yield of phenolics gets further enhanced by ultrasonication (593 v/s 522 μg GAE/g dw).

13.5.2.3 High Pressure Processing

In high hydrostatic pressure or high pressure processing (HPP), pressure up to 10,000 bar is applied to food instantly and uniformly (isostatic) followed by depressurization. Such pressurization and depressurization cycles can collapse intracellular vacuoles and damage cell membrane (Fellows 2005). Here also through cavitational action, cell permeability gets enhanced which further accelerates extraction process.

13.5.2.4 Pulsed Electric Field

Pulsed electric field (PEF) is a non-thermal technology which enhances mass transport by increasing permeability of cell membrane with less degradation of nutritional compounds (Lebovka et al. 2004; Janositz et al. 2011). PEF mainly works on the principle of electroporation of cell membranes where short pulses of high voltage electric fields cause increased permeability of cell membrane (Maskooki and Eshtiaghi 2012; Ignat et al. 2015); therefore, it is often used as a pretreatment for extraction, drying, freezing and salting in food matrix.

13.5.2.5 Microwave Assisted Extraction

Microwaving in pulsed form also creates cavitational force. Singh et al. (2011) optimized conditions for phenolic extraction from freeze-dried potato peels using RSM (Response Surface Methodology) by applying methanol concentration, power level and time as variables. They observed at 67.33% methanol content, 15 min of extraction time and 14.67% of power level, maximum concentration of phenolics can be obtained, i.e. 3.94 mg/g dry weight.

13.5.2.6 Enzyme Assisted Extraction

Enzyme assisted extraction alone or in combination of microwaving has been tried for variety of crops such as black carrot, citrus peels, beet roots, etc., however, reports on enzyme assisted extraction of potato peels are not available. Pectinolytic and cellulotic enzymes are generally used for this purpose. Commercial enzyme preparation viscozyme is a multienzyme complex and extensively used in research. The optimum pH and temperature of viscozyme activity is 3.3–5.5 and 25–55 °C. Viscozyme has been used for polyphenol extraction from *Citrus sinensis* (cv. Malta) peel (Nishad et al. 2019). They recommended processing conditions of 0.84% enzyme concentration, 30.94 mL/g solid to solvent ratio and 4.87 h extraction time for maximum polyphenol extraction using RSM. Enzyme assisted extraction procedures need to be explored more for commercial use.

13.5.2.7 Subcritical Fluid Extraction

Singh and Saldaña (2011) fractionated phenolics from potato peels by subcritical fluid extraction. They fractionated out eight different phenolics compounds, namely, gallic acid, chlorogenic acid, caffeic acid, protocatechuic acid, syringic acid, p-hydroxyl benzoic acid, ferulic acid and coumaric acid. In comparison to methanol, water was found to be a better solvent for total phenolics extraction. Higher yield of phenolics from potato peel can be obtained via subcritical fluid extraction without use of organic solvent (methanol and ethanol), since under subcritical conditions dielectric constant of water is equivalent to organic solvent (Ramos et al. 2002).

13.5.3 Glycoalkaloid Extraction

Health impacts of glycoalkaloids are controversial. In general, glycoalkaloids concentration should be less than 200 ppm, however, somewhat narrow safety concern has been reported by Morris and Lee (1984), i.e. 60 ppm. Since they are more concentrated in peels, therefore, care must be taken while incorporating peel or phenolics extract in any food product (Hossain et al. 2014; Maldonado et al. 2014). Water: acetic acid: sodium bisulphate (+kieselguhr) in 95:5:0.5 ratio has been used for glycoalkaloids extraction from potato peel for further quantification by HPLC (Raigond et al. 2016). Maldonado et al. (2014) found methanolic extract to have higher glycoalkaloids concentration than ethanolic extract of potato peels. Among three tested solvent mixtures, i.e. (1) 25% water, 70% methanol and 5% acetic acid; (2) 24% water, 67% ethanol and 9% acetic acid and (3) 46% water, 51% ethanol and 3% acetic acid; acidified methanol was found the best for extraction of glycoalkaloids as well as phenolics from potato peel. Further, fractionation has been standardized using a Sep Pak Vac 6 cc C18 cartridge at pH 7.0 and ethanol as eluent at 20% and 80% level for phenolics and glycoalkaloids extraction, respectively. Wu (2016) claimed that ultra-sound-assisted extraction of steroidal alkaloids (glycoalkaloids) has the potential for scale up and commercialization since it can be a potential ingredient for phyto-pharmaceutical industry.

13.6 Non-Food Uses of Potato Peel

Besides being used as a food for humans, numerous non-food uses of potato peel have also been explored, studied and practiced. Potato peel being a bio-degradable and renewable source of organic matter with essential nutrients opens a plethora of its applications. As potato is a commonly used low cost source of calories for major parts of the world, potato processing industries are continuously looking for novel applications of potato peel to increase their net profits. Potential applications of potato peel as a source of dietary supplements, organic matter, pharmaceutical ingredients, preservative, antioxidant, fibre and animal feed, etc., have been studied by various researchers (Tiwari et al. 2009; Parashar et al. 2014; Sepelev and Galoburda 2015; Akyol et al. 2016). This section elaborates on the possible applications of potato peel other than as human food and/or ingredient.

13.6.1 Nano-Crystal Synthesis

Raigond et al. (2018) synthesized nano-crystals from potato peel waste by acid hydrolysis of cellulosic fraction (alkaline extractable) of peels. Microscopic studies revealed size of potato peel nano-crystals ranged from 40 to 100 nm (length) and cellulose microcrystalline ranged from 4 to 20 nm (diameter). Both nanostructures can be considered safe for humans and the environment due to their bio-degradable nature and can be used as safe alternative of metallic nano-particles. However, like any other nano-particles, safety must be ensured before utilizing nano-particles in any food or non-food matrix.

13.6.2 For Enzyme Production

Potato peel extracts has been tried as a substrate for the production of extracellular enzymes by *Bacillus* in continuous culture (Mahmood et al. 1998). High content of starch (52 g/100 g of dry weight) makes it a good basis for fermentation (Arapoglou et al. 2009). As a source of low cost agro-industry waste material, potato peels were also utilized for the production of alpha-amylase and alkaline protease enzyme (Mukherjee et al. 2008, 2009). Among various lignocellulosic material such as peels of mango, apple, citrus, wheat straw, corn cob, rice husk; potato peel was found the most suitable matrix for the production of ß-mannanase, by *Bacillus amyloliquefaciens*. Enzyme ß-mannanase is extensively used in textile, coffee, oil and paper industry (Gubitz et al. 1996; Ademark et al. 1998).

13.6.3 Substrate for Fermentation

Potato peel waste can serve as a solid substrate for fermentation. By utilization of alkaline potato peel waste through fermentation, amylase production up to 26–40

SKB units/mL could be achieved by *Aspergillus foetidus* NRRL 337. Alcoholic fermentation of peel waste gave alcohol yields of up to 90% on a stoichiometric carbohydrates basis. Since, fermentable reducing sugar content is very low, 0.6 g/100 g of dry weight, fermentation is not practical and an initial hydrolysis of carbohydrates is necessary. This may be achieved by hydrolysis with various enzymes and/or acid, and fermented by *Saccharomyces cerevisiae* var. *bayanus* (Arapoglou et al. 2009, 2010). Studies showed that in comparison with enzymes, sulphuric acid exhibited more hydrolysis activity of dried potato samples for further fermentation (Guerra-Rodríguez et al. 2012).

Potato peels are also a good substrate for single cell protein production. *Scytalidium acidophilum* fungus, with a promising amino acid composition was produced on potato peel waste to be used in animal feed. Simultaneously, the BOD of the potato peel waste was reduced by 98%.

13.6.4 Bio-Fuel

With the continued depletion of fossil fuels, the search for better, environment friendly and renewable sources of energy is on a high. According to Arapoglou et al. (2010), potato peel contains sufficient quantities of starch, cellulose, hemicellulose, lignin and fermentable sugars to be used as source of bioethanol. They devised the whole set-up for production of ethanol using potato peel waste by enzymatic hydrolysis to enhance and release the fermentable sugar followed by fermentation using *S. cerevisiae var. bayanus* for production of ethanol at levels of 7.58 g/L which is approximately 96.5% of the theoretical maximum. In an attempt to determine ability of fermentation and ethanol production, they hydrolyzed potato peel waste using enzymes and/or acid, and fermentation by *Saccharomyces cerevisiae* var. *bayanus*. Hydrolysis with a combination of enzymes released 18.5 g/L sugars and produced 7.6 g/L ethanol after fermentation.

13.6.5 Bio-Gas

Potato peel has a noteworthy potential to be used as a source of renewable energy like bio-gas (Adeyosoye et al. 2010). Bio-gas production ability of potato peels by mesophilic anaerobic bacteria was reviewed by Gebrechristos and Chen (2018). For this, a bio-reactor with eco-friendly way of waste utilization can generate bio-manure, electricity and high quality of bio-gas, depending upon capacity of bio-reactor and amount of bio-waste, etc.

13.6.6 Animal Feed

Different food wastes are being increasingly explored as animal feeds since cost cutting of feed production is on a continuous rise. Potato peel is one of the prominent food wastes that could be used as alternative animal feed due to natural sources of energy and fibre with low levels of protein (Chimonyo 2017).

13.7 Conclusion

Across the globe, potato is considered the foremost horticultural crop to provide bulk to human diet. Technological intervention for pre- and post-harvest operations of potatoes is always welcomed and desired by growers, processors and policymakers. Recycling and utilization of waste generated of this huge resource is utmost important which otherwise is termed as 'zero value'. Use of peel can reduce the waste burden of environment and can cater to the food and pharmaceutical industries. However, high moisture content and adhered starch are the major constraints for storability of fresh potato peels. Therefore, provision of drying must be in place to avoid unnecessary bulk and microbial deterioration of peel. Safety concern while utilizing peels as whole or in extract form must be taken care of or fractionated in view point of glycoalkaloids content.

References

Abdel-Magied MM (1991) Effect of dietary fiber of potato peel on the rheological and organoleptic characteristics of biscuits. Egypt J Food Sci 19:293–300

Ademark P, Varga A, Medve J et al (1998) Softwood hemicellulose-degrading enzymes from Aspergillus niger: purification and properties of ß-mannanase. J Biotechnol 63:199–210

Adeyosoye OI, Adesokan IA, Afolabi KD et al (2010) Estimation of proximate composition and biogas production from in vitro gas fermentation of sweet potato (*Ipomoea batatas*) and wild cocoyam (*Colocasia esculenta*) peels. Afr J Environ Sci Technol 4:388–391

Akyol H, Riciputi Y, Capanoglu E et al (2016) Phenolic compounds in the potato and its byproducts: an overview. Int J Mol Sci 17:835–837

Al-Saikhan MS, Howard LR, Miller JC (1995) Antioxidant activity and total phenolics in different genotypes of potato (*Solanum tuberosum* L). J Food Sci 60:341–343

Al-Weshahy A, Rao AV (2009) Isolation and characterization of functional components from peel samples of six potatoes varieties growing in Ontario. Food Res Int 42:1062–1066

Al-Weshahy A, Rao VA (2012) Potato peel as a source of important phytochemical antioxidant nutraceuticals and their role in human health—a review. In: Rao V (ed) Phytochemicals as nutraceuticals—global approaches to their role in nutrition and health. InTech Publishers, Rijeka. http://www.intechopen.com/books/phytochemicals-as-nutraceuticals-globalapproaches-to-their-role-in-nutrition-and-health/potato-peel-as-a-source-of-important-phytochemicalantioxidant-nutraceuticals-and-their-role-in-huma

Amado IR, Franco D, Sánchez M et al (2014) Optimisation of antioxidant extraction from *Solanum tuberosum* potato peel waste by surface response methodology. Food Chem 165:290–299

Amanpour R (2015) Antibacterial effects of *Solanum tuberosum* peel ethanol extract in vitro. J HerbMed Pharmacol 4:45–48

Arapoglou D, Vlyssides A, Haidemenaki K et al (2009) Alternative ways for potato industries waste utilisation. In: Proceedings of the 11th International Conference on Environmental Science and Technology, Chania, Crete, Greece, 3–5 September 2009, pp 3–5

Arapoglou D, Varzakas T, Vlyssides A et al (2010) Ethanol production from potato peel waste (PPW). Waste Manag 30:1898–1902

Arora A, Camire ME (1994) Performance of potato peels in muffins and cookies. Food Res Int 27:15–22
Arun KB, Chandran J, Dhanya R et al (2015) A comparative evaluation of antioxidant and antidiabetic potential of peel from young and matured potato. Food Biosci 9:36–46
Azizi AF, Sethi S, Joshi A, Singh AM, Raigond R, Singh MK, Yadav RK (2020) Biochemical and functional attributes of raw and boiled potato flesh and peel powder. J Food Sci Technol. https://doi.org/10.1007/s13197-020-04424-3
Camire ME, Flint SI (1991) Thermal processing effects of dietary fiber composition and hydration capacity in corn meal, oat meal and potato peels. Cereal Chem 68:645–647
Camire ME, Zhao J, Violette DA (1993) In vitro binding of bile acids by extruded potato peels. J Agric Food Chem 41:2391–2394
Camire ME, Zhao J, Dougherty MP et al (1995) In vitro binding of benzopyrene by extruded potato peels. J Agric Food Chem 43:970–973
Chandalia M, Garg A, Lutjohann D et al (2000) Beneficial effect of high dietary fiber intake in patients with type 2 diabetes mellitus. N Engl J Med 342:1392–1398
Charmley E, Nelson D, Zvomuya F (2019) Nutrient cycling in the vegetable processing industry: utilization of potato by-products. Can J Soil Sci 86:621–629
Chimonyo M (2017) A review of the utility of potato by-products as a feed resource for smallholder pig production. Anim Feed Sci Technol 227:107–117
Curti E, Carini E, Diantom A et al (2016) The use of potato fibre to improve bread physico-chemical properties during storage. Food Chem 195:64–70
Dale MFB, Griffiths DW, Bain H (1998) Effect of bruising on the total glycoalkaloids and chlorogenic acid content of potato (*Solanum tuberosum*) tubers of five cultivars. J Sci Food Agric 77:499–505
De Sotillo R, Hadley M, Holm ET (1994a) Phenolics in aqueous potato peel extract: extraction, identification and degradation. J Food Sci 59:649–651
De Sotillo R, Hadley M, Holm ET (1994b) Potato peel waste: stability and antioxidant activity of a freeze- dried extract. J Food Sci 59:1031–1033
Dudek N, Maria MD, Horta P et al (2013) In vivo wound healing and antiulcer properties of white sweet potato (*Ipomoea batatas*). J Adv Res 4:411–415
Fellows PJ (2005) Processing using electric fields, high hydrostatic pressure, light or ultrasound. In: Food processing technology: principles and practice, 4th edn. Woodhead Publishing, Duxford, pp 210–227
Formanek Z, Lynch A, Galvin K et al (2003) Combined effects of irradiation and the use of natural antioxidants on the shelf-life stability of overwrapped minced beef. Meat Sci 63:433–440
Forsythe WA, Chenoweth WL, Bennink MR (1976) The effect of various dietary fibres on serum cholesterol and laxation in the rat. J Nutr 106:26–32
Friedman M (2004) Analysis of biologically active compounds in potatoes (*Solanum tuberosum*), tomatoes (*Lycopersicon esculentum*), and jimson weed (*Datura stramonium*) seeds. J Chromatogr A 1054:143–155
Gebrechristos HY, Chen W (2018) Utilization of potato peel as eco-friendly products: a review. Food Sci Nutr 6:1352–1356
Gubitz GM, Hayn M, Urbanz G, Steiner W (1996) Purification and properties of an acid ß-mannanase from Sclerotium rolfsii. J Biotechnol 45:165–172
Guerra-Rodríguez E, Portilla-Rivera OM, Ramírez J et al (2012) Modelling of the acid hydrolysis of potato (Solanum tuberosum) for fermentative purposes. Biomass Bioenergy 42:59–68
Harris PJ, Ferguson LR (1993) Dietary fibre: its composition and role in protection against colorectal cancer. Mutat Res 290:97–110
Hossain MB, Tiwari BK, Gangopadhyay N et al (2014) Ultrasonic extraction of steroidal alkaloids from potato peel waste. Ultrason Sonochem 21:1470–1476
Ignat A, Manzocco L, Brunton NP et al (2015) The effect of pulsed electric field pretreatments prior to deep-fat frying on quality aspects of potato fries innovative. Innov Food Sci Emerg Technol 29:65–69

Im HW, Suh BS, Lee SU et al (2008) Analysis of phenolic compounds by high-performance liquid chromatography and liquid chromatography/mass spectrometry in potato plant flowers, leaves, stems, and tubers and in home-processed potatoes. J Agric Food Chem 56:3341–3349

Janositz A, Noack AK, Knorr D (2011) Pulsed electric fields and their impact on the diffusion characteristics of potato slices. LWT- Food Sci Technol 44:1939–1945

Jeddou KB, Bouaziz F, Ellouzi SZ et al (2017) Improvement of texture and sensory properties of cakes by addition of potato peel powder with high level of dietary fiber and protein. Food Chem 217:668–677

Kähkönen MP, Hopia AI, Vuorela HJ et al (1999) Antioxidant activity of plant extracts containing phenolic compounds. J Agric Food Chem 47:3954–3962

Kanatt SR, Chander R, Radhakrishna P et al (2005) Potato peel extract—a natural antioxidant for retarding lipid peroxidation in radiation processed lamb meat. J Agric Food Chem 53:1499–1504

Kang HJ, Min SC (2010) Potato peel-based biopolymer film development using high-pressure homogenization, irradiation, and ultrasound. LWT- Food Sci Technol 43:903–909

Kannat S, Chander R, Radhakrishna P et al (2005) Potato peel extracts, a natural antioxidant for retarding lipid peroxidation in radiation processed lamb meat. J Agric Food Chem 53:1499–1504

Koduvayur Habeebullah SF, Nielsen NS, Jacobsen C (2010) Antioxidant activity of potato peel extracts in a fish-rapeseed oil mixture and in oil-in-water emulsions. J Am Oil Chem Soc 87:1319–1332

Lazarov K, Werman MJ (1996) Hypocholesterolemic effect of potato peels as a dietary fiber source. J Med Sci Res 24:581–582

Lebovka NI, Praporscic I, Vorobiev E (2004) Effect of moderate thermal and pulsed electric field treatments on textural properties of carrots, potatoes and apples. Innovative Food Sci Emerg Technol 5:9–16

Li CL, Yu SY, Lu Y (2019) Study on extraction of dietary fiber from potato peel by acid-base chemical method. IOP Conf Ser Earth Environ Sci. https://doi.org/10.1088/1755-1315/267/5/052028

Lisinska G, Leszczynski W (1987) Potato tubers as raw material for processing and nutrition. In: Lisinska G, Leszczynski W (eds) Potato science and technology. Elsevier Applied Science, London, pp 34–38

Mäder J, Rawel H, Kroh LW (2009) Composition of phenolic compounds and glycoalkaloids alpha-solanine and alpha-chaconine during commercial potato processing. J Agric Food Chem 57:6292–6297

Mahmood AU, Greenman J, Scragg AH (1998) Orange and potato peel extracts: analysis and use as *Bacillus* substrates for the production of extracellular enzymes in continuous culture. Enzym Microb Technol 22:130–137

Makris D, Boskou PG, Andrikopoulos NK (2007) Polyphenolic content and in vitro antioxidant characteristics of industry and other agri-food solid waste extracts. J Food Compos Anal 20:125–132

Maldonado AFS, Mudge E, Gänzle MG et al (2014) Extraction and fractionation of phenolic acids and glycoalkaloids from potato peels using acidified water/ethanol-based solvents. Food Res Int 65:27–34

Mansour EH, Khalil AH (2000) Evaluation of antioxidant activity of some plant extracts and their application to ground beef patties. Food Chem 69:135–141

Maskooki A, Eshtiaghi MN (2012) Impact of pulsed electric field on cell disintegration and mass transfer in sugar beet. Food Bioprod Process 90:377–384

Morris SC, Lee TH (1984) The toxicity and teratogenicity of Solanaceae glycoalkaloids, particularly those of potato (*Solanum tuberosum*). Food Technol Aust 36:118–124

Mukherjee KA, Adhikari H, Rai KS (2008) Production of alkaline protease by a thermophilic Bacillus subtilis under solid-state fermentation (SSF) condition using Imperata cylindrica grass

and potato peel as low-cost medium: characterization and application of enzyme in detergent formulation. Biochem Eng J 39:353–361
Mukherjee KA, Borahb M, Rai KS (2009) To study the influence of different components of fermentable substrates on induction of extracellular α- amylase synthesis by Bacillus subtilis DM-03 in solid-state fermentation and exploration of feasibility for inclusion of α- amylase in laundry detergent formulations. Biochem Eng J 43:149–156
Nara K, Miyoshi T, Honma T et al (2006) Antioxidative activity of bound-form phenolics in potato peel. Biosci Biotechnol Biochem 70:1489–1491
Nishad J, Saha S, Kaur C (2019) Enzyme- and ultrasound-assisted extractions of polyphenols from *Citrus sinensis* (cv. Malta) peel: a comparative study. J Food Process Preserv 43:e14046. https://doi.org/10.1111/jfpp.14046
Nostro A, Germanò MP, D'angelo V et al (2000) Extraction methods and bioautography for evaluation of medicinal plant antimicrobial activity. Lett Appl Microbiol 30:379–384
Onyechi UA, Judd PA, Ellis PR (1998) African plant foods rich in non-starch polysaccharides reduce postprandial blood glucose and insulin concentrations in healthy human subjects. Br J Nutr 80:419–428
Orr PH, Toma RB, Munson ST et al (1982) Sensory evaluation of breads containing various levels of potato peel. Am Potato J 59:605–612
Panda V, Sonkamble M, Patil S (2011) Wound healing activity of *Ipomoea batatas* tubers (sweet potato). Funct Foods Health Dis 10:403–415
Parashar S, Sharma H, Garg M (2014) Antimicrobial and antioxidant activities of fruits and vegetable peels: a review. J Pharmacogn Phytother 3:160–164
Pourcel L, Routaboul JM, Cheynier V et al (2007) Flavonoid oxidation in plants: from biochemical properties to physiological functions. Trends Plant Sci 12:29–36
Raigond P, Bandana JA et al (2016) Potato biochemistry: a laboratory manual. Technical Bulletin No. 102, ICAR-Central Potato Research Institute, Shimla, India, 84p
Raigond P, Sood A, Kalia A et al (2018) Antimicrobial activity of potato starch based active biodegradable nanocomposite films. Potato Res 62:69. https://doi.org/10.1007/s11540-018-9397-9
Ramos L, Kristenson EM, Brinkman UA (2002) Current use of pressurized liquid extraction and subcritical water extraction in environmental analysis. J Chromatogr A 975:3–29
Rowayshed G, Sharaf AM, El-Faham SY et al (2015) Utilization of potato peels extract as source of phytochemicals in biscuits. J Basic Appl Res Int 8:190–201
Sabeena Farvin KH, Grejsen HD et al (2012) Potato peel extract as a natural antioxidant in chilled storage of minced horse mackerel (*Trachurus trachurus*): effect on lipid and protein oxidation. Food Chem 131:843–851
Sahin S, Sumnu SG (2006) Physical properties of foods. Springer, New York
Samarin AM, Poorazarang H, Hematyar N et al (2012) Phenolics in potato peels: extraction and utilization as natural antioxidants. World Appl Sci J 18:191–195
Schieber A, Saldaña MDA (2009) Potato peels: a source of nutritionally and pharmacologically interesting compounds—a review. Food 3:23–29
Schieber A, Stinzing FC, Carle R (2001) By-products of plant food processing as a source of functional compounds-recent developments. Trends Food Sci Technol 12:401–413
Sepelev I, Galoburda R (2015) Industrial potato peel waste application in food production: a review. Res Rural Dev 1:130–136
Singh BP (2013) Final report of NAIP sub-project 'value chain on potato and potato products'. Central Potato Research Institute, Meerut
Singh PP, Saldaña MDA (2011) Subcritical water extraction of phenolic compounds from potato peel. Food Res Int 44:2452–2458
Singh A, Sabally K, Kubow S et al (2011) Microwave-assisted extraction of phenolic antioxidants from potato peels. Molecules 16(3):2218–2232
Singh B, Raigond P, Barwal S et al (2016) Glycoalkaloids in peels of Indian potatoes. Potato J 43:86–92

Thorat ID, Jagtap DD, Mohapatra D et al (2013) Antioxidants, their properties, uses in food products and their legal implications. Int J Food Stud 2:81–104

Tiwari BK, Valdramidis VP, O'Donnell CP et al (2009) Application of natural antimicrobials for food preservation. J Agric Food Chem 57:5987–6000

Toma RB, Orr PH, D'appolonia B et al (1979) Physical and chemical properties of potato peel as a source of dietary fiber in bread. J Food Sci 44:1403–1407

Wu D (2016) Recycle technology for potato peel waste processing: a review. Procedia Environ Sci 31:103–107

Zia-ur-Rehman, Habib F, Shah WH (2004) Utilization of potato peels extract as a natural antioxidant in soy bean oil. Food Chem 85:215–220

Nutritional Significance of Processed Potato Products

14

Arvind Kumar Jaiswal

Abstract

Globally, potato (*Solanum tuberosum* L.) shares a major part of our daily diet. Nowadays being a perishable crop and consumer preference its utilization is shifting towards the processing and value addition instead of fresh consumption. Due to frequent consumption, potatoes and its processed products have nutritional significance for human health. Potatoes are carbohydrate rich foods and frying makes them high energy and fat rich also. Potato tubers also contain plenty of dietary carbohydrates, proteins, vitamins, minerals, bioactive molecules, phytochemicals, anthocyanin, carotenoids, and metabolites having beneficial effects on human health and must be present in daily diet. Potato protein contains well-balanced amino acid composition with a higher content of essential amino acids and is one of the best sources of lysine among the vegetable and cereal sources. Among the processed products of potatoes, chips and French fries are most popular, although they are rich in fat and energy. Beside this, various frozen products such as mashed potatoes, hash brown and puffs, dehydrated potatoes, and starch are also available in the market. All these products are good source of carbohydrates, potassium, phosphorous, vitamin C, niacin, and many more. Recently, ICAR-CPRI Shimla has developed potato-based novel value-added products such as cookies, *halwa* premix, semolina, porridge which are rich in dietary fiber, protein, vitamin C, and potassium. The low sodium content and gluten-free composition makes these products also fit for celiac/wheat allergic and hypertensive population. This chapter provides the better understanding of nutritional significance of potato and its products. Besides this, the chapter also provides the overview of new dimensions of value addition in potatoes.

A. K. Jaiswal (✉)
ICAR-Central Potato Research Station, Jalandhar, Punjab, India
e-mail: jaiswal.arvind@live.com

P. Raigond et al. (eds.), *Potato*, https://doi.org/10.1007/978-981-15-7662-1_14

Keywords

Potato · Frozen products · Fried products · Cookies · Porridge · Halwa · Chips · Nutrition · Protein · Amino acids · Minerals · Vitamins

14.1 Introduction

Potato (*Solanum tuberosum* L.) is one of the most popular food commodities, globally sharing a major part of diet either as fresh potato or in the processed form. The total world potato production was estimated 0.388 billion tons in 2017 (FAOSTAT 2019) with 99.21 million tons share from China and 48.61 million tons from India. About 50% of potatoes grown worldwide are freshly consumed. The rest is processed into different potato food products and ingredients, feed for animals and birds, starch for industry, and as seed potato. The processed form includes fried, pre-fried and frozen, baked, dehydrated, and other miscellaneous products such as alcohol, organic acids, boiled and peeled, canned, and mashed. With the technological advancements, potato processing is growing rapidly and driven by the market. Globally, the potato utilization trend is also shifting towards the processing and value addition instead of fresh consumption. Presently, mainly frozen French fries and chips/crisps constitute a major proportion of value-added potato products market worldwide. The world's appetite for factory-made French fries has been put at more than 7 million tons a year (International Potato Center 2019). According to the IMARC report, in 2018, the global frozen finger chips market was worth US$ 20.4 billion with the compound annual growth rate (CAGR) of 3.7% during 2011–2018 (IMARC 2018). Dehydrated potato flakes, an emerging product shares a major proportion of the global market for its use in the preparation of mashed potato products, fabricated products, and as an ingredient in several fried snacks. Potato flour is also gaining popularity due to its use for binding of meat mixtures and as a thickener for gravies and soups. Due to fineness, bland taste, outstanding mouth-feel, viscosity, and clarity, potato starch has huge industrial importance for use in sauces, baked items, dough, and frozen desserts.

According to the IMARC Group, during the 2009–2016 global potato chips market has grown at a CAGR of around 4% and reached a market value worth US$ 26 billion and in the upcoming five years, it is expected to grow at a CGAR of 4.4%. In the year 2017, the global potato chips market accounted for US$ 28 billion (IMARC 2018). The main reason behind this growth is easy affordability and availability. North America and Europe represent the biggest market which is dominating the global chips market accounting for almost two-thirds of the total global demand. Especially the younger generation is the key driver for high market demand. Now the markets of India, China, and Russia are also witnessing promising growth rates. With the economic development, the incidence of many cardiovascular diseases, cancer, and diabetes is rapidly increasing worldwide. Consequently, the consumers are now inclining towards the fortified potato chips. Improved marketing channels are also responsible for continuously increasing CGAR. In a report by grand view research in 2016, the supermarkets/hypermarkets dominated the overall industry with a revenue share of approximately 76.0%, whereas convenience stores accounted for 15.3% of the U.S. potato chips market and the rest share was through other distribution channels such as service stations, drug stores, and online retail. In some countries especially in eastern Europe and Scandinavia, potatoes are used for the distillation of alcoholic beverages, such as vodka and akvavit. In some parts of Brazil, potato starch is used for the production of beer. The quality of any potato-based product severely depends upon the quality of raw material which is affected by various factors starting from the field to the storage of potatoes.

14.2 Nutritional Significance of Potato Products

The nutritional significance of any crop is highly dependent upon their inherent metabolic composition for which potato has a unique position. Starch is the major nutrient present in the potato which is one of the largest sources of carbohydrates in our daily diet. In fact, potato tubers contain plenty of dietary carbohydrates, proteins, vitamins, minerals, bioactive molecules, phytochemicals, anthocyanin, and carotenoids and metabolites which have beneficial effects on human health and must be present in daily diet (Burlingame et al. 2009; Ezekiel et al. 2013; Katan and Roos 2004; King and Slavin 2013; Love and Pavek 2008; Weaver and Marr 2013). Due to the high per capita consumption, among fruits and vegetables, potatoes are the third-largest contributor of total phenolic content to the diet (Song et al. 2010). If we look towards the contribution of potato as a source of the energy, potatoes yield more calories per unit area (216 MJ ha^{-1} day^{-1}) than any other major crop (Anderson 2010), such as corn (159 MJ ha^{-1} day^{-1}) and rice (121 MJ ha^{-1} day^{-1}).

In general, a major part of the potato is water (~80%) and the remaining part is solids (~20%) which remain after removal of water and may vary among the cultivars. This 20% part of the tuber is composed of about 17 g carbohydrates, 2 g proteins and the remaining part is composed of other ingredients. Potato also contains small proportion of resistant starch (Jansen et al. 2001) which provides different health benefits such as protection against colon cancer (Keenan et al. 2015; Liu and Xu 2008), improves glucose tolerance and insulin sensitivity (Keenan et al. 2015; Maki et al. 2012; Robertson et al. 2005), lowers plasma cholesterol (Dodevska et al. 2016; Gentile et al. 2015), triglyceride concentrations (Bindels et al. 2015), and increases satiety (Bodinham et al. 2010). Even though potato contains a very low amount of protein, it has excellent biological value (Waglay et al. 2014). Potato protein contains relatively well-balanced amino acid concentrations along with a higher content of essential amino acids compared to other proteins of plant-origin. In fact, potato is one of the best sources of lysine among the vegetable and cereal sources (Waglay et al. 2014). According to the USDA nutrient database, 100 g of potatoes provides about 4% of the RDA for calorie intake, 6% of RDA for total carbohydrate, 9% of RDA for dietary fiber, 4% of RDA for protein, 33% of RDA for vitamin C, 15% of RDA for vitamin B6, and 12% of RDA for potassium, 4% of

RDA for iron (USDA-SR Legacy 170032). Therefore, since potato is a stable part of the diet, nutrients present in the potato have a significant impact and dietary relevance than other foods eaten in sparse quantities.

14.2.1 Most Popular Processed Potato Products

14.2.1.1 Chips

Potato chips are one of the most popular snacks consumed worldwide, with a history of more than 165 years back. As per Statista, Inc, New York, data, the worldwide current revenue from potato chips are nearly US$41,873 million and the market is further expected to grow annually by CAGR of 3.8% in the next five years. At present, the global average per capita consumption of potato chips is around 0.6 kg annually. Furthermore, easy access to several brands and flavors of potato chips are some of the key factors promoting the growth of the market. Chips are often considered as "junk food," and leads to obesity, while many consumers perceive potatoes as high in calories and carbohydrates. But the reality is that potatoes have relatively low energy density and can contribute significant amounts of fiber and micronutrients, such as potassium, which are shortfall nutrients in the Western diet (Freedman and Keast 2011). With the rising health awareness, many big players also have introduced healthier varieties of potato chips without any added artificial flavors or colors, non-GMO ingredients, and also low oil chips such as baked, which are pushing the market. PepsiCo, Inc. (USA), Diamond Foods, Inc. (USA), Lorenz Bahlsen Snack-World Group (Germany), Herr Foods Inc. (USA), and Intersnack Group (Germany) are some of the key players leading the potato chip industry at the global level.

Nutritionally potato chips are fat-rich, high energy food (Table 14.1). Based on a 2000-calorie reference diet 100 g of plain salted potato chips can provide 27% of recommended daily value for energy, 58% of daily value for fat, 13% of daily value for protein, 17% of daily value for carbohydrate, 18% of daily value for dietary fiber, and 3.9 g of ash (USDA 2019). Out of the total energy, about 37% of energy comes from carbohydrates, more than 60% from fat, and about 3.3% from protein present in the chips.

Potato chips contain 3–11 g/100 g of saturated fat accounting for up to 55% of recommended daily value. The predominant fatty acids in potato chips include palmitic acid, stearic acid, myristic acid, and some part as lauric acid. Depending upon the quality of oil used during frying, 100 g chips can provide 9.8 g of monounsaturated fatty acids (MUFAs), 12.2 g of polyunsaturated fatty acids (PUFAs), 190 mg of omega-3 fatty acids, and 12 g of omega-6 fatty acids. Oleic acid constitutes the major part of monounsaturated fatty acid and linoleic acid (omega-6) as a major constituent of polyunsaturated fatty acid (USDA-SR Legacy 169677).

Protein and Amino Acids Potatoes are one of the highest per hectare protein supplying crops (Wu 2016). Fresh tubers of potato contain about 1.7–2.0% protein.

Table 14.1 Nutritive value of raw potato, popular fried products, and potato flour (per 100 g) (USDA 2019)

Parameters	Raw potato with skin[a]	Plain salted potato chips[b]	French fries fried in vegetable oil[c]	Fried sticks[d]	Potato flour[e]
Water (%)	77.0	1.24–2.44	34.6–45.2	2.2	6.5
Energy (kcal)	77	547	312	522	357
Total carbohydrate (%)	18.4	49.7–53.8	41.44	53.3	83.1
Dietary fiber (%)	2.2	2.2–3.4	2.8–4.3	3.4	5.9
Sugars (%)	0.8	0.3–0.4	0.18–0.83	0.22	3.5
Total protein (%)	2.0	5.6–6.6	2.63–4.2	6.7	6.9
Total fat (%)	0.1	30.7–38.39	11.9–16.6	34.4	0.3
Saturated fat (%)	0.0	3.4–11.0	2.3	8.9	0.1
Monounsaturated fat (%)	0.0	9.8–18.9	6.0	6.2	0.0
Polyunsaturated fat (%)	0.0	8.3–12.2	5.4	17.9	0.2
Total trans fatty acids (mg)	–	0–84	60	–	–
Total Omega-3 fatty acids (mg)	10.0	190.0	168.0	70.0	35.0
Total Omega-6 fatty acids (mg)	32.0	11980.0	2831.0	17802.0	~
Ash (%)	1.1	3.13–4.23	1.4–2.0	3.3	3.1
Calcium (mg)	12.0	12.0–31.0	12–23	18.0	65.0
Iron (mg)	0.8	1.0–1.5	0.63–1.07	2.3	1.4
Magnesium (mg)	23.0	49.0–81.0	25–40	64.0	65.0
Phosphorus (mg)	57.0	120–201	88–153	172.0	168.0
Potassium (mg)	421.0	983–1380	451–675	1237.0	1001.0
Sodium (mg)	6.0	235–633	115–363	633.0	55.0
Zinc (mg)	0.3	0.77–1.3	0.39–0.73	1.0	0.5
Copper (mg)	0.1	0.059–0.378	0.072–0.212	0.3	0.2
Manganese (mg)	0.2	0.34–0.57	0.183–0.343	0.4	0.3
Selenium (μg)	0.3	0.40–5.3	0.9–1.2	8.1	1.1
Vitamin A (IU)	2.0	0.0	0.0	0.0	0.0
Vitamin C (mg)	19.7	18.6–21.6	1.3–6.8	47.3	3.8
Vitamin E (α-Tocopherol) (mg)	0.0	6.7–10.5	1.7	9.1	0.3
Vitamin K (μg)	1.6	22–25	11.2	22.1	~
Thiamin (mg)	0.1	0.15–0.25	0.14–0.19	0.1	0.2
Riboflavin (mg)	0.0	0.08–0.12	0.03–0.06	0.1	0.1
Niacin (mg)	1.1	3.98–7.15	2.09–3.66	4.8	3.5
Vitamin B6 (mg)	0.3	0.40–0.64	0.28–0.44	0.3	0.8
Folate (μg)	16.0	29.0–75.0	25–64	40.0	25.0

(continued)

Table 14.1 (continued)

Parameters	Raw potato with skin[a]	Plain salted potato chips[b]	French fries fried in vegetable oil[c]	Fried sticks[d]	Potato flour[e]
Pantothenic acid (mg)	0.3	0.78–1.0	0.47–0.70	0.4	0.5

[a]Potatoes, raw, skin (USDA Standard Reference Legacy, 170032) (2019)
[b]Snacks, potato chips, plain, salted (USDA Standard Reference Legacy, 169677) (2019)
[c]Fast foods, potato, French fried in vegetable oil (USDA Standard Reference Legacy, 170698) (2019)
[d]Snacks, potato sticks (USDA Standard Reference Legacy, 168854) (2019)
[e]Potato flour (USDA Standard Reference Legacy, 168446)

Potato proteins are considered as high-quality proteins, equivalent to egg protein, lysozyme (Bártová and Bárta 2009; Ralet and Guéguen 2000) containing a high proportion of essential amino acids lysine, threonine, and sulfur-containing amino acids (He et al. 2013; Kamnerdpetch et al. 2007). Patatin is the major potato protein that represents approximately 40% of the soluble glycoproteins (Kärenlampi and White 2009). Along with nutritional properties patatin has been shown to possess various bioactivities, viz. antioxidant activity (Liu et al. 2003; Sun et al. 2013; Wang and Xiong 2005) and ACE inhibition (Makinen et al. 2008; Mäkinen et al. 2016).

Another protein, protease inhibitor also constitutes about 30–40% part of total tuber protein (Jørgensen et al. 2006; Pouvreau et al. 2001). Protease inhibitor has very low allergic response (Fu et al. 2002; He et al. 2013; Kärenlampi and White 2009) and several health benefits such as antimicrobial effects (Kim et al. 2005), antioxidant properties (Pihlanto et al. 2008), blood pressure and serum cholesterol regulation (Liyanage et al. 2008; Pihlanto et al. 2008), and anti-carcinogenic effect (Blanco-Aparicio et al. 1998). Moreover, all potato protein fractions are also rich in non-essential amino acid such as aspartic and glutamic acid.

Potato chips contain 5.6–6.6% protein which is more than 3 times of fresh potatoes. Denaturation is the prime reaction affecting the protein during frying. Most of the proteins denature at temperatures higher than 100 °C, which is desirable from the nutrition point of view, as the digestibility of denatured proteins is better than the native proteins. Therefore, the digestibility of potato chips is better than fresh potatoes. Whereas, some studies show that the soluble protein fraction decreases during tortilla chip preparation, accompanied by some unfavorable changes in protein digestibility (Vivas-Rodríguez et al. 1990). During frying, some undesirable changes also occur due to various reactions between amino acids and other components. The average losses of amino acids may reach up to 7%. Due to the presence of free primary amino group, lysine is the most reactive amino acid and forms lysino amino acids during frying by dehydration and condensation. As per USDA standard reference legacy database, fresh potatoes provide about 126 mg of lysine, whereas plain salted chips provide 424 mg lysine per 100 g weight (USDA-SR Legacy 170032). Potato chips contain 108 mg of tryptophan, whereas fresh potatoes contain 32 mg of tryptophan per 100 g weight (USDA-SR Legacy 169677).

The presence of the indole group in tryptophan increases the sensitivity towards heat-induced losses such as thermolysis and condensation. Likewise, the easy cleavability of the thiol group in cysteine is responsible for significantly higher losses. Fresh potatoes contain only 26 mg cystine, whereas potato chips can provide 89 mg cystine per 100 g serving. At frying temperatures free serine and cysteine bind with lysine to form lysinoalanine, a potentially moderately toxic compound. Potato chips contain 303 mg of serine which is 3.37 times higher than fresh potatoes (90 mg/100 g). At frying temperature methionine also undergoes various oxidation and lytic reactions. Potato chips contain 110 mg of methionine which is 3.33 times higher than fresh potatoes (33 mg/100 g). Other essential amino acids are 3.35–3.42 times higher than fresh potatoes. Potato chips can provide 153 mg of histidine, 283 mg isoleucine, 419 mg leucine, 310 mg phenylalanine, 253 mg threonine, and 392 mg valine per 100 g weight. Likewise, the non-essential amino acid content of potato chips is 3.34–3.42 times higher than fresh potatoes. Potato chips can provide 214 mg of alanine, 321 mg of arginine, 1706 mg of aspartic acid, 1170 mg of glutamic acid, 251 mg of glycine, 251 mg of proline, and 259 mg of tyrosine per 100 g weight (USDA-SR Legacy 169677).

Dietary Fibers Potatoes are a good source of dietary fibers. Intake of dietary fibers lowers the risk of coronary heart disease and improves digestive health. Kalita and Jayanty (2017) found that fiber level in potato tubers of different cultivars grown in the USA ranged from 1.75% to 2.59% with an average value of 2.14%. The average total dietary fiber in chips prepared by the continuous process was 2.5% with a range of 2.14–3.09% and 3.39% in kettle chips with a range of 3.02–3.71%. Total dietary fiber may increase after baking (Mullin and Smith 1991) with a relatively higher proportion of soluble dietary fibers. As per USDA SR legacy database, plain salted potato chips can provide 2.2–3.4% dietary fibers, whereas fresh potatoes with skin can provide up to 2.2% dietary fibers (USDA-SR Legacy 169677).

Vitamins Vitamin C has a major role in detoxifying reactive oxygen species, functioning as an electron donor and also serves as a cofactor for numerous enzymes. It is a well-known fact that raw potatoes are good source of vitamin C and potato chips have more dietary vitamin C due to the rapid dehydration of potato tubers during frying. Potato chips can provide 18.6–21.6 mg/100 g of vitamin C. Due to the heat sensitivity, the high temperature during frying results in a 30–50% loss in vitamin C (Miyoshi et al. 2006; Tian et al. 2016). In a study, tissue from SMP30/GNL knockout (KO) mice unable to synthesize vitamin C showed that administration of potato chips diet reduced the reactive oxygen species levels in the different parts of the body compared with ascorbic acid depleted diet-fed mice (Kondo et al. 2014). Potatoes are a readily bioavailable source of dietary vitamin C. A dietary study showed that vitamin C from mashed potatoes and potato chips was effectively absorbed in the intestine and transferred to the blood and hence increased plasma levels after consumption (Kondo et al. 2012).

Niacin (Vitamin B3) refers to the pyridine derivatives, nicotinamide or nicotinic acid. Most nicotinic acid exists in animals and plants in the form of nicotinamide, but it is absorbed in the form of nicotinic acid. Niacin is important for general good health and functions as a cofactor in various redox reactions that are primarily connected with photosynthesis and carbohydrate and fatty acid metabolism (Kwiterovich 1998; Wong and Kashyap 2000). At high concentration, niacin is reported to be beneficial for lowering cholesterol levels and may promote vasodilation and hypotension (Garg et al. 2017; Rolfe 2014; Siniawski et al. 2017). The current RDA values for niacin are 14 and 18 mg per day for women and men, respectively (RDA 2009). As per the USDA database, different variants of potato chips can provide up to 3.9–7.15 mg/100 g niacin.

Vitamin E is well-known most effective lipid-soluble, chain-breaking antioxidant, which protects cell membranes from peroxidative damage (Packer 1991). Vitamin E content of the potato chips depends upon the vitamin E content of frying oil used; therefore, vitamin E content of the chips is highly influenced by the oil. As per USDA Standard Reference database for potato, fresh potato chips can provide 6.7–10.5 mg vitamin E/100 g as alpha tocopherols.

Vitamin B6 (Pyridoxine) plays an important role in the body. It is needed to maintain the health of nerves, skin, and red blood cells. In coenzyme forms, it performs a wide variety of functions and is involved in more than 100 enzyme reactions, mostly concerned with protein metabolism (DRI 1998). Both pyridoxal 5′ phosphate and pyridoxamine 5′ phosphate are involved in amino acid metabolism. Pyridoxal 5′ phosphate is also involved in the metabolism of carbohydrates and lipids. Vitamin B6 is also involved in the biosynthesis of neurotransmitters and in maintaining normal levels of homocysteine, in gluconeogenesis and glycogenolysis, immune function and hemoglobin formation (Mackey et al. 2006). Potato chips provide 0.40–0.64 mg/100 g vitamin B6, whereas fresh potato can provide only 0.3 mg per 100 g serving (USDA-SR Legacy 169677). In a study on dietary vitamin B6 intake and its food sources in the young Korean population, it was found that the average daily intake of vitamin B6 was 1.57 mg/d in men and 1.44 mg/d in women. Potatoes were found as one of the major dietary sources of vitamin B6 in their diet. Potatoes contributed 4.10% of total vitamin B6 in 44.98% men population, whereas potato chips contributed 0.14% in 0.91% men (Cho and Kim 2005).

Folate is the generic name for 5,6,7,8-tetrahydrofolate (THF), a central compound in 1-carbon metabolism (Hanson and Gregory III 2011). Folate is involved in the synthesis of methionine, vitamin B5, thymidylate, purines, and for sulfur–iron cluster metabolism (da Silva et al. 2014; Tibbetts and Appling 2010). The presence of folate in lower concentrations is the main factor for making folate deficiency a nutritional problem worldwide. Moreover, insufficient uptake of folate is connected with cardiovascular diseases, anemia, and birth defects and increased risk for certain cancers (Blancquaert et al. 2010). The current RDA value for both men and women are 400 μg of dietary folate equivalents. It is very sensitive to oxidizing conditions or reducing agents during food processing operations (Delchier et al. 2014). Compared to unprocessed potatoes, processed potato products often contain relatively low amounts of folate (Goyer and Navarre 2009; Goyer and Sweek 2011). As per

USDA Standard Reference legacy data, fresh potatoes can provide 16 μg/100 g, whereas chips provide 29.0–75.0 μg folate/100 g weight. Pantothenic acid is a precursor of the amino acid β-alanine and a core component of coenzyme A. It is involved in various biosynthetic pathways that are connected with carbohydrate, amino acid, and fatty acid metabolism (Depeint et al. 2006). Potato chips can provide 0.78–1.0 mg pantothenic acid per 100 g of chips.

Minerals The predominant minerals found in potatoes are calcium, potassium, magnesium, and phosphorous. Potato is the major contributor of potassium in our diet providing about 18–20% RDA of potassium. Fresh potatoes can provide up to 6% RDA of iron, magnesium, and phosphorous; 2% of zinc and calcium. The content of these minerals varies due to genotypic differences (Andre et al. 2007; Randhawa et al. 1984) and various other factors such as soil, environmental interactions, and cultural practices (Lombardo et al. 2013). The mineral composition of potato is highly affected by processing conditions. Deep fat frying results in minimal changes in mineral content of the potato (Finglas and Faulks 1984). As per USDA Standard Reference legacy database, 100 g of plain salted potato chips can provide 17–31 mg of calcium; 1.02–1.49 mg of iron; 49–81 mg of magnesium; 120–201 mg of phosphorous; 983–1380 mg of potassium; 235–633 mg of sodium; 0.77–1.35 mg of zinc; 0.059–0.378 mg of copper and 0.335–0.57 mg of manganese. Whereas, raw potato with skin can provide 30 mg of calcium; 3.24 mg of iron; 23 mg of magnesium; 38 mg of phosphorous; 413 mg of potassium; 10 mg of sodium; 0.35 mg of zinc; 0.423 mg of copper; and 0.602 mg of manganese (USDA-SR Legacy 169677).

14.2.1.2 French Fries

French fries are among the highest viable potato products all around the world and are traditionally produced by cutting potato strips from fresh potatoes which are then deep fat fried. Jack Simplot in the USA who produced dried potatoes for the US army during World War II was the first to start the production of frozen French fries. In the developed regions a considerable part of potatoes, one to two thirds, is consumed in the form of French fries. In the French fry industry, five global players, three North American and two European, have an estimated share of about 75% of the global French fry production. At present three major kinds of French fries are produced at a commercial scale: (1) deep-frozen completely fried strips, which just require oven heating; (2) deep-frozen partially fried strips, which require additional frying before eating; and (3) refrigerated partially fried strips, which have a short shelf life and need additional frying (Lisinska and Leszczynski 1989).

Like potato chips, French fries are also fat-rich, high energy food (Table 14.1). A 100 g serving of French fries can provide 312 kcal of energy, 3.43 g protein, 14.73 g of fat, 41.44 g carbohydrates, 3.8 g total dietary fiber, 0.3 g sugars, and 1.85 g of ash. Depending upon the frying medium used during processing, French fries contain 2.34 g saturated fat, 5.97 g monounsaturated fatty acids, 5.40 g polyunsaturated fatty acids, 0.06 g trans fatty acids, 168 mg total omega-3 fatty acids, and 2.83 g of total omega-6 fatty acids (USDA-SR Legacy 170698). However, this composition can vary depending upon the quality of oil used, frying time and temperature.

Vitamins Vitamin content of French fries is highly influenced by the quality and vitamin content of the oil used during the processing. A 100 g serving of French fries can provide 1.3–6.8 mg of vitàmin C, 0.14–0.19 mg of thiamine, 0.03–0.06 mg of riboflavin, 2.06–3.66 mg of niacin, 0.47–0.7 mg of pantothenic acid, 0.28–0.44 mg of vitamin B6, 8–64 μg of total folate, and 6.4–16 μg of vitamin K (phylloquinone). Depending upon the tocopherol content of the frying oil, fries can provide 0.91–2.55 mg of alpha-tocopherol, 0.02-0.12 mg of beta-tocopherol, 1.03–8.4 mg of gamma-tocopherol, and 0.28–2.68 mg of delta-tocopherol (USDA-SR Legacy 170698). French fries are poor source of vitamin A, vitamin B12, and vitamin D.

Minerals The mineral content of the French fries varies with the minerals present in the potato used for the processing. Generally, a 100 g serving of French fries can provide 12–23 mg of calcium, 0.63–1.07 mg of iron, 25–40 mg of magnesium, 88–153 mg of phosphorus, 115–363 mg of sodium, 0.39–0.73 mg of zinc, 0.072–0.212 mg of copper, 0.183–0.343 mg of manganese, and 0.4–1.2 μg of selenium. French fries are a good source of potassium and can provide 451–675 mg of potassium per 100 g serving (USDA-SR Legacy 170698).

Protein and Amino Acids French fries contain 3.43 g of high-quality protein and provide 6.13% of RDA for adult males and 7.46% of RDA value for adult females (DRI 2006). French fries contain both essential and non-essential amino acids. Among the non-essential amino acids, a 100 g serving of French fries can provide 119 mg of alanine, 179 mg of arginine, 765 mg of aspartic acid, 33 mg of cysteine, 586 mg of glutamic acid, 103 mg glycine, 126 mg of proline, 125 mg of serine, and 87 mg of tyrosine. Among the essential amino acids, a 100 g serving of French fries can provide 66 mg of histidine, 120 mg of isoleucine, 188 mg of leucine, 225 mg of lysine, 58 mg of methionine, 206 mg of phenylalanine, 120 mg threonine, 38 mg tryptophan, and 178 mg of valine.

14.2.1.3 Sticks and Shreds

Potato sticks and shreds are deep-fried dehydrated products. Nutritionally sticks and shreds are similar to potato chips. A 100 g serving of potato sticks can provide 522 kcal of energy which is 26% of the daily value for a 2000 calorie reference diet. About 41% of this energy is contributed by the carbohydrates, 55% by fat and 4% by proteins present in the sticks. Potato sticks contain 53.3 g of total carbohydrates, 3.4 g of dietary fiber and fulfill the 18% and 14% of the daily value for 2000 calorie reference diet, respectively. Depending upon the variety of potato used and time-temperature combination of frying, potato sticks contain about 34.5% average fat content. About 26% of the fat is contributed by saturated fatty acids, 18% by monounsaturated fatty acids, and 52% by polyunsaturated fatty acids. This fat also contains 70 mg of omega-3 fatty acids and 17.8 g of omega-6 fatty acids (Table 14.1). However, this proportion may vary depending upon the quality of oil used during frying. Protein content may range between 6.5 and 6.8% (USDA-SR Legacy 168854).

Vitamins Like other fried products vitamin content of the sticks and shreds is highly affected by the quality of the oil used during frying. A 100 g serving of the sticks can provide 47.3 mg of vitamin C which is 79% of daily value for a 2000 calorie reference diet, 9.1 mg of vitamin E (46% of daily value), 22.1 μg of vitamin K (28% of daily value), 0.1 mg of thiamine (6% of daily value), 0.1 mg of riboflavin (7% of daily value), 4.8 mg of niacin (24% of daily value), 0.3 mg of vitamin B6 (16% of daily value), 40 μg of folate (10% of daily value), and 0.4 mg of pantothenic acid (4% of daily value).

Minerals A 100 g serving of potato sticks contains 18 mg of calcium, 2.3 mg of iron (13% of daily value), 64 mg of magnesium (16% of daily value), 172 mg of phosphorus (17% of daily value), 250 mg of sodium (10% of daily value), 1.0 mg zinc (7% of daily value), 0.3 mg copper (16% of daily value), 0.4 mg manganese (21% of daily value), and 8.1 μg selenium (12% of daily value). Potato sticks and *lachha* are rich sources of potassium (1237 mg) that can provide 35% of the daily value of a 2000 calorie reference diet (DRI 2006).

14.2.1.4 Frozen Products

Frozen potatoes are processed potato products which are highly convenient and flexible in terms of preparation time, containing different nutrients and vitamins in the preserved form with a long shelf life. Frozen potatoes and their different products are obtained via processing of fresh potatoes using a variety of advanced types of machinery at very low temperatures. These are available in various forms in the market, such as French fries, hash brown, shapes, mashed, battered/cooked, twice baked, topped/stuffed, and other frozen potatoes. These products are either consumed via quick-service restaurants (QSRs) or through retail stores at home by customers.

Unsalted French-Fried Potatoes

The global frozen potato market is projected to reach $74,403 million by 2025, registering a CAGR of 3.8% from 2018 to 2025 (Sable 2019). The key manufacturers in the Frozen French Fries include Ore-Ida, Cascadian Farm Organic, Alexia Foods, Trader Joe's Fan, Checkers & Rally's, Arby's IP Holder, McCain, and Kroger. Being fried in oil, French fries are generally high-calorie food and provide some beneficial nutrients such as minerals and vitamins. Nutritionally French fries mainly consist of carbohydrates, fat, and some protein. Reports show that from a nutritional point of view baked fries are better than deep-fried fries.

As per the standard reference data available on food data central, 100 g of unprepared frozen French fries contains 63–70 g water, 150 kcal energy which is 8% of the daily value for 2000 cal reference diet (Table 14.2). About 66% of total calories are contributed by carbohydrates, 28% by fat, and 6% by proteins. Frozen fries contain 24–25% carbohydrates, 7–9% of the total carbohydrates are contributed by dietary fibers and about 70–75% by starch and remaining part is contributed by simple sugars. The fat content of the frozen fries' ranges between 3.3 and 6.1%, about 19–20% of the total fat is contributed by saturated fatty acids, 63–65% by monounsaturated fatty acids, 6–7% by polyunsaturated fatty acids. Moreover, frozen French fries also contain 18 mg of total omega-3 fatty acids and 247 mg of total omega-6 fatty acids per 100 g serving. This composition may vary depending upon the quality of oil used during frying. Depending upon the variety used frozen French fries contain 1.98–2.41% high-quality protein.

Frozen French fries are good sources of vitamin C, provide 10.5–24.5 mg of vitamin C per 100 g serving which may provide 30% of the daily value for a 2000 cal reference diet. Depending upon the quality of oil used frozen French fries can provide 0.05–0.16 mg of alpha-tocopherol, 0.1–0.6 mg of gamma-tocopherol, and 0.24–0.76 mg of delta-tocopherol. Besides this, a 100 g serving of the frozen French fries can provide 0.64–1.28 mg of thiamine, 0.03–0.103 mg of riboflavin, 1.86–2.24 mg of niacin, 0.394–0.577 mg of pantothenic acid, 0.157–0.203 mg vitamin B6, 21–42 μg of total folate, and 1.6–2.8 μg of vitamin K.

Frozen French fries are good source of potassium containing 385–433 mg of potassium (12% of daily value), 8–11 mg calcium, 0.54–0.77 mg iron, 20–23 mg magnesium, 70–103 mg phosphorus, 21–23 mg sodium, 0.33–0.37 mg of zinc, 0.71–0.127 mg copper, and 0.143–0.179 mg manganese per 100 g serving (USDA-SR Legacy 170523).

Potato Wedges

Potato wedges are wedge-shaped cut potatoes coated with some seasoning and herbs. Potato wedges contain 68–69% moisture and provide 123–129 kcal energy (Table 14.2). About 73% of the energy value is contributed by carbohydrates, 16% by fats and 8.7% by proteins. Frozen wedges contain 25.5% carbohydrates, 7% of the total carbohydrate is contributed by dietary fibers (8% of daily value). The fat content of the frozen wedge ranges between 2.0% and 2.5%, about 22–23% of the total fat is contributed by saturated fatty acids, 68% by monounsaturated fatty acids, and remaining by polyunsaturated fatty acids. Frozen wedges also contain 102 mg of total omega-6 fatty acids per 100 g serving. Depending upon the variety used frozen French fries contain 2.7% high-quality protein which provides 5% of the daily value for a 2000 cal reference diet. A 100 g serving of frozen wedges can provide 11.2 mg of vitamin C (19% of daily value), 0.1 mg of thiamine, 1.5 mg of niacin, and 0.4 mg vitamin B6. Among the minerals, frozen wedges can provide 15 mg calcium, 0.1 mg iron, 19 mg magnesium, 87 mg phosphorus, 394 mg potassium, 49 mg sodium, and 0.4 mg of zinc (USDA-SR Legacy 168443).

Potato Puffs

According to standard reference legacy data, potato puffs contain 62.4% water and provide 175 kcal of energy. About 59% of the calories are contributed by carbohydrates, 37% by the fats, and remaining by proteins. Potato puffs contain

Table 14.2 Nutritive value of popular potato-based frozen products (per 100 g) (USDA 2019)

Parameters	Unsalted French fries[a]	Potato wedges[b]	Potato puffs[c]	Hash brown potatoes[d]	Whole boiled potatoes[e]
Water (g)	63.16–69.93	68.3	61.51–63.93	78.85	82.80
Energy (kcal)	147	129	178	82	65
Energy (kJ)	613	539	746	343	272
Protein (g)	1.98–2.41	2.7	1.62–2.25	2.06	1.98
Total lipid (fat) (g)	3.31–6.12	2.2	7.37–10.54	0.62	0.13
Ash (g)	1.24–2.02	1.34	1.29–1.91	0.75	0.57
Carbohydrate, by difference (g)	24.81	25.46	24.80	17.72	14.52
Fiber, total dietary (g)	1.5–2.4	2	2.30	1.40	1.40
Calcium (mg)	8–11	15	9–15	10.00	7.00
Iron (mg)	0.54–0.77	0.7	0.4–0.61	0.98	0.84
Magnesium (mg)	20–23	19	13–20	11.00	11.00
Phosphorus (mg)	70–103	87	49–88	47.00	26.00
Potassium (mg)	385–433	394	190–354	285.00	287.00
Sodium (mg)	23	49	257–483	22.00	20.00
Zinc (mg)	0.33–0.37	0.37	0.18–0.35	0.21	0.25
Copper (mg)	0.071–0.127	–	0.049–0.095	0.10	0.08
Manganese (mg)	0.143–0.179	–	0.093–0.174	0.15	0.19
Vitamin C (mg)	10.5–24.5	11.2	6.90	8.20	9.40
Thiamin (mg)	0.064–0.128	0.1	0.18	0.10	0.10
Riboflavin (mg)	0.03–0.103	0.04	0.06	0.01	0.03
Niacin (mg)	1.86–2.24	1.54	1.82	1.66	1.33
Pantothenic acid (mg)	0.394–0.577	–	0.58	0.32	0.28
Vitamin B-6 (mg)	0.157–0.203	0.35	0.19	0.09	0.20
Folate, total (μg)	21–42	–	17.00	4.00	8.00
Vitamin E (alpha-tocopherol) (mg)	0.05–0.16	–	0.11	0.00	0.00
Carotene, beta (μg)	–	–	3.00	0.00	0.00
Vitamin A (IU)	–	–	4.00	0.00	0.00
Lutein + zeaxanthin (μg)	–	–	15.00	0.00	0.00
Vitamin K (phylloquinone) (μg)	1.6–2.8	–	2.50	0.00	0.00
Fatty acids, total saturated (g)	0.94	0.55	1.42	0.16	0.03
Fatty acids, total monounsaturated (g)	–	1.451	3.38	0.01	0.00
Fatty acids, total polyunsaturated (g)	0.265	0.102	3.14	0.27	0.06

(continued)

Table 14.2 (continued)

Parameters	Unsalted French fries[a]	Potato wedges[b]	Potato puffs[c]	Hash brown potatoes[d]	Whole boiled potatoes[e]
Fatty acids, total trans	–	0	0.06	0.00	0.00
Tryptophan	–	–	0.02	0.03	0.03
Threonine	–	–	0.08	0.09	0.07
Isoleucine	–	–	0.08	0.09	0.08
Leucine	–	–	0.14	0.12	0.12
Lysine	–	–	0.13	0.11	0.12
Methionine	–	–	0.03	0.02	0.03
Cystine	–	–	0.04	0.01	0.03
Phenylalanine	–	–	0.09	0.09	0.09
Tyrosine	–	–	0.07	0.05	0.07
Valine	–	–	0.12	0.11	0.11
Arginine	–	–	0.12	0.10	0.09
Histidine	–	–	0.04	0.04	0.04
Alanine	–	–	0.09	0.07	0.06
Aspartic acid	–	–	0.35	0.48	0.48
Glutamic acid	–	–	0.31	0.32	0.33
Glycine	–	–	0.08	0.08	0.06
Proline	–	–	0.09	0.07	0.07
Serine	–	–	0.10	0.07	0.09

[a]Potatoes, French fried, all types, salt not added in processing, frozen, as purchased (USDA Standard Reference Legacy, 170523) (2019)
[b]Potato wedges, frozen (USDA Standard Reference Legacy, 168443) (2019)
[c]Potato puffs, frozen, unprepared (USDA Standard Reference Legacy, 170047) (2019)
[d]Potatoes, hash brown, frozen, plain, unprepared (USDA Standard Reference Legacy, 170043) (2019)
[e]Potatoes, frozen, whole, cooked, boiled, drained, without salt (USDA Standard Reference Legacy, 170049 (2019)

24.8–26% carbohydrates, 7.3% of the total carbohydrates are contributed by dietary fibers, 76.15% by the starch and remaining by the simple sugars (Table 14.2). The fat content ranges between 7.37 and 10.54% which is equivalent to 12% of daily value based on a 2000 cal reference diet. About 16–20% of the total fat is contributed by the saturated fat, 38% by the monounsaturated fatty acids, and 36% by polyunsaturated fatty acids. Among the omega fatty acids, omega-3 fatty acid content ranges between 21 and 25 mg and omega 6 fatty acid content between 360 and 365 mg per 100 g of frozen unprepared potato puff. The protein content of the puff ranges between 1.62 and 2.25% which contains 371 mg of aspartic acid, 320 mg glutamic acid, 151 mg leucine, 133 mg lysine, 128 mg arginine, 122 mg valine, and 100 mg of serine amino acids. A 100 g serving of the frozen potato puffs can provide 4 IU of vitamin A, 3 μg of beta carotene, 15 μg lutein + zeaxanthin, 7.5 mg of vitamin C,

0.2 mg of alpha-tocopherol, 1.1 mg of gamma-tocopherol, 1.1 mg of delta-tocopherol, 3.4 mg vitamin K, 0.1 mg thiamine, 1.4 mg niacin, 0.1 mg vitamin B6, 35 μg folate, and 0.3 mg of pantothenic acid. As compared to fresh potato and frozen products puffs contain a relatively higher amount of sodium (257–483 mg/100 g). Among the other minerals, potato puffs contain 9–15 mg calcium, 0.4–0.6 mg of iron, 13–20 mg of magnesium, 49–88 mg of phosphorus, 190–354 mg of potassium, 0.18–0.35 mg of zinc, 0.049–0.095 mg of copper, 0.093–0.174 mg of manganese (USDA-SR Legacy 170047).

Hash Brown Potatoes

Frozen hash browned potatoes are prepared from mature, sound, white potatoes that are washed, peeled, sorted, and trimmed to assure a wholesome product. The potatoes so prepared are blanched, may or may not be fried, and are shredded or diced or chopped and frozen. As per USDA standard reference database, unprepared frozen plain hash brown potatoes contain 82 kcal energy, about 86% of the total energy is contributed by carbohydrates, 6.3% energy by fats, and remaining by proteins (Table 14.2). The total carbohydrate content of the hash browned potato ranges between 17 and 8% and about 8% of the carbohydrates are contributed by the dietary fibers. The total fat content is 0.6% and 33% of the total fat is contributed by the saturated fat and 50% by the polyunsaturated fats. Moreover, hash browned potatoes also contain 64 mg omega-3 fatty acids and 201 mg omega-6 fatty acids. The protein content ranges between 2.0 and 2.5% which can provide 4% of daily value based on a 2000 cal reference diet. This protein is rich in aspartic (478 mg) and glutamic acid (322 mg). Among the other amino acids, 100 g serving of hash browned potatoes can provide 124 mg of leucine, 110 mg of lysine, 105 mg of valine, 98 mg of arginine, 94 mg of threonine, 89 mg isoleucine, 88 mg phenylalanine, 75 mg glycine, 74 mg serine, 71 mg alanine, 66 mg of proline, and 52 mg of tyrosine. Besides this, other amino acids are present in small amounts. Like other potato-based frozen products, frozen hash browned potatoes are good source of vitamin C containing 8.2 mg of vitamin C per 100 g serving. Among the other vitamins, 100 g serving can provide 1.7 mg of niacin, 0.3 mg pantothenic acid, and thiamine, vitamin B6, and folate in small amounts. Hash browned potatoes are rich in potassium (285 mg/100 g) and low in sodium (22 mg/100 g). Ash content of the hash browned potatoes ranges between 0.7 and 0.8%, a 100 g serving can provide 10 mg of calcium, 1.0 mg of iron, 11 mg of magnesium, 47 mg of phosphorus, 0.2 mg zinc, 0.1 mg copper and manganese (USDA-SR Legacy 170043).

Whole, Boiled Potatoes

Boiled potatoes contain 82–83% moisture and 65 kcal of energy (Table 14.2). About 90% of energy is contributed by carbohydrates, 1.6% by fats, and 8.4% by proteins. Boiled potatoes contain 14.5% total carbohydrates and 1.4% dietary fiber. Boiled potatoes contain almost nil fat content (0.1%) in the form of mono and polyunsaturated fat. Boiled potatoes are poor sources of proteins (2.0%) but the quality of protein is similar to animal proteins. A 100 g serving of boiled frozen potatoes can provide 31 mg of tryptophan, 72 mg threonine, 80 mg isoleucine, 119 mg leucine,

120 mg lysine, 31 mg methionine, 25 mg cysteine, 88 mg phenylalanine, 73 mg tyrosine, 111 mg valine, 91 mg arginine, 43 mg histidine, 61 mg alanine, 59 mg glycine, 71 mg proline, and 86 mg serine. Potato proteins are rich in aspartic (483 mg) and glutamic acids (331 mg). Like other potato products, boiled potatoes are rich in vitamin C containing 9.4 mg per 100 g serving which is equivalent to 16% of the daily value on a 2000 cal reference diet. A 100 g serving of the boiled potato can provide 0.1 mg thiamine, 1.3 mg niacin, 0.2 mg vitamin B6, 8 μg total folate, and 0.3 mg pantothenic acid. Boiled potato contains 0.6% ash and is rich in potassium (287 mg/100 g). Among the other minerals, major minerals are calcium (7 mg), iron (0.8 mg), magnesium (11 mg), phosphorous (26 mg), and sodium (20 mg/100 g) (USDA-SR Legacy 170049).

14.2.2 Indian Cuisines

Wheat-based porridge and semolina are very popular food products in several countries that are consumed as a part of their breakfast, lunch, or dinner, frequently with added milk and sugar or salt for enhancing acceptability. Semolina is also used for the preparation of many south Indian dishes such as *idli*, *dhokla*, and *upma*. They are easy to digest and rich in fibers also. Besides, wheat-based porridge and semolina have various other nutritional benefits. Similarly, more than 80% of baked products such as bread, biscuits, cookies, cakes are prepared from the flour-containing wheat as a major ingredient. In addition, the various health issues are also associated with the consumption of wheat. The major drawbacks of wheat-based products are allergic reactions, gastrointestinal disorders, and celiac disease. In most cases, wheat protein gluten is the major allergen responsible for various types of allergic reactions (Czaja-Bulsa 2015; Shan et al. 2002; Uhde et al. 2016). In such cases, gluten withdrawal from the diet is the best remedy (Biesiekierski et al. 2011; Sapone et al. 2012; Verdu et al. 2009; Wahnschaffe et al. 2007; Wahnschaffe et al. 2001). As an alternative of wheat-based products, recently Indian Council of Agricultural Research Institute, Central Potato Research Institute, Shimla, India has developed some potato-based alternatives of the above-mentioned food products which are devoid of gluten and rich in fibers.

14.2.2.1 Potato Porridge and Semolina

Potato-based porridge and semolina are similar to wheat porridge and semolina except for their source of origin. The process of cooking is the same as for wheat-based porridge or bulgur and semolina. Potato porridge and semolina contain about 6.5–7.5% moisture and 350–370 kcal energy (Fig. 14.1a, b). Total carbohydrate content ranges between 81 and 85%. These products are rich in fibers containing about 9.0–10.7% dietary fibers which are almost equivalent to fiber content present in wheat-based porridge and bulgur (12.5%) (USDA SR legacy data). The fat content of the potato-based porridge and semolina is very low (0.15–0.20%), whereas wheat bulgur contains 1.33% fat and the major proportion of the fat is contributed by the polyunsaturated fatty acids. The protein content ranges between

Fig. 14.1 (**a**) Potato porridge and (**b**) semolina

7 and 8.3% with the amino acid score of 103 which is 50% higher than the amino acid score of wheat bulgur. This is due to the presence of all essential amino acids. Among the essential amino acids, leucine is present in highest (0.42%) amount followed by lysine, valine, isoleucine, phenylalanine, histidine, tryptophan, and methionine. Among the non-essential amino acids, aspartic acid (1.25%) and glutamic acid (0.95%) are present in the highest amount. This shows that the proteins from potato-based porridge and semolina are of better quality as compared to the protein from wheat-based products. These products are a very good source of vitamin B complex, except for vitamin B12, and vitamin C. A 100 g serving of potato porridge can provide 7–9% of dietary reference intake of vitamin C for adult males and females (DRI 2006). Moreover, potato porridge and semolina are good sources of minerals. A 100 g uncooked serving of potato porridge can provide 784.7 mg potassium, 50–60 mg of sodium, 62–70 mg calcium, 4.5–6.75 mg of iron. Moreover, potato porridge and semolina are good sources of polyphenolic compounds, anthocyanins, and other antioxidants. Unlike fried products, these products are free from acrylamide and other toxic compounds that are formed during the Maillard reaction.

14.2.2.2 Potato Cookies

Cookies are one of the popular bakery products worldwide. In most of the cookies, refined wheat flour is used as one of the major ingredients in the recipe. Besides some millets, maize, oats-based cookies are also available in the market. At present more than fifty thousand variants of the cookies are available in the market. Recently, ICAR-Central Potato Research Institute, Shimla has standardized and developed the process for potato-based cookies which are made from whole potato flour (Fig. 14.2). As these cookies are devoid of wheat, barley or rye flour, therefore, these cookies are free from gluten. Nutritionally potato cookies contain 4.0–4.55% high-quality complete protein rich in aspartic acid, glutamic acid, leucine, and lysine. The fat content of the cookies ranges between 23 and 26%, about 55% of fat is contributed by saturated fatty acids, 35% by monounsaturated fatty acids, and 7% by polyunsaturated fatty acids. The protein and fat content of the potato cookies

Fig. 14.2 Potato cookies

Fig. 14.3 Potato halwa premix

are almost like normal wheat flour-based cookies. The dietary fiber content of the cookies ranges between 6.2 and 6.7% and total carbohydrates ranges between 65 and 68%. Potato cookies are rich in total folate (13.3 μg/100 g) and vitamin C (2.027 mg/100 g) and tocopherols (α and γ). A 100 g pack of potato cookies can provide an average calcium content of 85.1 mg, magnesium 32.57 mg, phosphorus 85.12 mg, potassium 674.18 mg, sodium 322.94 mg, and iron 2.10 mg.

14.2.2.3 Potato Halwa Premix

In India, halwa is a very popular dish and consumed as dessert with or without added dry fruits. Traditionally, potato halwa is prepared from freshly boiled potatoes and the cooking involves a lengthy process. The shelf life of traditionally prepared potato halwa is very short. ICAR-CPRI, Shimla has standardized the process for potato halwa premix having a shelf life of up to 4 months at ambient temperature conditions (Fig. 14.3). Being prepared from whole potatoes the potato halwa premix contains a relatively higher amount of total dietary fiber (3.53%) as compared to halwa prepared from boiled and peeled potatoes. Other advantages over the traditionally prepared potato halwa are low fat, gluten-free, suitability for celiac and wheat allergic population. Halwa premix is prepared from potato solids along with sugar, milk solids, dry fruits, flavor, and some other additives in a very specific proportion.

Whole potatoes of any shape, size, and variety can also be used. Moreover, partially damaged or cold-stored potatoes can also be utilized for premix. From a nutritional standpoint, uncooked potato halwa premix contains more than 8% protein, 10–11% fat, 74% total carbohydrates and provides 424 kcal energy per 100 g premix. Potato halwa premix is rich in total folate (14 μg), vitamin C, vitamin K, lutein (0.97 μg), and total carotenoids (2.32 μg). Potato halwa premix is a rich source of potassium (466 mg), calcium (155 mg), and phosphorous (137 mg/100 g premix). Other minerals are available in small amounts.

14.3 Conclusion/Future Direction

A very large proportion of the population in India still thinks that processed food is unhealthy and a source of ailments. However, with the increasing number of working women, higher disposable income, and Western influence on young Indians, the number of potato product eaters is increasing very rapidly. Rapid urbanization and improvements in living standards have led to phenomenal growth in potato processing in recent past. Nowadays to improve the nutritive value of potato, various strategies such as supplementation, fortification, and the development of nutritionally enhanced crop cultivars are being practiced and all have been shown to be effective. However, from an economic point of view bio-fortification through breeding is the most stable and viable strategy for enhancing the nutritional status of potato and people.

Potatoes are nutrient-dense food. It is well established that potatoes are a rich source of high-value protein, essential vitamins, minerals, and trace elements. Beside this, potato also contains some anti-nutritional factors such as glycoalkaloids in permissible limits. It is well known that potato contains high-quality complete protein which is composed of a high proportion of lysine, threonine, tryptophan, and methionine an essential amino acid often lacking in cereal and vegetable crops. Because of their amino acid composition, potato proteins have been recognized to be nutritionally equivalent to animal protein, lysozyme. Several health benefits such as lower allergic response, antimicrobial, antioxidant properties, and the regulatory effect on blood pressure and blood serum cholesterol-lowering effect and anticarcinogenic behavior also have been recognized in the potato proteins. The nutritionally beneficial effect of their high starch content has also been recognized, which supplies the body with energy and dietary fiber. The fatty acid profile of lipids isolated from potatoes is especially nutritionally advantageous because the essential part of all fatty acids is formed by unsaturated fatty acids with one to three double bonds, mainly oleic, linoleic, and linolenic acids. Potatoes are best known as a source of vitamin C and potassium and serves as a less expensive source of vitamin C. In potatoes, major minerals include potassium, phosphorus, calcium, and magnesium which are nutritionally essential major minerals for the human body. Besides potato also contains nutritionally essential minor and trace minerals such as iron, copper, selenium, and sulfur. Along with these, potatoes also contain various phytochemicals such as polyphenols, flavonols, anthocyanins, phenolic acids, carotenoids, polyamines, glycoalkaloids, tocopherols, calystegines, and sesquiterpenes. Up to some concentrations, these glycoalkaloids have a beneficial effect on human health, but their intake at higher concentration has a harmful effect on the body such as anti-proliferative activity. In Indian potato varieties glycoalkaloid content is within the permissible limit (Singh et al. 2016).

Depending upon the processing conditions applied during processing, some changes (nutrient losses or improved digestibility) occur in the nutritional profile of the finished product. Some processing operations such as frying, in the making of French fries and crisps, may cause up to 80–90% loss in nutrients. Being rich in aspartic acid and the presence of reducing sugars there is the possibility for the formation of acrylamide during frying and baking. In general, the acrylamide is formed during prolonged heating or cooking at high temperatures (>120 °C) which is very common in frying, baking, and roasting. However, the loss of these nutrients can be minimized by proper control of the processing parameters.

References

Anderson P (2010) The contribution of potatoes to global food security. Director General International Potato Center September 10

Andre CM, Ghislain M, Bertin P, Oufir M, Del Rosario Herrera M, Hoffmann L, Hausman J-F, Larondelle Y, Evers D (2007) Andean potato cultivars (*Solanum tuberosum* L.) as a source of antioxidant and mineral micronutrients. J Agric Food Chem 55(2):366–378

Bártová V, Bárta J (2009) Chemical composition and nutritional value of protein concentrates isolated from potato (*Solanum tuberosum* L.) fruit juice by precipitation with ethanol or ferric chloride. J Agric Food Chem 57(19):9028–9034

Biesiekierski JR, Newnham ED, Irving PM, Barrett JS, Haines M, Doecke JD, Shepherd SJ, Muir JG, Gibson PR (2011) Gluten causes gastrointestinal symptoms in subjects without celiac disease: a double-blind randomized placebo-controlled trial. Am J Gastroenterol 106(3):508

Bindels LB, Walter J, Ramer-Tait AE (2015) Resistant starches for the management of metabolic diseases. Curr Opin Clin Nutr Metab Care 18(6):559

Blanco-Aparicio C, Molina MA, Fernández-Salas E, Frazier ML, Mas JM, Querol E, Avilés FX, De Llorens R (1998) Potato carboxypeptidase inhibitor, a T-knot protein, is an epidermal growth factor antagonist that inhibits tumor cell growth. J Biol Chem 273(20):12370–12377

Blancquaert D, Storozhenko S, Loizeau K, De Steur H, De Brouwer V, Viaene J, Ravanel S, Rebeille F, Lambert W, Van Der Straeten D (2010) Folates and folic acid: from fundamental research toward sustainable health. Crit Rev Plant Sci 29(1):14–35

Bodinham CL, Frost GS, Robertson MD (2010) Acute ingestion of resistant starch reduces food intake in healthy adults. Br J Nutr 103(6):917–922

Burlingame B, Mouillé B, Charrondiere R (2009) Nutrients, bioactive non-nutrients and anti-nutrients in potatoes. J Food Comp Anal 22(6):494–502

Cho Y-O, Kim B-Y (2005) Vitamin B6 intake by Koreans should be based on sufficient amount and a variety of food sources. Nutrition 21(11–12):1113–1119

Czaja-Bulsa G (2015) Non coeliac gluten sensitivity–A new disease with gluten intolerance. Clin Nutr 34(2):189–194

Da Silva RP, Kelly KB, Al Rajabi A, Jacobs RL (2014) Novel insights on interactions between folate and lipid metabolism. Biofactors 40(3):277–283

Delchier N, Ringling C, Maingonnat J-F, Rychlik M, Renard CM (2014) Mechanisms of folate losses during processing: diffusion vs. heat degradation. Food Chem 157:439–447

Depeint F, Bruce WR, Shangari N, Mehta R, O'Brien PJ (2006) Mitochondrial function and toxicity: role of the B vitamin family on mitochondrial energy metabolism. Chem-Biol Interact 163(1-2):94–112

Dodevska MS, Sobajic SS, Djordjevic PB, Dimitrijevic-Sreckovic VS, Spasojevic-Kalimanovska VV, Djordjevic BI (2016) Effects of total fibre or resistant starch-rich diets within lifestyle intervention in obese prediabetic adults. Eur J Nutr 55(1):127–137

DRI (1998) Dietary reference intakes for thiamin, riboflavin, niacin, vitamin B6, folate, vitamin B12, pantothenic acid, biotin, and choline. National Academies Press, Washington

DRI (2006) Dietary reference intakes: the essential guide to nutrient requirements. National Academies Press, Washington

Ezekiel R, Singh N, Sharma S, Kaur A (2013) Beneficial phytochemicals in potato—a review. Food Res Int 50(2):487–496

FAOSTAT (2019) FAO, Rome, Italy

Fast Foods, Potato, French Fried in Vegetable Oil (Standard Reference Legacy, 170698) (2019) U.S. Department of Agriculture, Agricultural Research Service, Food Data Central. https://fdc.nal.usda.gov/fdc-app.html#/food-details/170698/nutrients

Finglas PM, Faulks RM (1984) Nutritional composition of UK retail potatoes, both raw and cooked. J Sci Food Agric 35(12):1347–1356

Freedman MR, Keast DR (2011) White potatoes, including French fries, contribute shortfall nutrients to children's and adolescents' diets. Nutr Res 31(4):270–277

Fu T-J, Abbott UR, Hatzos C (2002) Digestibility of food allergens and nonallergenic proteins in simulated gastric fluid and simulated intestinal fluid a comparative study. J Agric Food Chem 50 (24):7154–7160

Garg A, Sharma A, Krishnamoorthy P, Garg J, Virmani D, Sharma T, Stefanini G, Kostis JB, Mukherjee D, Sikorskaya E (2017) Role of niacin in current clinical practice: a systematic review. Am J Med 130(2):173–187

Gentile CL, Ward E, Holst JJ, Astrup A, Ormsbee MJ, Connelly S, Arciero PJ (2015) Resistant starch and protein intake enhances fat oxidation and feelings of fullness in lean and overweight/obese women. Nutr J 14(1):113

Goyer A, Navarre DA (2009) Folate is higher in developmentally younger potato tubers. J Sci Food Agric 89(4):579–583

Goyer A, Sweek K (2011) Genetic diversity of thiamin and folate in primitive cultivated and wild potato (*Solanum*) species. J Agric Food Chem 59(24):13072–13080

Hanson AD, Gregory JF III (2011) Folate biosynthesis, turnover, and transport in plants. Annu Review Plant Biol 62:105–125

He T, Spelbrink RE, Witteman BJ, Giuseppin ML (2013) Digestion kinetics of potato protein isolates in vitro and in vivo. Int J Food Sci Nutr 64(7):787–793

IMARC (2018) Potato starch market: global industry trends, share, size, growth, opportunity and forecast 2018–2023

International Potato Center (2019) CIP potato processing and uses. https://cipotato.org/crops/potato/potato-processing-uses/. Accessed 13/11/12 2019

Jansen G, Flamme W, Schüler K, Vandrey M (2001) Tuber and starch quality of wild and cultivated potato species and cultivars. Pot Res 44(2):137–146

Jørgensen M, Bauw G, Welinder KG (2006) Molecular properties and activities of tuber proteins from starch potato cv. Kuras. J Agric Food Chem 54(25):9389–9397

Kalita D, Jayanty SS (2017) Nutrient composition of continuous and kettle cooked potato chips from three potato cultivars. Curr Res Nutr Food Sci J 5(2):75–88

Kamnerdpetch C, Weiss M, Kasper C, Scheper T (2007) An improvement of potato pulp protein hydrolyzation process by the combination of protease enzyme systems. Enzym Microb Technol 40(4):508–514

Kärenlampi SO, White PJ (2009) Potato proteins, lipids, and minerals. In: Advances in potato chemistry and technology. Elsevier, Amsterdam, pp 99–125

Katan MB, Roos NM (2004) Promises and problems of functional foods. Crit Rev Food Sci Nutr 44 (5):369–377

Keenan MJ, Zhou J, Hegsted M, Pelkman C, Durham HA, Coulon DB, Martin RJ (2015) Role of resistant starch in improving gut health, adiposity, and insulin resistance. Adv Nutr 6 (2):198–205

Kim J-Y, Park S-C, Kim M-H, Lim H-T, Park Y, Hahm K-S (2005) Antimicrobial activity studies on a trypsin–chymotrypsin protease inhibitor obtained from potato. Biochem Bioph Res Co 330 (3):921–927

King JC, Slavin JL (2013) White potatoes, human health, and dietary guidance. Adv Nutr 4 (3):393S–401S

Kondo Y, Higashi C, Iwama M, Ishihara K, Handa S, Mugita H, Maruyama N, Koga H, Ishigami A (2012) Bioavailability of vitamin C from mashed potatoes and potato chips after oral administration in healthy Japanese men. Br J Nutr 107(6):885–892

Kondo Y, Sakuma R, Ichisawa M, Ishihara K, Kubo M, Handa S, Mugita H, Maruyama N, Koga H, Ishigami A (2014) Potato chip intake increases ascorbic acid levels and decreases reactive oxygen species in SMP30/GNL knockout mouse tissues. J Agric Food Chem 62(38):9286–9295

Kwiterovich PO (1998) State-of-the-art update and review: clinical trials of lipid-lowering agents. Am J Cardiol 82(12):3U–17U

Lisinska G, Leszczynski W (1989) Potato science and technology. Springer, England

Liu R, Xu G (2008) Effects of resistant starch on colonic preneoplastic aberrant crypt foci in rats. Food Chem Toxicol 46(8):2672–2679

Liu Y-W, Han C-H, Lee M-H, Hsu F-L, Hou W-C (2003) Patatin, the tuber storage protein of potato (*Solanum tuberosum* L.), exhibits antioxidant activity in vitro. J Agric Food Chem 51 (15):4389–4393

Liyanage R, Han K-H, Watanabe S, Shimada K-I, Sekikawa M, Ohba K, Tokuji Y, Ohnishi M, Shibayama S, Nakamori T (2008) Potato and soy peptide diets modulate lipid metabolism in rats. Biosci Biotechnol Biochem 72(4):943–950

Lombardo S, Pandino G, Mauromicale G (2013) The influence of growing environment on the antioxidant and mineral content of "early" crop potato. J Food Comp Anal 32(1):28–35

Love S, Pavek J (2008) Positioning the potato as a primary food source of vitamin C. Am J Pot Res 85(4):277–285

Mackey A, Davis S, Gregory J III (2006) Vitamin B6. In: Shils ME, Shike M (eds) Modern nutrition in health and disease, 10th edn. Lippincott Williams & Wilkins, Baltimore, pp 452–461

Maki KC, Pelkman CL, Finocchiaro ET, Kelley KM, Lawless AL, Schild AL, Rains TM (2012) Resistant starch from high-amylose maize increases insulin sensitivity in overweight and obese men. J Nutr 142(4):717–723

Makinen S, Kelloniemi J, Pihlanto A, MAkinen K, Korhonen H, Hopia A, Valkonen JP (2008) Inhibition of angiotensin converting enzyme I caused by autolysis of potato proteins by enzymatic activities confined to different parts of the potato tuber. J Agric Food Chem 56 (21):9875–9883

Mäkinen S, Streng T, Larsen LB, Laine A, Pihlanto A (2016) Angiotensin I-converting enzyme inhibitory and antihypertensive properties of potato and rapeseed protein-derived peptides. J Funct Foods 25:160–173

Miyoshi T, Ishihara K, Nakamura K, Koga H (2006) Antioxidative compounds in potato chips. J Jpn Cooking Sci Soc 39(5):277–282. https://doi.org/10.11402/cookeryscience1995.39.5_277

Mullin WJ, Smith JM (1991) Dietary fiber in raw and cooked potatoes. J Food Comp Anal 4 (2):100–106

Packer L (1991) Protective role of vitamin E in biological systems. Am J Clin Nutr 53(4):1050S–1055S

Pihlanto A, Akkanen S, Korhonen HJ (2008) ACE-inhibitory and antioxidant properties of potato (*Solanum tuberosum*). Food Chem 109(1):104–112

Potato Puffs, Frozen, Unprepared (Standard Reference Legacy, 170047) (2019) U.S. Department of Agriculture, Agricultural Research Service, Food Data Central. https://fdc.nal.usda.gov/fdc-app.html#/food-details/170047/nutrients

Potato wedges, frozen (Standard Reference Legacy, 168443) (2019) U.S. Department of Agriculture, Agricultural Research Service, Food Data Central. https://fdc.nal.usda.gov/fdc-app.html#/food-details/168443/nutrients

Potatoes, French Fried, All Types, Salt not Added in Processing, Frozen, as Purchased (Standard Reference Legacy, 170523) (2019) U.S. Department of Agriculture, Agricultural Research Service, Food Data Central. https://fdc.nal.usda.gov/fdc-app.html#/food-details/170523/nutrients

Potatoes, Frozen, Whole, Cooked, Boiled, Drained, Without Salt (Standard Reference Legacy, 170049) (2019) U.S. Department of Agriculture, Agricultural Research Service, Food Data Central. https://fdc.nal.usda.gov/fdc-app.html#/food-details/170049/nutrients

Potatoes, Hash Brown, Frozen, Plain, Unprepared (Standard Reference Legacy, 170043) (2019) U.S. Department of Agriculture, Agricultural Research Service, Food Data Central. https://fdc.nal.usda.gov/fdc-app.html#/food-details/170043/nutrients

Potatoes, Raw, Skin (Standard Reference Legacy, 170032) (2019) U.S. Department of Agriculture, Agricultural Research Service, Food Data Central. https://fdc.nal.usda.gov/fdc-app.html#/food-details/170032/nutrients

Pouvreau L, Gruppen H, Piersma S, Van den Broek L, Van Koningsveld G, Voragen A (2001) Relative abundance and inhibitory distribution of protease inhibitors in potato juice from cv. Elkana. J Agric Food Chem 49(6):2864–2874

Ralet M-C, Guéguen J (2000) Fractionation of potato proteins: solubility, thermal coagulation and emulsifying properties. LWT-Food Sci Technol 33(5):380–387

Randhawa K, Sandhu K, Kaur G, Singh D (1984) Studies of the evaluation of different genotypes of potato (*Solanum tuberosum* L.) for yield and mineral contents. Plant Foods Hum Nutr 34 (4):239–242

RDA (2009) Nutrient requirements and recommended dietary allowances for Indians. National Institute of Nutrition, Indian Council of Medical Research, Jamai-Osmania

Robertson MD, Bickerton AS, Dennis AL, Vidal H, Frayn KN (2005) Insulin-sensitizing effects of dietary resistant starch and effects on skeletal muscle and adipose tissue metabolism. The Am J Clin Nutr 82(3):559–567

Rolfe HM (2014) A review of nicotinamide: treatment of skin diseases and potential side effects. J Cosmet Dermatol 13(4):324–328

Sable K (2019) Frozen potato market by product (French Fries, hash brown, shapes, mashed, sweet potatoes/yam, battered/cooked, twice baked, topped/stuff, and others) and end user (residential and commercial): global opportunity analysis and industry forecast, 2018–2025 UK

Sapone A, Bai JC, Ciacci C, Dolinsek J, Green PH, Hadjivassiliou M, Kaukinen K, Rostami K, Sanders DS, Schumann M (2012) Spectrum of gluten-related disorders: consensus on new nomenclature and classification. BMC Med 10(1):13

Shan L, Molberg Ø, Parrot I, Hausch F, Filiz F, Gray GM, Sollid LM, Khosla C (2002) Structural basis for gluten intolerance in celiac sprue. Science 297(5590):2275–2279

Singh B, Raigond P, Barwal S et al (2016) Glycoalkaloids in peels of Indian potatoes. Potato J 43:86–92

Siniawski D, Badimon JJ, Masson W (2017) Role of Niacin in Cardiovascular Prevention: The Debate Continues. Am J Med 130(8):e345

Snacks, Potato Chips, Plain, Salted (Standard Reference Legacy, 169677) (2019) U.S. Department of Agriculture, Agricultural Research Service, Food Data Central. https://fdc.nal.usda.gov/fdc-app.html#/food-details/169677/nutrients

Snacks, Potato Sticks (Standard Reference Legacy, 168854) (2019) U.S. Department of Agriculture, Agricultural Research Service, Food Data Central. https://fdc.nal.usda.gov/fdc-app.html#/food-details/168854/nutrients

Song W, Derito CM, Liu MK, He X, Dong M, Liu RH (2010) Cellular antioxidant activity of common vegetables. J Agric Food Chem 58(11):6621–6629
Sun Y, Jiang L, Wei D (2013) Partial characterization, in vitro antioxidant and antiproliferative activities of patatin purified from potato fruit juice. Food Funct 4(10):1502–1511
Tian J, Chen J, Lv F, Chen S, Chen J, Liu D, Ye X (2016) Domestic cooking methods affect the phytochemical composition and antioxidant activity of purple-fleshed potatoes. Food Chem 197:1264–1270
Tibbetts AS, Appling DR (2010) Compartmentalization of mammalian folate-mediated one-carbon metabolism. Annu Rev Nutr 30:57–81
Uhde M, Ajamian M, Caio G, De Giorgio R, Indart A, Green PH, Verna EC, Volta U, Alaedini A (2016) Intestinal cell damage and systemic immune activation in individuals reporting sensitivity to wheat in the absence of coeliac disease. Gut 65(12):1930–1937
US Department of Agriculture, Agricultural Research Service (2019) Food Data Central. https://fdc.nal.usda.gov
Verdu EF, Armstrong D, Murray JA (2009) Between celiac disease and irritable bowel syndrome: the "no man's land" of gluten sensitivity. Am J Gastroenterol 104(6):1587
Vivas-Rodríguez N, Serna-Saldivar S, Waniska R, Rooney L (1990) Effect of tortilla chip preparation on the protein fractions of quality protein maize, regular maize and sorghum. J Cereal Sci 12 (3):289–296
Waglay A, Karboune S, Alli I (2014) Potato protein isolates: recovery and characterization of their properties. Food Chem 142:373–382
Wahnschaffe U, Ullrich R, Riecken E, Schulzke J (2001) Celiac disease–like abnormalities in a subgroup of patients with irritable bowel syndrome. Gastroenterol 121(6):1329–1338
Wahnschaffe U, Schulzke JD, Zeitz M, Ullrich R (2007) Predictors of clinical response to gluten-free diet in patients diagnosed with diarrhea-predominant irritable bowel syndrome. Clin Gastroenterol Hepatol 5(7):844–850
Wang LL, Xiong YL (2005) Inhibition of lipid oxidation in cooked beef patties by hydrolyzed potato protein is related to its reducing and radical scavenging ability. J Agric Food Chem 53 (23):9186–9192
Weaver C, Marr ET (2013) White vegetables: a forgotten source of nutrients: purdue roundtable executive summary. Adv Nutr 4(3):318S–326S
Wong ND, Kashyap ML (2000) Cholesterol and lipids. In: Wong ND, Black HR, Gardin JM (eds) Preventive cardiology. McGraw Hill Professional, New York, pp 165–194
Wu D (2016) Recycle technology for potato peel waste processing: a review. Procedia Environ Sci 31:103–107

Potato Probiotics for Human Health

15

Dharmendra Kumar, Som Dutt, Pinky Raigond, Sushil Sudhakar Changan, Milan Kumar Lal, Devender Sharma, and Brajesh Singh

Abstract

The microorganisms which are beneficial for human as well as animal health when consumed in the proper amount are known as probiotics. Probiotics control the metabolism of the host. Probiotics are beneficial in improving lactose metabolism, reduction of antimicrobial activities, reduction of gastrointestinal infection, controlling serum cholesterol, have anticarcinogenic properties, antidiarrhoeal properties, and improve inflammatory bowel diseases. The consumer is mainly dependent on the probiotics which are available in the market in the form of fermented dairy products. However, there are two problems regarding the consumption of the fermented dairy product such as lactose intolerance and increase in cholesterol content. In current scenario demand for vegetarian probiotics products has increased. Potato is currently the third most important food crop which is consumed globally. The advantages of using potato along with their low cost are the presence of low fat, high on carbohydrates, high quality of proteins, fibre, vitamins, cocktails of minerals, and also the presence of other health-promoting compounds such as phytonutrients, carotenoids, flavonoids, and chlorogenic acids. The major advantage of fermented potato beverages are for those who are allergic to dairy products and are vegetarian.

Keywords

Probiotics · Potato · Microorganisms · Lactic acid bacteria

D. Kumar (✉) · S. Dutt · P. Raigond · S. S. Changan · M. K. Lal · B. Singh
ICAR-Central Potato Research Institute, Shimla, Himachal Pradesh, India
e-mail: dharmendra851@gmail.com

D. Sharma
ICAR-Vivekananda Parvatiya Krishi Anusandhan Sansthan, Almora, Uttarakhand, India

P. Raigond et al. (eds.), *Potato*, https://doi.org/10.1007/978-981-15-7662-1_15

15.1 Introduction

In modern society, the consumers are more aware of food and health benefitting compounds which led to an increase in the interest of healthy, functional, and fermented beverages in the present time. However, the idea of healthy food is not new. Hippocrates wrote 2400 years ago "Let food be the medicine and medicine be the food". Healthy food means functional foods that exert a beneficial effect in addition to the traditional nutrition and has a specific body function. There are many well-known functional foods which contain or prepared with bioactive compounds like phytochemicals, dietary fibre, oligosaccharides, antioxidants, and active friendly bacteria (Graf et al. 2015). The vegetables, fruits, and milk products based probiotic beverages are getting attention these days because of their nutritional status, besides being delicious gut microbiota, they form a diverse resident microbial community that is essential for health. The composition and abundance of probiotics varies in different regions of the gut. In the human gut, it plays an important role in maintaining the health and metabolism. The key factor for shaping microbiota composition is diet, which influences the gut community function. Probiotics are live microorganisms that cope with constructive effects on the host in the correct and adequate amounts. Probiotics are such foods which confer health benefits if administered in adequate quantity and are selectively microbial, viable, and dietary supplements that help to maintain good metabolism in body (Perricone et al. 2015; Bibi et al. 2019).

Probiotic's impact on human health is being progressively increasing, since they are being promoted by health and medical professionals. Functional foods with both probiotics and prebiotics are referred to as synbiotics. Prebiotics are outlined as food supplements that are non-digestible by the host, however, improve the health by selectively stimulating the expansion and/or activities of selected enteric microorganism. Most prebiotics are non-digestible carbohydrates with completely different variety of sugar moieties from two to several hundred. Some examples are galactose, fructooligosaccharides, lactulose, and resistant starch (Sanders et al. 2005; De Vrese and Schrezenmeir 2008; Bajramagic et al. 2019).

The availability of probiotic bacteria differs in a variety of dietary supplements, drugs, and food products. In ancient times most of the probiotics used are of dairy-based products. However, in modern society, the demand for probiotics has been increased due to its nutritional implications. The dairy products satisfy nutritional requirements by the use of lactic acid bacteria (LAB). Due to the increase in the trend of vegetarian lifestyle the challenging issue of lactose intolerance is the main hurdle. Moreover, the demand for low fat and low cholesterol foods has created a huge demand for new taste with non-dairy probiotic products. Thus, there is a huge demand for non-dairy probiotic foods, viz., coconut milk, soy products, nutrition bars, fruit drinks, cereals based products, vegetables, and fruit juices. Fresh fruits and vegetables are considered healthy products as it does not contain any dairy allergens that might prevent its usage by certain segments of the population. It provides ideal substrates for probiotics since they contain minerals, vitamins, antioxidants and fibres. Moreover, it also has a pleasing taste profile to all age groups and they are perceived as being healthy and refreshing (Rivera-Espinoza and Gallardo-Navarro 2010; Silanikove et al. 2015; Beals 2019; Nguyen et al. 2019).

Potato is the world's fourth most significant food crop after rice, wheat, and maize. India is the second largest potato producer (52 million tonnes in 2018–19). India produces 12% of the world's potatoes from 7% of the global area. Moreover, potato processing is an important part of the agro-processing sector in India. The rapid urbanization and enhancement in the living standard of the population have created tremendous growth in potato processing in recent years. Being carbohydrate and nutrient-rich, potato is considered as an important staple food. It is also rich in complex carbohydrates, dietary fibre, resistant starch, vitamins, polyphenols, and minerals. Potato also contains phenolics such as chlorogenic acid and anthocyanin. In potato, the anthocyanins are present in stable form due to its acetylated form as compared to other food (Kook et al. 2019; Bibi et al. 2019; Singh et al. 2020). The beneficial effect on gut microbiota is due to the dietary fibre and resistant starch content in potato. Polyphenols act as antioxidant and anti-inflammatory constituents which are known to have prebiotic effects. It was also reported that intake of purple sweet potato anthocyanins, flavanol-enriched cocoa powder, or high-cocoa flavanol drink promote *Bifidobacteria* and *Lactobacilli* in gut microbiota. These studies show that phytonutrients associated with potato consumption can have profound effects on the gut microbiota (Pradeep and Reddy 2010).

15.2 Interactions Between Probiotics, Prebiotics, and Synbiotics

Functional foods with both probiotics and prebiotics are referred to as synbiotics. Most prebiotics commonly used presently are possibly non-digestible carbohydrates consisting of two to hundreds of moieties like lactulose, galactose, fructooligosaccharides, and resistant starch. The idea of synbiotics is comparatively new and there are not much studies of interactions between probiotics and prebiotics. The prebiotics may also act as the starter where it may influence the expansion and survival of probiotic by influencing the expansion and metabolites. Interaction in vivo can be favoured by associate adaptation of the probiotic to the prebiotic. By adapting prebiotics metabolism to a substrate simultaneously with the probiotic may end up in a competitive advantage for the probiotic. However, most of the studies have been done with fully grown probiotic and not on the associated adaptation of probiotics and prebiotics (Nomoto 2005; De Vrese and Schrezenmeir 2008; Roberfroid et al. 2010; Wang and Wan 2019; Nguyen et al. 2019).

The oral feeding of external microorganisms is a mechanism to extend the number of useful microorganisms inside the gut. Addition of prebiotics is another mechanism (Parvez et al. 2006; Patel et al. 2014); these prebiotics act as food supplements that are non-digestible by the host, however, they improve health by selective stimulation of the expansion and/or activities of selected enteric microorganism (Andersson et al. 2001; Salmerón 2017; Battistini et al. 2018). This increase in gut microorganism might, in turn, inhibit the growth of pathogenic species.

Prebiotics are not hydrolyzed or absorbed within the intestine, however, they work as substrates for gut microorganism. The probiotics are often combined with a prebiotic to make what is referred to as a dependent. The colonization by associate exogenous probiotics might increase by synchronic administration of a prebiotic that is utilized by the probiotic within the enteric tract (Sanders et al. 2005; De Vrese and Schrezenmeir 2008; Nikolaou et al. 2020). To date, no organized clinical trials in humans have tested prebiotics or synbiotics for hindrance or treatment of enteric disorders. Probiotics enhance regulation of intestinal microbial homeostasis, antimicrobial activity, immune modulation, and pathogen exclusion (Shah 2000; Mohammadi et al. 2017; Anwar et al. 2019) (Table 15.1).

15.3 Microorganisms for Probiotics

The most distinguished probiotic microorganisms are LAB and *Bifidobacteria* species that are famous "generally recognized as safe" (GRAS) in the food industry. The strains of *Lactobacillus plantarum*, *Lactobacillus lactis*, *Bifidobacterium infantis*, *Lactobacillus reuteri*, *Lactobacillus acidophilus*, *Lactobacillus rhamnosus*, *Lactobacillus johnsonii*, *Lactobacillus casei*, *Lactobacillus delbrueckii subsp. bulgaricus*, *Pediococcus acidilactici*, *Lactobacillus paracasei*, *Bifidobacterium lactis*, *Bifidobacterium longum*, *Enterococcus*, and *Lactobacillus thermophilus* additionally as *Escherichia coli*, *Propionibacterium freudenreichii*, *Saccharomyces cerevisiae*, *Bifidobacterium brevis*, and some fungus genus have been regularly used as industrial starter cultures (Sanders et al. 2005; Parvez et al. 2006; Lahtinen et al. 2006; Farnworth et al. 2007; Perricone et al. 2015; Salmerón 2017). In addition to those, *Streptococcus thermophilus*, *Pediococcus*, *Enterococcus francium*, and *Leuconostoc* species might be used as probiotics. Probiotic bacteria primarily used in fermented foods and dairy farm products play a predominant role as carriers of probiotics (Bernet et al. 1993; Kumura et al. 2004; Yoon et al. 2006). To deliver these microorganisms with efficiency, the survival of the probiotic bacterium in non-fermented food products and non-dairy merchandise is being studied (Kaila et al. 1995; Sheehan et al. 2007; Pereira et al. 2011; Nguyen et al. 2019).

Almost for two decades, probiotic (health-promoting) microorganisms are being increasingly incorporated in varieties of food products, particularly in fermented milk. Safety, practical and technological characteristics, has to be considered for the selection method of probiotic microorganisms. The selection criteria for probiotics theoretically include functional, safety, and technological aspects. Several reports have defined the safety criteria of probiotics (Adams 1999; Heenan et al. 2004; Canani et al. 2007; Collado et al. 2008; Ranadheera et al. 2010). There were many debates regarding the current successful strains linked to the human origin of probiotics. It has been suggested that the probiotic strain may function better in a similar environmental condition (human GI-tract) as where it was initially isolated in case of human. The specification for the safety aspects of probiotics includes:

Table 15.1 Effect of probiotic microorganisms on health

Disease	Strains	Health impact	References
Immunity	*L. casei Shirota, L. rhamnosus, L. acidophilus, Bf. lactis, Bacillus circulans, Lactobacillus plantarum*	Positive impact on human health by enhancement of the level of immune reactive cells	Homayouni et al. (2012), Nguyen et al. (2019)
Dental problems	*Lactobacillus, Bifidobacterium*	Decrease dental problems	Seminario-Amez et al. (2017)
Antibiotic therapy	*L. salivarius, L. acidophilus, L. johnsonii, Enterococcus mundtii, Lactobacillus plantarum, Lactobacillus brevis, Lactobacillus strains, Bifidobacterium strains*	Minimize the disruptive effect of antibiotics to normal bacterial flora	Sanders et al. (1996, 2005)
Cancer	*Bifidobacterium sp., L. casei Shirota, L. acidophilus, Propionibacterium sp., L. rhamnosus*	Positive impact on human health. Detoxify the ingested carcinogens	Sanders et al. (1996, 2005)
Diarrhoea	*L. rhamnosus, L. casei, Bf. lactis, Bf. bifidum, S. thermophilus, Lactobacillus casei*	Competition with pathogenic bacteria on epithelial cells	Parvez et al. (2006)
Inflammatory bowel diseases	*E. coli, Saccharomyces boulardii, Bifidobacterium longum, B. breve, B infantis, Lactobacillus casei, L. plantarum, L. acidophilus, L. delbrueckii subsp. bulgaricus, and Streptococcus salivarius subsp. thermophilus*	Positive impact on human health. Bowel syndrome and inflammatory bowel diseases are reduced	Ventura and Perozzi (2011)
Anaemia	*Lactobacillus lactobacilli*	Increase the expression of iron transporters in the caecum due to production of propionic acid	Balamurugan et al. (2010)
Hypertension	*L. rhamnosus, L. lactis*	Positive impact on human health by reducing blood pressure	Sanders et al. (1996, 2005)
Pancreatitis	*L. rhamnosus GG, Bf. lactis BB–12*	Positive impact on human health by decreasing the occurrence of pancreatic infection	Pezzilli and Fantini (2006)
Urinary tract infection	*Lactobacillus rhamnosus, Lactobacillus reuteri, L. acidophilus*	Decrease urinary tract diseases	Anukam et al. (2009)

(continued)

Table 15.1 (continued)

Disease	Strains	Health impact	References
Hyper cholesterolemia and cardiovascular diseases	*Lactobacillus spp., Bifidobacterium spp., Enterococcus faecium, Lactobacillus plantarum, Propionibacterium freudenreichii, Lactobacillus plantarum*	Reduced dietary cholesterol absorption	Sanders et al. (2005), Homayouni et al. (2012), Phan et al. (2017)
Lactose intolerance	*L. rhamnosus, L. plantarum, L. delbrueckii, Bf. lactis, Lactobacillus acidophilus*	Digestion of lactose	Sanders et al. (1996, 2005)

The table is adopted from Panghal et al. (2018)

1. The strain of probiotics for human use should ideally derived from human origin.
2. The strains are isolated from healthy human GI-tract.
3. The isolated strain of probiotics should have a history of being no-pathogenic.
4. The probiotics strains should not have a history of infectious disease like infective endocarditis of GI disorder.
5. The probiotics should not be involved in the deconjugation in the gall salts (bile salt deconjugation—the process in which dehydroxylation would lead to negative attributes within the tiny bowel).
6. They should not carry any transmissible antibiotic resistance genes.

The importance and adequacy of probiotics required to be established through in vitro methods and the results of these studies need to be reflected in human studies under control condition. While choosing a desirable probiotic strain, many aspects of practicality need to be considered:

1. Human digestive juice and acid tolerance.
2. Tolerance to bile pigment (this is necessary for the survival within the bowel).
3. It should have persistence results within the human GI-tract and also adhere to the animal tissues.
4. It should not have any pro-inflammatory results and immunostimulation.
5. Antagonistic activity against pathogens like *Helicobacter pylori*, *Salmonella sp.*, *Listeria monocytogenes*.
6. Anticarcinogenic and antimutagenic properties (Mitsuoka 1992; Magge and Lembo 2012; Kubota et al. 2020).

It was reported in the clinical trial that the probiotic strain generally disappears from the GI-tract within the few weeks when the uptake is interrupted. However, the role of probiotic persistence within the human GI-tract system has been questioned. The temporary persistence of many edible probiotics strains showed enhanced probability of their useful functions in the GI-tract which plays a significant role. Even supposing a probiotic strain fulfils the mandatory safety and purposeful criteria, the aspects associated with probiotic production and process also need to be discussed. Many technological aspects have to be thought of while choosing probiotics (Johansson et al. 1993; Silanikove et al. 2015; Naghizadeh Raeisi et al. 2018), like

1. Virus resistance
2. Sensible sensory protein
3. Viability
4. Enhanced stability within the product and also through storage

In selecting probiotic bacterium for industrial production, safety, purpose, and technological characteristics need to be assessed. Thus, the safety of probiotic bacterium in terms of infectivity in immune-compromised animal models, metabolic activities (D-Lactate, bile salt deconjugation), antibiotic resistance profiles, haemolytic activity, poisonous substance production, genetic and pathological side effects, need to be ascertained (post-market) (Mattila-Sandholm et al. 2002; Mohammadi et al. 2011).

Good viability and activity of probiotics are important for optimum usage. However, many reports have shown that the non-viable probiotics should have a useful effect like immune modulation and substance binding within the host (Ouwehand et al. 2002; Blandino et al. 2008; Zheng et al. 2015, 2017). Thus, it should be desirable that the probiotic strain must grow well throughout the initial production phase (to get high enough numbers within the product) and they may not essentially have to retain the sensible viability throughout the storage. The purposeful properties include the viability and stability of the cells inside a food matrix, persistence activity in the alimentary tract, species and strain characteristics, daily dose, fermentation technology, storage conditions, and accessibility of prebiotics (Brown and Valiere 2004; Ranadheera et al. 2010).

Indian probiotic trade is still in its infancy stage though it is evolving at a gradual pace with expectations for tremendous growth soon. India is rising as a serious probiotic market in the long run with an annual growth rate of 6%. Till 2015, major players in Indian probiotic trade were Amul, Mother Dairy farm, Yakult, Danone, and Nestle, besides few minor players in operation in several regions as per their capacities. With their advent, the Indian probiotic market turnover is anticipated to exceed $ 20 million by the year 2020. Probiotics in India typically involves milk and soured milk product sharing 62% and 38% of the market share, respectively. Fruits and vegetables, being delicious and good sources of nutrients, are however not linked with the development of probiotic beverages, though in terms of production of fruits and vegetables in the world India secures the second position (Hajela et al. 2014).

Presently, the probiotic microorganisms are offered in several food products, dietary supplements, and medicines. Most probiotic products are dairy farm based since their inception. The dairy farm could be an alimental substrate that satisfies the nutritionary necessities of fastidious work. But, increasing trend of the vegetarian lifestyle, problems of genetic disorders, low steroid alcohol foods demand, and low fat have created a growing demand for non-dairy products (Champagne et al. 2005; Rivera-Espinoza and Gallardo-Navarro 2010; Granato et al. 2010; Panghal et al. 2017). These include non-dairy probiotic foods, like nutrition bars, coconut milk, soy product, fruit drinks, and cereal-based products. Similarly, there is an increasing interest in the development of drinks primarily based on probiotic products and fruits and vegetables that are suitable and may give ideal substrates for probiotics, due to presence of minerals, vitamins, antioxidants, and fibres and not containing any dairy allergens. The fermented beverages and functional foods have acceptability for all age groups since they are perceived as healthy and refreshing taste. In the development of the latest probiotic beverages, the choice of the probiotic organism is the main challenge for food industries. The selection of organisms depends on survival in the intestinal region, treatments and storage, viability throughout the process, overall healthy impact on gut, and finally human health (Mattila-Sandholm et al. 2002; Chiu et al. 2007; Ventura and Perozzi 2011). The initial step in the selection of probiotic strains for commercialization hence is assessed on the following parameters:

1. The stability of probiotics during storage

Due to the storage and viability of the probiotic organisms, the adequacy of a variety of probiotic microorganism is challenging at the time of consumption. The food additives and ingredients like sugars, sweeteners, salts, aroma compounds, natural or artificial colouring agents, other natural or artificial ingredients, acid preservatives, enzymes, and nitrites may inhibit the expansion of probiotics. The LABs are typically strong probiotic organisms than *Bifidobacterium* and may survive harsh conditions of storage and process. LABs are immune to low pH scale, so preferred for many food applications compared to different probiotic organisms (Mättö et al. 2006; Tripathi and Giri 2014).

2. Sensory criteria

The addition of probiotic microorganism in the production of non-dairy probiotic products leads to change in some sensory criteria. The aroma, flavour, and nature of the probiotic products are altered by the addition of probiotic organisms due to the nature completely different from various metabolic parts like lactic acid and other metabolites in living cells having different spp. throughout fermentation and during storage. In probiotic food preparation, the probiotic microorganism ferments macromolecules of the vegetables, fruit, cereals, and legumes and produces gases and alcohol content by fermentation. The major challenge for the probiotic market is product satisfaction by the customers. Thus, the presence of the probiotic culture in food should not adversely affect product sensory properties (Champagne et al. 2005; Katina et al. 2007; Mohammadi et al. 2011; Min et al. 2019).

15.4 Advantages of Using Non-Dairy Probiotics

The dairy products are the most popular probiotic microorganism since they provide suitable habitat for a probiotic microorganism that supports their growth and viability. However, with the rise in consumer diet preferences throughout the world, there is an increasing demand for other probiotic products. There have been huge interest and demand among the vegetarians and lactose intolerance customers for non-dairy products. According to The National Institute of Diabetes and Digestive and Kidney Diseases (NIDDK) of the USA National Institutes of Health, 75% of the world population is lactose intolerant. However, the non-dairy probiotic food products are not in abundance since they need to meet the consumer's expectancy for health benefits. On the other hand, dairy allergens are also preventing some consumers to consume milk-based probiotics, but fruits and vegetables are free from milk allergens, disaccharide, and cholesterin and may act as appropriate substrates (Luckow and Delahunty 2004; Granato et al. 2010; Esaiassen et al. 2017).

A large variety of fermented lactic acid acid-based products are available and also consumed throughout the world, but these are ancient products which have not been explored for their potential probiotics characteristics. In some cases, change of state before consumption ends up in the killing of probiotic microorganisms, e.g. in Indian food products like idli, dosa, dhokla, etc. Presently, several commercially available non-dairy probiotics are being popularized (Johansson et al. 1993; Agrawal et al. 2000; Sridevi et al. 2010; Patel et al. 2014).

The major symptoms of lactose intolerance are abdominal pain and distension, borborygmi, flatus, and diarrhoea which are induced by dairy products. However, it may be associated with primary and secondary genetic defects. The biological mechanism of lactose absorption is established and several other investigations related to genetics, scrutiny, and physiological tests already exist. The lactose intolerance depends not solely on the expression of lactose, but also the dose of lactose, microorganism, epithelial duct motility, little enteral microorganism overgrowth, and sensitivity of the canal to the generation of gas and alternative fermentation product of lactose digestion. Treatment of lactose intolerance includes the lactose-reduced diet and protein replacement. This is often effective if symptoms are solely associated with dairy products; but, lactose intolerance may be a part of a wider intolerance to variably absorbed, possible Fermentable Oligo-, Di-, Monosaccharides And Polyols (FODMAPs). This often affects patients with irritable bowel syndrome (IBS) and they solely not need restriction of lactose intake however, additionally a coffee FODMAP diet is offered to enhance epithelial duct complaints. The semi-permanent effects of a dairy-free, low FODMAPs diet on biological health and also the feculent microbiome though have not been well studied (Muthukumarasamy and Holley 2006; Farnworth et al. 2007; Keddari et al. 2016).

The fermentation of vegetables has been familiar since ancient times and these offer an appropriate media to deliver probiotics. However, low incubation temperature of vegetable fermentation may be a limitation for the introduction of the standard *L. acidophilus* and *Bifidobacterium* probiotic bacterium. Probiotic of *L. casei*, *L. plantarum*, and *L. rhamnosus* are often custom-made for the vegetable during fermentation. Nonetheless, once the temperature is adjusted at 37 °C, the probiotic bacterium grows quickly in plant-based substrates. To develop the new probiotic vegetable product, several studies are being done. The quality of juice as a substrate for the assembly of probiotic food with *Bifidobacterium* strains has been investigated. The study showed that *Bifidobacteria* was capable of getting fermentation activities in juice without any nutrient supplementation.

Yoon et al. (2005) studied the quality of juice for the assembly of a probiotic product by *L. plantarum*, *L. acidophilus*, *L. delbrueckii*, and *L. casei* and tested the quality of cabbage to supply probiotic cabbage juice and advised that cabbage juice supports the viability of probiotics and functions as a healthy drink. The potential of red beets as the substrate for the assembly of probiotic beet juice by four strains of the lactic acid bacterium was well documented for growth and lactic acid production. The viability of assorted *Bifidobacteria* in kimchi was investigated under various conditions and results showed the appropriate levels of probiotics in kimchi. Similarly, sauerkraut-type foods like hard cabbage, potato, carrots, onions, and cucumbers supported dairy product fermentation by *L. plantarum* and may act as good probiotic carriers. However, ancient ways of production may lead to inactivation of the probiotic cultures and therefore the use of probiotics in hard vegetables would need low temperature storage of the products (Champagne et al. 2005; Di Cagno et al. 2008).

The vegetables are a good source of sugars, vitamins, minerals, and other health-promoting compounds. These vegetables are enriched with phytochemicals and phytonutrients and due to the absence of disaccharide; these probiotic merchandises are similar to temperaments for the lactose intolerant. Cabbage, carrot root, tomato, beet, onion, ginger, peanuts, etc., are utilized in probiotics. The LABs like *L. plantarum*, *L. acidophilus*, *B. longum*, and *L. casei* are often utilized for vegetable-based probiotic foods. Lactic acid fermentation improves the nutritional value of products for the viability of probiotic bacteria (Mishra et al. 2015; Panghal et al. 2017).

15.5 Potatoes' Significance as Probiotic

More than a billion people worldwide eat potato and global total crop production exceeds 300 million metrics tons. India is the second largest producer of potato in the world after China. In 2018–19, the country produced 52.59 million tons of potato from 2.18 million hectares with an average productivity of 24.08 t/ha. India produces 11% of the world's potatoes from 7% of the global area. About 68.5% (16.9 mt) of the potato output, equivalent to 14.9 kg per capita, goes to domestic "table" consumption since potatoes are mostly used as an unprocessed vegetable. Out of the country's per capita annual potato production of about 21.5 kg, the consumption of processed quality potatoes is only 1.5 kg and they are typically not used as animal feed (Rana 2011; Ranadheera et al. 2010).

However, using potato juice as a medium for the production of probiotic beverages is being assessed in terms of the quality as the carrier for probiotic bacterium and sensory satisfaction of the consumers. The customers are quite attentive to the link between the physiological state and food intake, significantly from naturally derived foods like fruits, vegetables, and probiotic dairy farm products. Several kinds of food products containing probiotic bacteria have been introduced into the markets including non-dairy products, like chocolate, cereals, beverages, fruits, and vegetable products. The potato juices contain useful nutrients which may act as a perfect medium for probiotics (Abadias et al. 2008).

Accordingly, any sort of probiotic fruit and vegetable products are developed and marketed, like fruits and vegetable juices, dried fruits, hard vegetables, and desserts. Potato juice has qualitative nutrients and may have acceptability by all age and economic groups and possesses excellent potential in developing probiotic potato beverages. A major development in the area of healthy foods relates to foods containing probiotics and prebiotics that enhance health-promoting microorganism flora within the gut. Potato probiotic beverages may be helpful particularly to consumers who cannot consume dairy farm products and also for those who have lactose intolerance, hypercholesterolaemia, and strictly vegetarian people. There is growing scientific proof to support the idea that the upkeep of healthy gut microflora might offer protection against epithelial duct disorders together with epithelial duct infections, inflammatory internal organ diseases, and even cancer (Klaenhammer and Kleeman 1981; Agerholm-Larsen et al. 2000).

The utilization of probiotic microorganism cultures stimulates the expansion of most popular microorganisms, crowds out probably harmful bacteria, and reinforces the body's natural defence mechanisms. A lot of proofs exists on the positive effects of probiotics on human health. However, this has sometimes been inconsistent in the human populations (Sanz et al. 2006; Canani et al. 2007; Katina et al. 2007; Baldassarre et al. 2018).

15.6 Mechanisms of Probiotics

There are several planned mechanisms using which probiotics could benefit the host from internal disorders. On such direct exclusion, the mechanism includes a competition for nutrition/strict hindrance, adhesion to the epithelium, and secretion of antimicrobial compounds. Similarly, non-competitive mechanisms such as immune exclusion, host antimicrobial peptides, epithelial barrier function exist. Probiotic microorganisms produce a range of inhibitory substances such as organic acids, peroxide, and bacteriocins. These compounds inhibit the population of unwanted microbes in the gut. Recent reports suggest that stimulation of specific and nonspecific immunity could also be another mechanism by that probiotics may provide immunity against enteric diseases (Kaila et al. 1995; Gopal et al. 2001; Song et al. 2015; Imane and Amel 2018; Fijan et al. 2019).

15.7 Challenges in Probiotic Industry

Fruits, vegetables, cereals, and legume products represent a promising carrier for probiotic microorganisms with sensible nutraceutical parts. However, some limitations might preclude non-dairy probiotics production at the economic level which includes sensory traits, overall acceptance, and most vital being the survival of probiotics during the supply chain. The viability of probiotics through completely different acidic/basic/salt medium thus has to be ascertained. The selected microorganisms have to have the ability to tolerate bile and acid within the human gut, acceptability towards individual animal tissue cell; sensible growth characteristics, non-pathogenicity, GRAS (Generally considered Safe) listing, and sensible impact on the individual health. The application of probiotic cultures in non-dairy products and environments represents a big challenge. The probiotic viability within the food matrix depends on factors like pH, storage temperature, presence of competitor microorganisms, and inhibitors (e.g. NaCl). In merchandise like probiotic-containing baby foods or confectionery (e.g. chocolate) the formulation must maintain the activity and viability of the probiotic for extended periods. The storage temperature, commonly used for non-dairy products like cereals, drinks, chocolate, etc., may or may not be suitable for stability in potato juice. Hence, it is suggested to use probiotic encapsulation technology to confirm the viability and stability of probiotic cultures. The stable probiotic-containing baby food formulations and confectioneries have been developed and marketed. The probiotic microbes are stable with microencapsulation to preserve them from prejudicial processes and storage factors like high acidity and low pH. The advanced technologies, e.g. growth-promoting food ingredients, gas barrier packaging materials, antioxidants, and storage atmosphere modification have enabled microorganisms to survive under different environments. Hence there is the market potential for non-dairy probiotics like vegetable-based products, fruit-based products, cereal and legumes-based products, confectionery products, and breakfast cereals.

15.8 Conclusion

Potato is a rich source of phytonutrients and bioactive food components which include resistant starch, dietary fibre, anthocyanins, and polyphenols. Potato consumption improves the composition of gut microbiota. Development of the probiotic product depends on pH, gas content, salt concentration, the temperature of process and storage, shelf-life, and range of viable microorganisms throughout consumption. The use of probiotic microorganisms in the non-dairy sector though is difficult, is the need of time and emphasizes on exploring the possibility of using non-dairy probiotics like potatoes.

The future of probiotics depends on rigorous scientific studies that are needed for chemical drug entities, together with irregular, placebo-controlled, double-blind studies, pharmacokinetic studies (i.e., Dose-dependent efficaciousness, absorption, distribution, metabolism, excretion, and length of effect), and multicenter trials to ascertain the reliability of these probiotics, particularly while developing non-dairy probiotics including vegetables like potatoes.

References

Abadias M, Usall J, Anguera M et al (2008) Microbiological quality of fresh, minimally-processed fruit and vegetables, and sprouts from retail establishments. Int J Food Microbiol 123:121–129. https://doi.org/10.1016/j.ijfoodmicro.2007.12.013

Adams MR (1999) Safety of industrial lactic acid bacteria. J Biotechnol 68:171–178. https://doi.org/10.1016/S0168-1656(98)00198-9

Agerholm-Larsen L, Raben A, Haulrik N et al (2000) Effect of 8 week intake of probiotic milk products on risk factors for cardiovascular diseases. Eur J Clin Nutr 54:288–297. https://doi.org/10.1038/sj.ejcn.1600937

Agrawal R, Rati ER, Vijayendra SVN et al (2000) Flavour profile of idli batter prepared from defined microbial starter cultures. World J Microbiol Biotechnol 16:687–690. https://doi.org/10.1023/A:1008939807778

Andersson H, Asp NG, Bruce Å et al (2001) Health effects of probiotics and prebiotics a literature review on human studies. Scand J Nutr 45:58–75. https://doi.org/10.3402/fnr.v45i0.1790

Anukam KC, Hayes K, Summers K, Reid G (2009) Case Report Probiotic Lactobacillus rhamnosus GR-1 and Lactobacillus reuteri RC-14 May Help Downregulate TNF-Alpha, IL-6, IL-8, IL-10 and IL-12 (p70) in the Neurogenic Bladder of Spinal Cord Injured Patient with Urinary Tract Infections: A Two-Case Study. Adv Urol 2009:680363. https://doi.org/10.1155/2009/680363

Anwar A, Siddiqui R, Shah MR, Khan NA (2019) Antidiabetic drugs and their nanoconjugates repurposed as novel antimicrobial agents against Acanthamoeba castellanii. J Microbiol Biotechnol 29:713–720. https://doi.org/10.4014/jmb/1903.03009

Bajramagic S, Hodzic E, Mulabdic A et al (2019) Usage of probiotics and its clinical significance at surgically treated patients suffering from Colorectal Carcinoma. Med Arch 73:316–320. https://doi.org/10.5455/medarh.2019.73.316-320

Balamurugan R, Mary RR, Chittaranjan S et al (2010) Low levels of faecal lactobacilli in women with iron-deficiency anaemia in south India. Br J Nutr 104:931–934. https://doi.org/10.1017/S0007114510001637

Baldassarre ME, Palladino V, Amoruso A et al (2018) Rationale of probiotic supplementation during pregnancy and neonatal period. Nutrients 10(11):1693

Battistini C, Gullón B, Ichimura ES et al (2018) Development and characterization of an innovative synbiotic fermented beverage based on vegetable soybean. Braz J Microbiol 49:303–309. https://doi.org/10.1016/j.bjm.2017.08.006

Beals KA (2019) Potatoes, nutrition and health. Am J Potato Res 96:102–110

Bernet MF, Brassart D, Neeser JR, Servin AL (1993) Adhesion of human bifidobacterial strains to cultured human intestinal epithelial cells and inhibition of enteropathogen-cell interactions. Appl Environ Microbiol 59:4121–4128. https://doi.org/10.1128/aem.59.12.4121-4128.1993

Bibi S, Navarre DA, Sun X et al (2019) Beneficial effect of potato consumption on gut microbiota and intestinal epithelial health. Am J Potato Res 96:170–176

Blandino G, Fazio D, Di Marco R (2008) Probiotics: overview of microbiological and immunological characteristics. Expert Rev Anti Infect Ther 6:497–508. https://doi.org/10.1586/14787210.6.4.497

Brown AC, Valiere A (2004) Probiotics and medical nutrition therapy. Nutr Clin Care 7:56–68

Canani RB, Cirillo P, Terrin G et al (2007) Probiotics for treatment of acute diarrhoea in children: randomised clinical trial of five different preparations. Br Med J 335:340–342. https://doi.org/10.1136/bmj.39272.581736.55

Champagne CP, Gardner NJ, Roy D (2005) Challenges in the addition of probiotic cultures to foods. Crit Rev Food Sci Nutr 45:61–84

Chiu H-H, Tsai C-C, Hsih H-Y, Tsen H-Y (2007) Screening from pickled vegetables the potential probiotic strains of lactic acid bacteria able to inhibit the Salmonella invasion in mice. J Appl Microbiol 104(2):605–612. https://doi.org/10.1111/j.1365-2672.2007.03573.x

Collado MC, Meriluoto J, Salminen S (2008) Adhesion and aggregation properties of probiotic and pathogen strains. Eur Food Res Technol 226:1065–1073. https://doi.org/10.1007/s00217-007-0632-x

De Vrese M, Schrezenmeir J (2008) Probiotics, prebiotics, and synbiotics. Adv Biochem Eng Biotechnol 111:1–66

Di Cagno R, Surico RF, Siragusa S et al (2008) Selection and use of autochthonous mixed starter for lactic acid fermentation of carrots, French beans or marrows. Int J Food Microbiol 127:220–228. https://doi.org/10.1016/j.ijfoodmicro.2008.07.010

Esaiassen E, Hjerde E, Cavanagh JP et al (2017) Bifidobacterium bacteremia: clinical characteristics and a genomic approach to assess pathogenicity. J Clin Microbiol 55:2234–2248. https://doi.org/10.1128/JCM.00150-17

Farnworth ER, Mainville I, Desjardins MP et al (2007) Growth of probiotic bacteria and bifidobacteria in a soy yogurt formulation. Int J Food Microbiol 116:174–181. https://doi.org/10.1016/j.ijfoodmicro.2006.12.015

Fijan S, Frauwallner A, Langerholc T et al (2019) Efficacy of using probiotics with antagonistic activity against pathogens of wound infections: an integrative review of literature. Biomed Res Int 2019:7585486. https://doi.org/10.1155/2019/7585486

Gopal PK, Prasad J, Smart J, Gill HS (2001) In vitro adherence properties of Lactobacillus rhamnosus DR20 and Bifidobacterium lactis DR10 strains and their antagonistic activity against an enterotoxigenic Escherichia coli. Int J Food Microbiol 67:207–216. https://doi.org/10.1016/S0168-1605(01)00440-8

Graf D, Di Cagno R, Fåk F et al (2015) Contribution of diet to the composition of the human gut microbiota. Microb Ecol Heal Dis 26:26164. https://doi.org/10.3402/mehd.v26.26164

Granato D, Branco GF, Nazzaro F et al (2010) Functional foods and nondairy probiotic food development: trends, concepts, and products. Compr Rev Food Sci Food Saf 9:292–302. https://doi.org/10.1111/j.1541-4337.2010.00110.x

Hajela N, Balakrish Nair G, Ramakrishna BS, Ganguly NK (2014) Probiotic foods: can their increasing use in India ameliorate the burden of chronic lifestyle disorders? Indian J Med Res 139(1):19–26

Heenan CN, Adams MC, Hosken RW, Fleet GH (2004) Survival and sensory acceptability of probiotic microorganisms in a nonfermented frozen vegetarian dessert. LWT - Food Sci Technol 37:461–466. https://doi.org/10.1016/j.lwt.2003.11.001

Homayouni A, Payahoo L, Azizi A (2012) Effects of probiotics on lipid profile: a review. Am J Food Technol 7:251–265

Imane HA, Amel D (2018) Characterization and screening of the potential probiotic lactic acid bacteria and bifidobacterium strains isolated of different biotopes. Med J Nutrition Metab 11:145–173. https://doi.org/10.3233/MNM-17191

Johansson ML, Molin G, Jeppsson B et al (1993) Administration of different Lactobacillus strains in fermented oatmeal soup: in vivo colonization of human intestinal mucosa and effect on the indigenous flora. Appl Environ Microbiol 59:15–20. https://doi.org/10.1128/aem.59.1.15-20.1993

Kaila M, Isolauri E, Saxelin M et al (1995) Viable versus inactivated lactobacillus strain GG in acute rotavirus diarrhoea. Arch Dis Child 72:51–53. https://doi.org/10.1136/adc.72.1.51

Katina K, Liukkonen KH, Kaukovirta-Norja A et al (2007) Fermentation-induced changes in the nutritional value of native or germinated rye. J Cereal Sci 46:348–355. https://doi.org/10.1016/j.jcs.2007.07.006

Keddari S, Benaoum N, Boufadi Y et al (2016) Antioxidant Activity and in vitro Fermentation of Dietary Fiber Extracts from Durum Wheat Bran. J Food Nutr Res 4:508–514. https://doi.org/10.12691/jfnr-4-8-4

Klaenhammer TR, Kleeman EG (1981) Growth characteristics, bile sensitivity and freeze damage in colonial variants of Lactobacillus acidophilus. Appl Environ Microbiol 41:1461–1467. https://doi.org/10.1128/aem.41.6.1461-1467.1981

Kook S-Y, Chung E-C, Lee Y et al (2019) Isolation and characterization of five novel probiotic strains from Korean infant and children faeces. PLoS One 14:e0223913. https://doi.org/10.1371/journal.pone.0223913

Kubota M, Ito K, Tomimoto K et al (2020) Lactobacillus reuteri DSM 17938 and magnesium oxide in children with functional chronic constipation: A double-blind and randomized clinical trial. Nutrients 12(1):225. https://doi.org/10.3390/nu12010225

Kumura H, Tanoue Y, Tsukahara M et al (2004) Screening of dairy yeast strains for probiotic applications. J Dairy Sci 87:4050–4056. https://doi.org/10.3168/jds.S0022-0302(04)73546-8

Lahtinen SJ, Gueimonde M, Ouwehand AC et al (2006) Comparison of four methods to enumerate probiotic bifidobacteria in a fermented food product. Food Microbiol 23:571–577. https://doi.org/10.1016/j.fm.2005.09.001

Luckow T, Delahunty C (2004) Which juice is "healthier"? A consumer study of probiotic non-dairy juice drinks. Food Qual Prefer 15:751–759. https://doi.org/10.1016/j.foodqual.2003.12.007

Magge S, Lembo A (2012) Low-FODMAP diet for treatment of irritable bowel syndrome. Gastroenterol Hepatol 8:739

Mattila-Sandholm T, Myllärinen P, Crittenden R et al (2002) Technological challenges for future probiotic foods. Int Dairy J 12:173–182. https://doi.org/10.1016/S0958-6946(01)00099-1

Mättö J, Alakomi HL, Vaari A et al (2006) Influence of processing conditions on Bifidobacterium animalis subsp. lactis functionality with a special focus on acid tolerance and factors affecting it. Int Dairy J 16:1029–1037. https://doi.org/10.1016/j.idairyj.2005.10.014

Min M, Bunt CR, Mason SL, Hussain MA (2019) Non-dairy probiotic food products: an emerging group of functional foods. Crit Rev Food Sci Nutr 59:2626–2641. https://doi.org/10.1080/10408398.2018.1462760

Mishra V, Shah C, Mokashe N et al (2015) Probiotics as potential antioxidants: a systematic review. J Agric Food Chem 63:3615–3626. https://doi.org/10.1021/jf506326t

Mitsuoka T (1992) The human gastrointestinal tract. In: The lactic acid bacteria, vol 1. Springer, Boston, pp 69–114

Mohammadi R, Mortazavian AM, Khosrokhavar R, Da Cruz AG (2011) Probiotic ice cream: viability of probiotic bacteria and sensory properties. Ann Microbiol 61:411–424

Mohammadi R, Yousefi M, Sarlak Z et al (2017) Influence of commercial culture composition and cow milk to soy milk ratio on the biochemical, microbiological, and sensory characteristics of a probiotic fermented composite drink. Food Sci Biotechnol 26:749–757. https://doi.org/10.1007/s10068-017-0097-z

Muthukumarasamy P, Holley RA (2006) Microbiological and sensory quality of dry fermented sausages containing alginate-microencapsulated Lactobacillus reuteri. Int J Food Microbiol 111:164–169. https://doi.org/10.1016/j.ijfoodmicro.2006.04.036

Naghizadeh Raeisi S, Ghoddusi HB, Juncker Boll E et al (2018) Antimicrobial susceptibility of bifidobacteria from probiotic milk products and determination of the genetic basis of tetracycline resistance in Enterococcus species after in vitro conjugation with Bifidobacterium animalis subsp. lactis. Food Control 94:205–211. https://doi.org/10.1016/j.foodcont.2018.07.016

Nguyen BT, Bujna E, Fekete N et al (2019) Probiotic beverage from pineapple juice fermented with Lactobacillus and Bifidobacterium strains. Front Nutr 6:54. https://doi.org/10.3389/fnut.2019.00054

Nikolaou E, Kamilari E, Savkov D et al (2020) Intestinal microbiome analysis demonstrates azithromycin post-treatment effects improve when combined with lactulose. World J Pediatr 16:168–176. https://doi.org/10.1007/s12519-019-00315-6

Nomoto K (2005) Prevention of infections by probiotics. J Biosci Bioeng 100:583–592. https://doi.org/10.1263/jbb.100.583

Ouwehand AC, Salminen S, Isolauri E (2002) Probiotics: an overview of beneficial effects. Antonie van Leeuwenhoek, Int J Gen Mol Microbiol 82:279–289. https://doi.org/10.1023/A:1020620607611

Panghal A, Virkar K, Kumar V et al (2017) Development of probiotic beetroot drink. Curr Res Nutr Food Sci 5:257–262. https://doi.org/10.12944/CRNFSJ.5.3.10

Panghal A, Janghu S, Virkar K et al (2018) Potential non-dairy probiotic products – A healthy approach. Food Biosci 21:80–89

Parvez S, Malik KA, Ah Kang S, Kim HY (2006) Probiotics and their fermented food products are beneficial for health. J Appl Microbiol 100:1171–1185

Patel A, Prajapati JB, Holst O, Ljungh A (2014) Determining probiotic potential of exopolysaccharide producing lactic acid bacteria isolated from vegetables and traditional Indian fermented food products. Food Biosci 5:27–33. https://doi.org/10.1016/j.fbio.2013.10.002

Pereira ALF, Maciel TC, Rodrigues S (2011) Probiotic beverage from cashew apple juice fermented with Lactobacillus casei. Food Res Int 44:1276–1283. https://doi.org/10.1016/j.foodres.2010.11.035

Perricone M, Bevilacqua A, Altieri C et al (2015) Challenges for the production of probiotic fruit juices. Beverages 1:95–103. https://doi.org/10.3390/beverages1020095

Pezzilli R, Fantini L (2006) Probiotics and severe acute pancreatitis. JOP 7:92–93

Phan YTN, Tang MT, Tran TTM et al (2017) Diversity of lactic acid bacteria in vegetable-based and meat-based fermented foods produced in the central region of Vietnam. AIMS Microbiol 3:61–70. https://doi.org/10.3934/microbiol.2017.1.61

Pradeep P, Reddy OVS (2010) High gravity fermentation of sugarcane molasses to produce ethanol: effect of nutrients. Indian J Microbiol 50:82–87. https://doi.org/10.1007/s12088-010-0006-0

Rana RK (2011) The Indian potato-processing industry. Outlook Agric 40(3):237–243

Ranadheera RDCS, Baines SK, Adams MC (2010) Importance of food in probiotic efficacy. Food Res Int 43:1–7

Rivera-Espinoza Y, Gallardo-Navarro Y (2010) Non-dairy probiotic products. Food Microbiol 27:1–11

Roberfroid M, Gibson GR, Hoyles L et al (2010) Prebiotic effects: metabolic and health benefits. Br J Nutr 104:S1–S63

Salmerón I (2017) Fermented cereal beverages: from probiotic, prebiotic and synbiotic towards nanoscience designed healthy drinks. Lett Appl Microbiol 65:114–124. https://doi.org/10.1111/lam.12740

Sanders ME, Walker DC, Walker KM et al (1996) Performance of commercial cultures in fluid milk applications. J Dairy Sci 79:943–955. https://doi.org/10.3168/jds.S0022-0302(96)76445-7

Sanders ME, Tompkins T, Heimbach JT, Kolida S (2005) Weight of evidence needed to substantiate a health effect for probiotics and prebiotics: regulatory considerations in Canada, E.U., and U.S. Eur J Nutr 44:303–310

Sanz ML, Côté GL, Gibson GR, Rastall RA (2006) Influence of glycosidic linkages and molecular weight on the fermentation of maltose-based oligosaccharides by human gut bacteria. J Agric Food Chem 54:9779–9784. https://doi.org/10.1021/jf061894v

Seminario-Amez M, López-López J, Estrugo-Devesa A et al (2017) Probiotics and oral health: a systematic review. Med Oral Patol Oral Cir Bucal 22:e282–e288

Shah NP (2000) Probiotic bacteria: selective enumeration and survival in dairy foods. J Dairy Sci 83:894–907. https://doi.org/10.3168/jds.S0022-0302(00)74953-8

Sheehan VM, Ross P, Fitzgerald GF (2007) Assessing the acid tolerance and the technological robustness of probiotic cultures for fortification in fruit juices. Innov Food Sci Emerg Technol 8:279–284. https://doi.org/10.1016/j.ifset.2007.01.007

Silanikove N, Leitner G, Merin U (2015) The interrelationships between lactose intolerance and the modern dairy industry: global perspectives in evolutional and historical backgrounds. Nutrients 7:7312–7331

Singh A, Raigond P, Lal MK et al (2020) Effect of cooking methods on glycemic index and in vitro bioaccessibility of potato (Solanum tuberosum L.) carbohydrates. LWT-Food Sci Technol 127:109363. https://doi.org/10.1016/j.lwt.2020.109363

Song M, Yun B, Moon JH et al (2015) Characterization of selected lactobacillus strains for use as probiotics. Korean J Food Sci Anim Resour 35:551–556. https://doi.org/10.5851/kosfa.2015.35.4.551

Sridevi J, Halami PM, Vijayendra SVN (2010) Selection of starter cultures for idli batter fermentation and their effect on quality of idlis. J Food Sci Technol 47:557–563. https://doi.org/10.1007/s13197-010-0101-6

Tripathi MK, Giri SK (2014) Probiotic functional foods: survival of probiotics during processing and storage. J Funct Foods 9:225–241

Ventura M, Perozzi G (2011) Introduction to the special issue Probiotic bacteria and human gut microbiota. Genes Nutr 6:203–204

Wang L, Wan Y-JY (2019) The role of gut microbiota in liver disease development and treatment. Liver Res 3:3–18. https://doi.org/10.1016/j.livres.2019.02.001

Yoon KY, Woodams EE, Hang YD (2005) Fermentation of beet juice by beneficial lactic acid bacteria. LWT - Food Sci Technol 38:73–75. https://doi.org/10.1016/j.lwt.2004.04.008

Yoon KY, Woodams EE, Hang YD (2006) Production of probiotic cabbage juice by lactic acid bacteria. Bioresour Technol 97:1427–1430. https://doi.org/10.1016/j.biortech.2005.06.018

Zheng X, Chu H, Cong Y et al (2015) Self-reported lactose intolerance in clinic patients with functional gastrointestinal symptoms: prevalence, risk factors, and impact on food choices. Neurogastroenterol Motil 27:1138–1146. https://doi.org/10.1111/nmo.12602

Zheng M, Zhang R, Tian X et al (2017) Assessing the risk of probiotic dietary supplements in the context of antibiotic resistance. Front Microbiol 8:908. https://doi.org/10.3389/fmicb.2017.00908

The manufacturer's authorised representative in the EU is Springer Nature Customer Service Centre GmbH, Europaplatz 3, 69115 Heidelberg, Germany. If you have any concerns regarding our products, please contact ProductSafety@springernature.com

Printed and bound by CPI Group (UK) Ltd, Croydon, CR0 4YY
30/09/2026
02225332-0009